AF770080

Environmental Impact: Assessment and Analysis

Environmental Impact: Assessment and Analysis

Edited by **Emma Layer**

SYRAWOOD
PUBLISHING HOUSE
New York

Published by Syrawood Publishing House,
750 Third Avenue, 9th Floor,
New York, NY 10017, USA
www.syrawoodpublishinghouse.com

Environmental Impact: Assessment and Analysis
Edited by Emma Layer

© 2016 Syrawood Publishing House

International Standard Book Number: 978-1-68286-203-2 (Hardback)

This book contains information obtained from authentic and highly regarded sources. Copyright for all individual chapters remain with the respective authors as indicated. All chapters are published with permission under the Creative Commons Attribution License or equivalent. A wide variety of references are listed. Permission and sources are indicated; for detailed attributions, please refer to the permissions page and list of contributors. Reasonable efforts have been made to publish reliable data and information, but the authors, editors and publisher cannot assume any responsibility for the validity of all materials or the consequences of their use.

The publisher's policy is to use permanent paper from mills that operate a sustainable forestry policy. Furthermore, the publisher ensures that the text paper and cover boards used have met acceptable environmental accreditation standards.

Trademark Notice: Registered trademark of products or corporate names are used only for explanation and identification without intent to infringe.

Printed in China.

Contents

Preface

Environmental impact is the study of various effects caused by human activities and other factors, both negative and positive, on the ecosystem and environment of the Earth. It emphasises on evaluating and understanding the implications of different human interferences and actions. The significant topics compiled in this book are ecosystem services, food security, sustainable energy use, biodiversity conservation, etc. The extensive content of this book provides the readers with a thorough understanding of the subject.

This book is a comprehensive compilation of works of different researchers from varied parts of the world. It includes valuable experiences of the researchers with the sole objective of providing the readers (learners) with a proper knowledge of the concerned field. This book will be beneficial in evoking inspiration and enhancing the knowledge of the interested readers.

In the end, I would like to extend my heartiest thanks to the authors who worked with great determination on their chapters. I also appreciate the publisher's support in the course of the book. I would also like to deeply acknowledge my family who stood by me as a source of inspiration during the project.

Editor

1

The effectiveness of integrated farm management, organic farming and agri-environment schemes for conserving biodiversity in temperate Europe - A systematic map

Nicola P Randall* and Katy L James

Abstract

Background: Agriculture is the dominant land use throughout much of Europe. Changes to farming practices have led to concerns about negative impacts on biodiversity, and current agricultural policy has an emphasis towards conservation. The objective of this study was to investigate and describe the nature and coverage of research pertaining to the effectiveness of integrated farm management, organic farming and agri-environment schemes as interventions for conserving biodiversity in temperate Europe.
Systematic mapping methodology was adapted from social sciences, and used to create a searchable database of relevant research.

Methods: Searches were made of 10 electronic databases containing peer reviewed journals, PhD theses, conference proceedings and organisational reports. Web searches for relevant research were also made. The title and abstracts of results were examined for relevance. Studies were included when published in English, when an intervention was applied to increase biodiversity or species diversity on farmland, and where there was a measured effect on study organism(s). Correlative and manipulative studies from temperate Europe were included. The research was incorporated into a searchable database (systematic map) and key wording used to describe, categorise and code studies.

Results: The searches identified 83,590 records. Following removal of duplicates and the application of inclusion criteria, 743 references were coded for the final systematic map database. Most of the studies reported were from Western Europe, particularly from the UK. Invertebrates were the most commonly studied organism followed by plants and birds, and field margins were the most commonly studied biotope.

Conclusions: The systematic map describes the scope of research on the topic. It can be used to inform future primary research, or research synthesis and evaluation methods such as systematic review. Areas for which there appear to be evidence gaps, and so may have potential for further primary research, are highlighted. They include the effectiveness of agri-environment options under different farming systems and in providing for amphibians and reptiles. Implications for the development of future systematic maps are discussed, including the question of how to incorporate study quality appraisal. The development of a Collaboration for Environmental Evidence systematic mapping methods group will address some of these issues.

Keywords: Farmland biodiversity, Conservation, Systematic map, Systematic review, Agri-environment, Organic farming, Integrated farm management, Evidence-based policy

* Correspondence: nrandall@harper-adams.ac.uk
Harper Adams University College, Newport, Shropshire TF10 8NB, UK

Background

Agriculture is the dominant land use throughout much of Western Europe [1]. It covers about 47% of total European Union territory [2]. Since the 1940s, farming has become increasingly intensive, resulting in the widespread decline of many farmland species. Declines in farmland birds have been particularly well documented [3,4] but also for other taxa, including mammals, arthropods and flowering plants [5,6]. Factors believed to lead to losses include an increase in the use of pesticides and fertilisers, loss of non-crop biotopes, a change from spring to autumn cropping, land drainage, increased stocking densities, switch from hay to silage and an increase in landscape homogeneity resulting from regional specialisation of farming practices [1,7-9]. With an increasing world population one of the major challenges for the future will be how best to meet the need to increase food production whilst minimising the harm to biodiversity [10].

Concerns about harm to biodiversity and the environment have led to a greater emphasis on environmental protection and biodiversity conservation in agricultural policy. This has included new agri-environment schemes that provide payment for management that is designed to be environmentally beneficial [7]. Following a European discussion paper in 1985, which addressed the environmental impact of agriculture [11], agri-environment schemes (AES) have been available to European Union member states under the Common Agricultural Policy [12], and agri-environment is a key European Union policy with 22.2 billion Euros allocated to schemes by member states for the period 2007 to 2013 [13]. However, these schemes are subject to change, and vary between different countries.

A number of studies have examined whether AES conserve or increase biodiversity [12,14 for example], but results appear variable and may be affected by a number of factors. For example AES may provide only some of the resources needed by target organisms, or provision may be inadequately located [15]. Landscape structure may also affect the success of farmland biodiversity initiatives [16]. Changes to future funding initiatives could be made in order to address these issues [15]. The European Commission has recently been carrying out consultations on how best to reform the Common Agricultural Policy, including how best to use future incentives for farmers to deal with environmental concerns such as biodiversity [12].

Wider concerns about the sustainability, environmental impacts and health effects of intensive farming have led to an increased demand for organic food, and development of alternative systems such as integrated farming. Integrated farming aims to minimise the use of pesticides and fertilisers through better targeting, and integration with cultural control of weeds, pests and diseases [7].

The European Commission consider integrated crop management to be of particular importance when combining modern farming with the capacity to ensure good environmental outcome [2]. Both organic and integrated farming may have positive benefits for biodiversity [17-20], although results vary between studies and species.

For example, Hole *et al.* [21] used data from Europe, New Zealand, the US and Canada to suggest that organic farming increased biodiversity at every level of the food chain, yet a meta-analysis by Bengstsson *et al.* [22] found that results varied between studies, and between organism groups and landscapes. Fuller *et al.* [23] suggested that the variation between taxa is because organic farms may be too small and isolated to affect species with large spatial needs. Furthermore, a recent study showed that the benefits of organic farming to wildlife and increases in biodiversity are much lower than previously thought [24].

In their recent consultation on reform of the Common Agricultural Policy, the European Commission identified the need to find out how best to protect maintain and enhance farmland habitats and biodiversity throughout Europe [2].

The primary objective of this study was to describe and 'map' research pertaining to 'The effectiveness of integrated farm management, organic farming and agri-environment schemes for conserving biodiversity in temperate Europe (defined here as European countries west of Russia, but not those south of France, Switzerland, Austria, Hungary and Romania) with an emphasis on birds and invertebrates. The resulting database can be used to identify knowledge gaps to inform future primary research, other research synthesis methods such as synopses of evidence, and synthesis and evaluation methods such as systematic review.

Systematic maps and their use in social sciences

Systematic mapping is used in social sciences to provide an overview of evidence for broad research questions. Systematic mapping like systematic review, is a robust, repeatable and transparent scientific method used to identify and categorise available literature on a particular topic. It uses established searching protocols and has rigorous criteria for inclusion or exclusion.

Unlike systematic review, however, systematic maps are not used to attempt to answer a question, and do not usually attempt full data extraction or critical appraisal of the relevant research. Instead, they identify and describe the nature, volume and characteristics of research in a chosen field [25,26].

Systematic maps can be used to review broad topic areas [27] and categorise the literature according to topic area, study design, and other factors of relevance to the reviewers [26].

The information from systematic maps can be used as a tool for decision makers or to identify where studies are similar enough for robust syntheses such as systematic review [26]. Systematic maps can also identify knowledge gaps to inform primary research [27,28]. This differs from scoping reviews, which are usually preliminary assessments of the size and scope of literature in a topic area, and are not usually considered final outputs in their own right [26].

Systematic mapping as a tool for environmental scientists

To date, the adaptation of systematic mapping methodology for environmental evidence is in the early stages and there is no formal methodology for applying it to environmental sciences.

We adapted methodology for systematic mapping from the Social Care Institute for Excellence [27]. Figure 1 provides an overview of the process. A searchable database is created to show the available research on a topic, and this together with an associated summary report forms the systematic map.

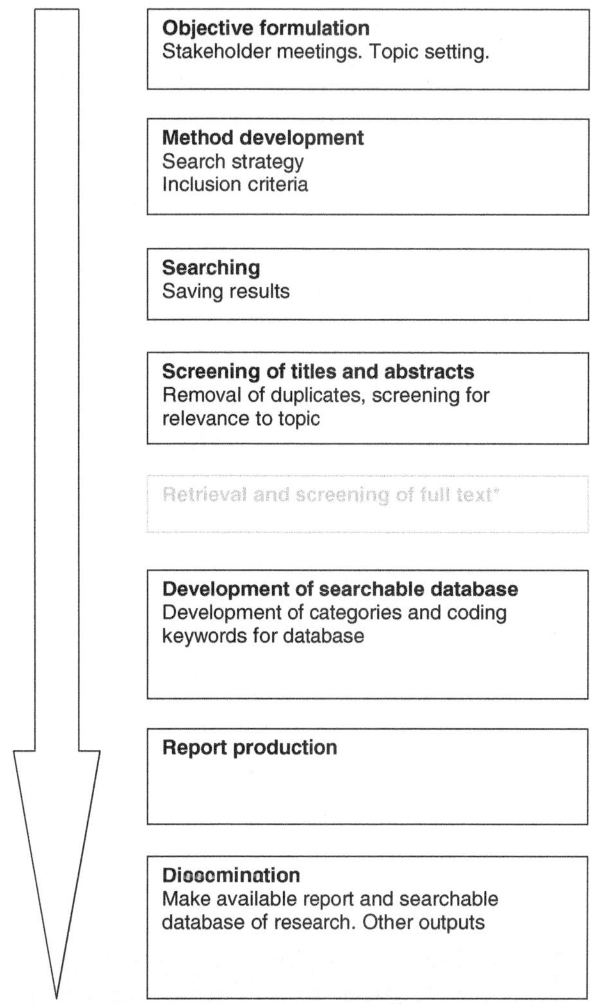

Figure 1 Stages of mapping (adapted from Clapton, Rutter and Sharif, 2009 [27]). *For this article full text articles were not retrieved or screened.

Objective of the map

Objective formulation

The question of how different farming systems compare in their effects on biodiversity and other environmental impacts was recently identified by policy makers and scientists as one of the top 100 policy-relevant ecological questions in the UK: 'How do farming systems such as conventional, integrate farm management and organic compare in terms of their effects on biodiversity and other environmental impacts?' The need to learn from Agri-environment Schemes to optimise future biodiversity gain and ecological benefits was also prioritised [29].

The need for a systematic review to answer these questions was discussed by a stakeholder group comprising; the UK Department for Environment Food and Rural Affairs (Defra), Natural England, Countryside Council for Wales, Farming and Wildlife Advisory Group, Linking Environment and Farming (LEAF) and individual farmer representatives. Feedback from meetings was incorporated into the finalised protocol to shape the review question and search strategy.

Stakeholders agreed that the study should focus on 'temperate' Europe, rather than solely on the UK, because these countries have similar farming systems, and a significant amount of money is invested into agri-environment schemes throughout the European Union [13,30]. The questions of how best to produce food sustainably, including the protection, maintenance and enhancement of farmland biodiversity are included in the European Commissions recent consultation into the Common Agricultural Policy [2], confirming the need to understand the benefits to biodiversity to ensure value for money, and to further improve the schemes.

A systematic review search strategy was established in 2008. A particular emphasis was placed on birds and invertebrates at the request of stakeholders, but the searches captured a large volume of relevant but disparate literature. When, in 2010, funding and resources became available to further the research, it became evident that the topic area was too broad for systematic review synthesis and alternative methods of presenting the data were considered. Subsequently the question posed by the group was formulated into a primary objective for a systematic map instead of a systematic review.

Primary objective

The study aimed to identify and systematically map published and unpublished research to address the following primary objective:

- To describe and map research that investigates the effectiveness of integrated farm management, organic farming and agri-environment schemes as interventions for conserving biodiversity in temperate Europe.

Secondary objectives

The secondary objectives were to:

- Describe how the research into these farm-scale interventions on biodiversity varies between different regions, farming-systems, agri-environmental options/ farmland biotopes and types of organism.

- Provide an overview of published research and grey literature in the subject area for use by practitioners, policymakers, researchers and the public.

- Provide a map that is searchable by topic

- Inform future research synthesis and evaluation methods such as systematic review and synopses of evidence

- Identify knowledge gaps in order to inform future primary research

Methods

The methods used in the development of the systematic map were adapted from Collaboration for Environmental Evidence (CEE) Systematic Review Guidelines [31], from existing systematic map reports [32,33] and Social Care Institute for Excellence (SCIE) Guidelines [25,27].

Searches

A comprehensive search of several information sources attempted to achieve an un-biased sample of both published and grey literature. There were two search phases, the first in 2008 and the second search in 2010, where the search was updated, and new databases were searched.

The following online literature databases were searched to identify relevant literature:

1) ISI Web of Knowledge involving the following products: ISI Web of Science; ISI Proceedings
2) Science Direct
3) Blackwell Synergy
4) Ingenta Connect
5) Index to Theses Online
6) CAB Abstracts
7) Centre for Ecology and Hydrology online database
8) Conservation Evidence.com
9) UK Department for Environment food and Rural Affairs online
10) NERC Open Research Archive

In addition, web searches were performed using the search engines: WWW.ALLTHEWEB.COM and HTTP://SCHOLAR.GOOGLE.COM. The first 50 hits (.doc .txt . xls and .pdf documents where this can be separated) from each data source were examined for appropriate data. No further links from the captured website were followed unless to a document/pdf file.

Database and repository searches were conducted in English language. A wildcard (*) was used to pick up multiple word endings (eg. Farm* would pick up farms, farming etc.). Birds and invertebrates were specifically searched for as potential farmland biodiversity indicators at the request of the original stakeholder group. The following terms were used in combination to search the online databases/web searches:

Farm* AND biodiversity
Organic AND biodiversity
Farm* AND (diversity or abundance) not fish
Agri-environment*
(Farmland or farming) AND bird*
(Farm* AND invertebrate*) not fish
Agri-environment* AND biodiversity

Where wildcards were not permitted extra searches were carried out to cover multiple word endings. 'Integrated farm management' was not searched for as a phrase, as any searches with farm* would have identified these references. Brackets were only used where databases allowed.

The results of the searches were imported into EndNote X2^{TM} library files. A separate file was made for each search made. When searching was complete, all the database libraries were incorporated into one new library, and the number of references captured was recorded. Using the automatic function in the EndNote X2^{TM} software, any duplicates were removed. The library was then searched for references relevant to the topic.

Study inclusion criteria

The inclusion criteria were applied by one reviewer to all studies at title and abstract level. For a random 20% of studies, and where it was not clear whether a study met the inclusion criteria, the study was screened by a second reviewer and any differences resolved through discussion. Reviewers discussed the procedures to ensure a consistent understanding of the criteria at both the screening and coding stages.

Relevant subject(s): Studies on some aspect of biodiversity or species diversity on farmland were included, irrespective of scale. All biotopes were included except for ponds, farm woodland and orchards. The scope of temperate Europe was defined as all European countries west of Russia, but not those south of France, Switzerland Austria, Hungary and Romania.

Types of intervention: Interventions that aimed to increase biodiversity on farmland, or some element of it, were considered: these included organic farming and integrated farm management, named agri-environment schemes, and specific options and biotopes that could form part of agri-environment, organic or integrated farming schemes.

Types of comparator: Conventional farming or a conventionally farmed plot.

Types of outcome: Effects on any organism or organisms excluding fungi, bacteria and agricultural pests. Any measure of biodiversity was considered. Differences or similarities in biodiversity on different farm types, and differences or similarities in biodiversity following management changes on farmland were considered.

Types of study: Any type of study that investigated biodiversity on farmland was considered, including correlative studies and manipulative studies but not pure ecology studies, statistical models, or review papers. Systematic maps, lacking strict critical appraisal, often include studies that would not be included in systematic reviews [27]. Only primary research studies were incorporated into the final systematic map.

Language: Studies published in English.

Date: No date restrictions were applied, but searches were limited according to the scope of the individual databases.

There was a high volume of irrelevant references present in the intial library (for example large numbers of medical papers referring to biodiversity of the gut were present) so some filters were made of keywords in the title to aid exclusion of irrelevant references. The titles of all records filtered by this method were visually checked to ensure that no potentially relevant references were excluded. Where there was doubt about relevance, records were removed from the exclusion list and passed to the next inclusion stage. The references were then hand searched for relevance, first by title, and then by title and abstract. Due to the high volume of relevant literature captured, searching was taken to the abstract phase only, full text was not retrieved. The resulting references were used to formulate a database.

Definitions

Definitions are important when both searching and screening [33]. When creating the searchable database clear definitions help to overcome ambiguity and improve rigour. The operational definitions and clarification of the categories used for the systematic map are detailed in an additional file (Additional File 1).

Recording the search and screening

A record of each search was made to enable a re-run of the search if needed.

The following data were recorded: date of search, database name, search term, number of hits, date limits of the database searched (*e.g.* 1999-2006), and notes. For each stage of the screening process the number of references obtained and excluded was recorded.

Coding for the systematic map

Key wording was used to describe, categorise and code studies. Systematic map reports can present data by country of origin, publication date, intervention type etc. Coding should be determined on a review-by-review basis [27].

Keywords were generated from the primary question, expert knowledge in the subject area and topics reported in the studies.

A wide range of organisms were studied in the publications. Individual species were not recorded as this would have further expanded and complicated the database. Instead organisms were categorised as general groups: invertebrate, plant, bird, mammal, reptile or amphibian. Invertebrates were further categorised into four functional groups (identified as important for farmland systems by the stakeholder workshop participants): pollinators, natural enemies, soil organisms and invertebrate food for farmland birds. Some functional groups were further divided. For example, pollinators have been subdivided into Lepidoptera (butterflies), Bombus spp. (bumblebees), Apoidae (solitary bees and unspecified bees) and Syrphidae (hoverflies). A subcategory 'plants of conservation concern' was also made where study authors had identified that the plants included in the research were of particular local or national conservation interest and/or that receive funding through agri-environment schemes.

The articles were coded under the following categories: author, full reference, date of publication, publication type, length of time of study, country/ies of study, spatial scale of experiment, farming system, farming prescription/intervention, biotope, organism, outcome investigated, and experimental design. Studies were defined as either correlative (where biodiversity outcomes had been compared between different biotopes as found) or manipulative (where artificial treatments were applied to test for effects on biodiversity). Studies could appear under more than one category where relevant. For example butterflies were categorised as invertebrate, pollinator, and as butterfly.

The coding was carefully moderated. All included articles were coded by one reviewer, and scrutinised by a second reviewer. Coding keywords were revised to resolve any inconsistencies between reviewers or in response to search findings. Any articles already coded, were re-coded using the revised keywords. A sample of coded articles was cross checked by a third independent reviewer to confirm understanding of coding.

The included references were exported from the EndNoteX2^{TM} library to a Microsoft excel workbook where the coding was applied. Details of the keyword coding used for each category included in the systematic map database are given in an additional file (Additional File 2).

Systematic map database

A searchable map (database) was created to describe the scope of the research and identify apparent knowledge gaps. It can be used for further analysis as it enables both simple numerical accounts of the frequencies in each category, and more complex cross tablulations. The searchable database is included as a Microsoft Access file (Additional File 3) and details of how to search the database are provided separately (Additional File 4).

The systematic map data can be ordered by topic area, publication date, intervention type, organism, country of study etc. Cross tabulations can be generated from the map to allow reviewers to find correlations and trends and identify areas where there are no research outputs available. For example, interactions between farming system and agri-environment scheme can be displayed or used to examine whether there is bias towards an individual agri-environment scheme and its effect on a particular organism.

Results

Review descriptive statistics

The searches identified 83,590 records. Of these, 31,371 were found in the 2008 searches, and 52,219 by the 2010 searches. The 2010 searches were a combination of updates to the 2008 searches, and completes searches of some databases hence the higher number of hits in 2010. The number of records generated for specific searches are detailed in an additional file (Additional file 5). The records were screened for inclusion in the systematic map as illustrated in Figure 2.

A total of 743 articles were judged to have met the inclusion criteria based on title and abstract, and included in the final systematic map. The list of articles is provided in an additional file (Additional file 6). Each article was coded for the database. An online version of the systematic map is available as an electronic database, searchable by keyword, and can be found as an additional file (Additional file 3). Six articles were found that were either meta-analyses or systematic reviews. These were not included in the systematic map database, but have been included as an additional file (Additional File 7).

Of the 743 records included in the map, 388 articles reported on correlative studies and 339 manipulative. Sixteen records contained a combination of correlative and manipulative studies presented in the same article. These were predominantly PhD theses.

Overall, the majority of articles were journal papers (73%, $n = 539$), followed by conference papers (16%, $n = 121$) and theses (6%, $n = 47$). The remaining articles (5%) were reported within books or were organisation reports and final reports for the UK Department for Environment Food and Rural Affairs. Where information from one study (although not necessarily the same data) was presented in one or more articles, for example a thesis, conference paper and refereed journal, they were classified as 'multiple outputs of one study' (total = 116). All outputs found on any study were included within the database. Any known multiple outputs were marked as such.

The earliest article recorded was in 1962, and there was a substantial increase in the number of articles found year by year up to the present day (Figure 3).

Country of origin

Whilst the country of origin was not mentioned in many abstracts ($n = 181$) most studies were carried out in the UK ($n = 280$). Most of the remaining studies originated from Western European countries (particularly Germany, Switzerland, the Netherlands) when compared with the Eastern European countries included in this study (eg. Czech Republic, Slovakia, Latvia) (Figure 4). No studies were found for Romania, the Ukraine, Moldovia, or Belaruse.

Scale

Most studies were performed across a variety of spatial scales. The most common combination was 'within-field' 'multiple farm'. I.e. the study compared specific within-field interventions (such as field margin) with other interventions or crops, and evaluated results across a number of farms. Many of these studies could also be considered 'landscape scale' as studies were conducted on multiple farms in different regions of the UK. Very few articles ($n = 12$) reported studies made at the 'whole farm scale', i.e. investigating the impact of a named intervention (such as agri-environment scheme implementation) on biodiversity across the whole farm. A breakdown of the spatial scales of studies is given in Table 1. Some abstracts ($n = 24$) did not mention a spatial scale at all, and in other cases it was possible to discern some aspects of spatial scale, but not others (for example, experiments may have been described as

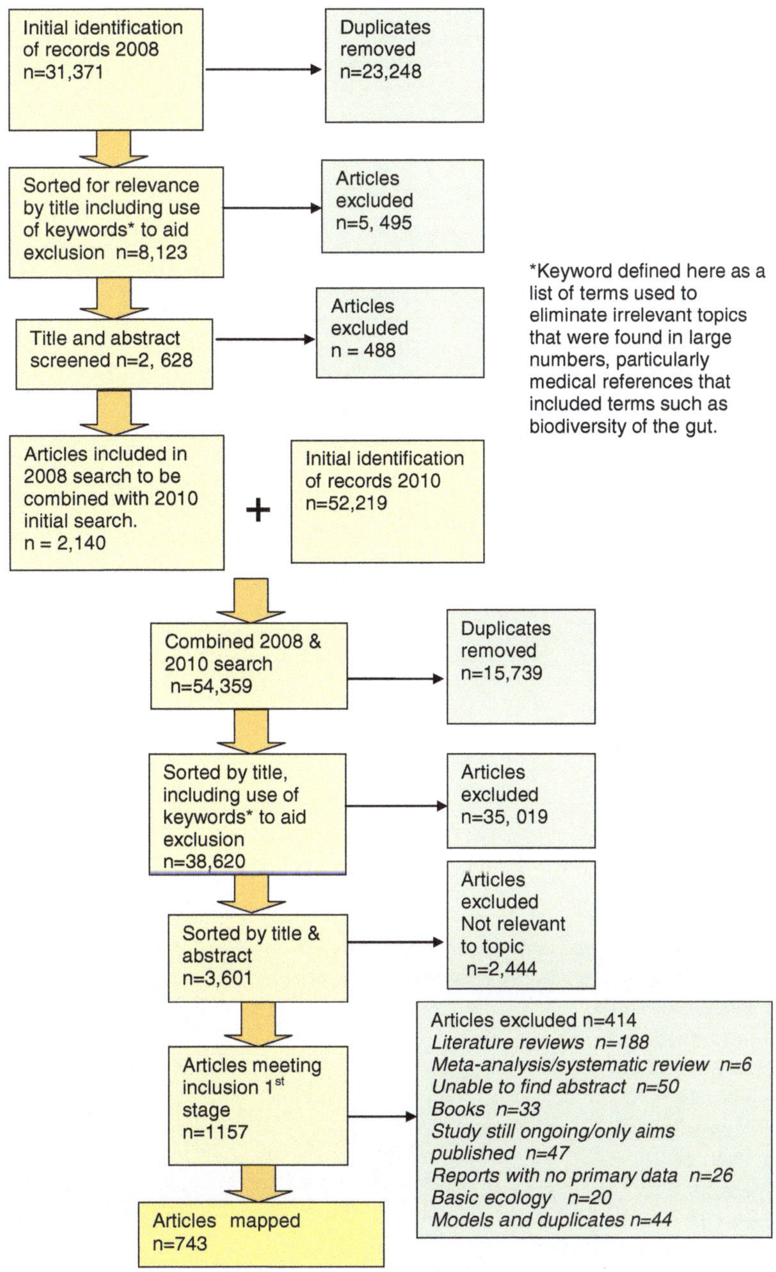

Figure 2 Literature included and excluded at each stage of the systematic mapping process.

'within field', but no mention given of whether this was replicated within one or multiple farms).

Period of study

Most abstracts (*n* = *396*) did not mention the period of time over which the work took place. Of the remaining, most studies (*50%*) were between 1 and 3 years duration, 19% were less than a year in duration, 10% took place over 4 years, 12% were carried out over 5 to 9 years, and 9% were carried out over 10 years or more.

Organisms and functional groups

Invertebrates were the most commonly studied organisms (*n* = *422*). Followed by plants (*n* = *243*) and birds (*n* = *226*). There were very few studies on mammals and less on reptiles or amphibians (Figure 5). As specific searches were made for invertebrates and for birds, but not for other types of organisms, this is likely to have resulted in bias. Freshwater biotopes were not included in the searches and were screened out, which is likely to have had an impact on organisms specifically associated with them.

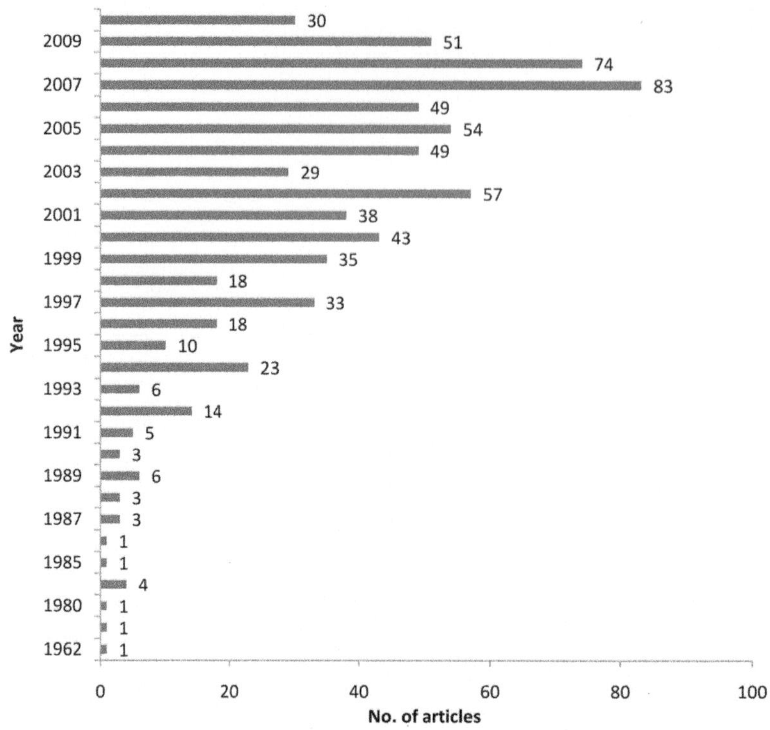

Figure 3 Number of articles by year. The apparent increase will reflect (at least in part) bias in dates available for the databases searched.

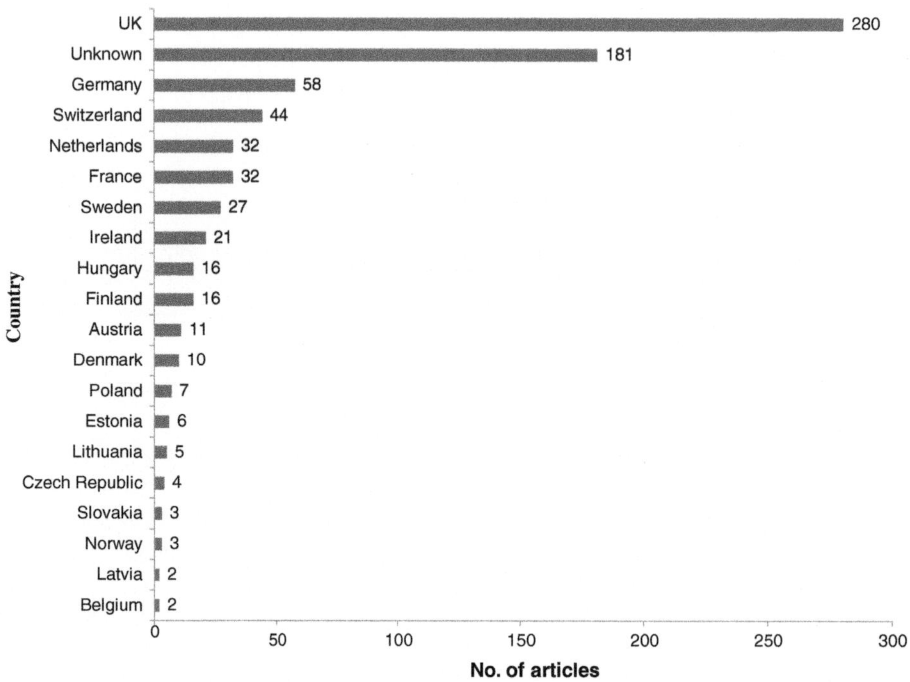

Figure 4 Country of Origin of study. The total number of publications is shown at the end of each column. The overall total is more than 743 because some studies were performed in more than one country. The bias towards UK studies is likely to reflect (at least in part) the English language searching and coding strategy used.

Table 1 The number of articles at each spatial scale recorded

Scale	*No. of articles*
Landscape	206
Multiple farm	350
Single farm	98
Whole farm	12
Within field	680

The total number of articles exceeds 743 because some studies were performed on multiple farms at landscape scale, within field. Some abstracts did not include details of spatial scale so are not included in the totals.

Correlative and manipulative studies were found in similar numbers of articles for most groups, with the exception of birds, which were most commonly investigated in correlative studies (145 correlative; 75 manipulative; 6 combined correlative and manipulative). Articles reporting plants studies were slightly biased towards manipulative research (135 manipulative; 103 correlative; 5 combined correlative and manipulative) (Figure 5).

Four functional groups were selected to be investigated further to investigate the scope of evidence relating to these subgroups. This also provided an indication of the breakdown of the invertebrate composition, the most commonly studied taxanomic group. Pollinators ($n = 128$) and natural enemies ($n = 127$) were the most studied functional groups. There were fewer study articles for soil organisms ($n = 58$) and invertebrate food for farmland birds ($n = 46$). There were similar numbers of correlative and manipulative studies for each functional group. Pollinator studies were mainly on Lepidoptera (butterflies) (52% of articles) and Bombus sp. (bumblebees) (34%). Studies on Apoidae (unspecified bees) (21%), Syrphidae (hoverflies) (16%), Apis sp. (honey bees) (5%) and Apoidae (solitary bees) (5%) made up the remainder. Natural enemy studies were mainly of Araneae (spiders) (71% of articles) and Carabidae (ground beetles) (48%) and to a lesser extent on Staphylinidae (rove beetles) (22%) and Syrphidae (hoverflies) (7%). Lumbricadae (earthworms) were the most commonly studied soil inverebrate ($n = 35$) followed by Collembola (springtails) ($n = 14$). The remaining 9 articles either investigated other soil invertebrates or the organism was not mentioned in the abstract.

Only 17 articles reported investigations of plants which were of particular conservation interest either in the country or locality of the study due to rarity or declines. These were categorised separately within the database as 'plants of conservation concern' as some receive conservation funding through the Common Agricultural Policy.

Out of the 743 articles, 698 abstracts reported the outcome measured, most commonly abundance. This was followed by diversity, richness, density, composition, reproductive success, survival and distribution. Figure 6 shows the number of times each outcome was recorded.

Farming system and biodiversity

Studies investigating farming system and biodiversity encompassed a wide variety of subject area, scales and design. For example, multiple farm studies sometimes compared biodiversity between matched pairs of different farming systems (such as organic and non-organic) [34], whereas others investigated changes in biodiversity on farms during conversion from conventional to organic or integrated farming systems [19,35 for example]. Other studies looked at the impact on biodiversity of management of biotopes such as field margins [36] or farm practices such as fertilisation [37] in specific farming systems.

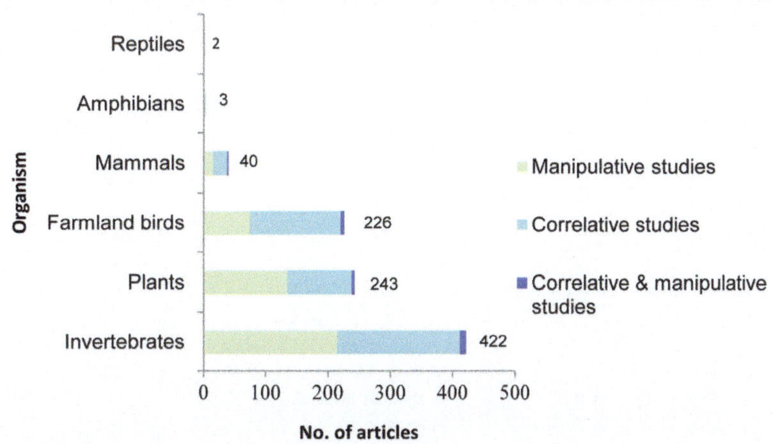

Figure 5 Number of articles for each type of organism. Total number of articles for each is presented at the end of each column. The total number of articles exceeds 743 because more than one organism may have been investigated in a study.

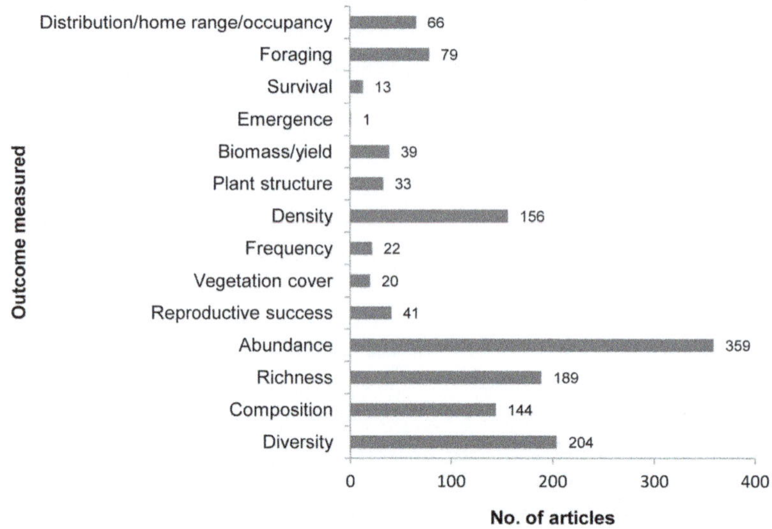

Figure 6 Number of times each outcome is reported.

Organic & integrated farming systems

The number of studies of organic farming increased from the late 1990s. In contrast, those of integrated farming stayed at a low level throughout, although numbers did increase slightly in the 1990s before declining again (Figure 7).

Most studies of organic farming were in Western Europe, especially the UK and Germany (UK *n = 31 articles*; Germany *n = 22*; Sweden, Netherlands and Switzerland *n = 8* each; Denmark *n = 7*; Austria *n = 6*; France *n = 4*; Ireland and Finland *n = 1*). Studies from Eastern Europe were from Estonia (*n = 3*), Slovakia (*n = 2*), Poland, Hungary & Czech Republic (*n = 1* each).

The incidence was similar for integrated farm management with 36 articles reporting studies from Western Europe (UK *n = 20*; Germany *n = 9*; Switzerland *n = 6*; Finland and Netherlands *n = 1*) and 3 Eastern European countries (Estonia, Lithuania and Slovakia *n = 1* each).

More correlative studies were found for organic farming than for integrated farming but the number of manipulative studies was similar. Most comparisons of organic and conventional farming were correlative, whereas those of integrated and conventional farming were manipulative. Eight articles reported studies comparing organic and integrated farming systems (Table 2).

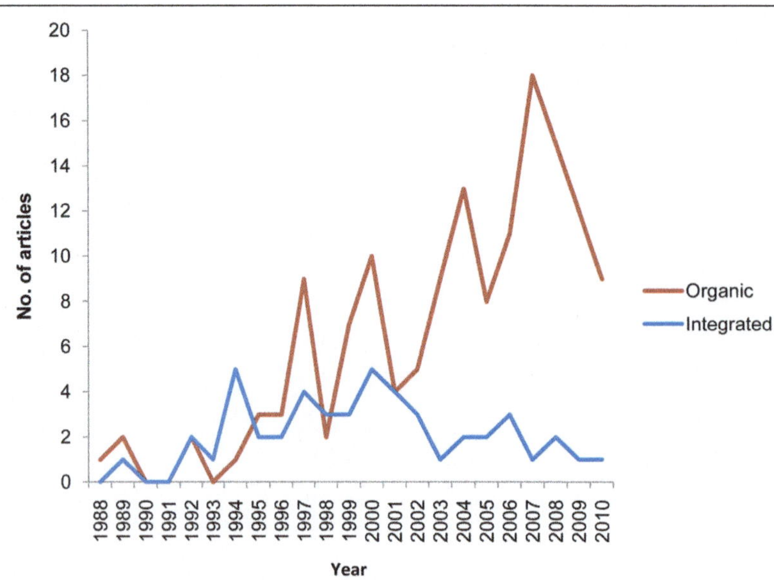

Figure 7 Number of articles for of organic and integrated farming by year.

Table 2 The number of articles reporting correlative and manipulative studies of organic, conventional and integrated farming systems

	Correlative	*Manipulative*	*Correlative and manipulative*	*Total*
Organic	117	27	2	144
Conventional	112	39	2	151
Integrated	20	28	0	48
Organic versus conventional farm management	89	14	2	103
Organic versus conventional versus integrated farm management	11	4	0	15
Integrated versus conventional farm management	3	17	0	20
Organic versus integrated farm management	6	2	0	8

Table 3 summarises the number of articles for organism and functional group in organic, conventional and integrated farming systems. Invertebrates were the most studied group for all systems followed by plants and birds. Few studies examined mammals and plants of conservation concern and no studies were found for amphibians or reptiles. Invertebrates are considered important to organic and integrated farming, particularly natural enemies which are often used as a form of pest control. It was therefore unsurprising to find that natural enemies were the most studied invertebrate functional group. Soil organisms are also important in these systems for maintaining healthy soils and this is reflected in the number of articles captured. Natural enemies (n = 8) and soil organisms (n = 7) were also the most studied functional groups in comparisons of integrated and conventional farm management. In contrast, when organic and converntional farming systems were compared, plants (n = 36) and birds (n = 24) were more commonly reported than any invertebrate functional group. Pollinator studies were reported far more frequently in comparisons between organic and conventional farming systems (n = 13) than in integrated farming systems (n = 2). There were very few studies of farming system and invertebrate food for farmland birds.

Other farming systems studied

Other farming systems, captured and included in the map included conservation tillage (Table 4) and extensive grassland management, in particular "extensive" grazing systems. These can be used as agri-environment options, or be a component of specific farming systems. For example, non-inversion tillage is practiced in integrated farm management [38].

Invertebrates, particularly soil organisms, were the focus of the tillage studies. One of the aims of conservation tillage is to reduce soil organism mortality, and so improve soil condition. Conservation tillage can also indirectly benefit other organisms by increasing prey availability (invertebrate food for birds for example).

Invertebrates, particularly pollinators, were also the most commonly reported study organisms in experiments investigating grassland (Table 5) and grazing (Table 6) systems. The number of studies investigating plants and birds in the two systems was similar. Few studies investigated mammals and none were found for amphibians or reptiles.

Impact of agri-environment scheme

The abstracts of 64 articles stated that they were comparing farms under agri-environment schemes with those not under them. The actual prescriptions may or may not have been mentioned. These were predominantly manipulative studies (86%). There were only 3 articles for Eastern Europe (Czech Republic n = 1 and Estonia n = 2). The others were: UK (*n = 29*); Switzerland (*n = 13*); Netherlands (*n = 8*); Ireland (*n = 6*); Austria (*n = 2*); France and Germany (*n = 1each*). Two articles did not give a country in the abstract, although from personal knowledge, one was UK. Invertebrates (*n = 33*), plants (*n = 31*) and farmland birds (*n = 26*) were the most commonly studied groups. Few studies investigated mammals and none amphibians or reptiles (Figure 8). Pollinators were the most common functional group, with half on butterflies (*n = 10*) and the others on unspecified bees (*n = 7*), bumblebees (n = 3) and hoverflies (*n = 2*). Figure 9 shows the number of articles for specific management options. Field margins (*n = 17*), Meadows (*n = 10*) including Ecological Compensation Area Meadows (specific to Switzerland *n = 9*) and hedgerows (*n = 11*) were the most common. It may be that more evidence is needed to test the efficacy of management options. Some options are for specific organisms eg. Lapwing option, whereas others are not. Table 7 gives the group studied for each option.

Biotope/agri-environment options

Twenty six types of biotope intervention were identified, and recorded in the database. Arable crop and grass field, however, were usually used either as a comparator (in [39] for example, bumblebee numbers on experimental field margins were compared to those of the arable crop), or associated with other interventions such as

Table 3 Number of articles for farming system and type of organism

	Invertebrates	*Plants*	*Farmland birds*	*Mammals*	*Plants of conservation concern*	*Pollinators*	*Natural enemies*	*Soil organisms*	*Invertebrate food for farmland birds*
Organic	88	50	28	7	6	18	30	17	3
Conventional	95	46	30	8	4	16	32	21	6
Integrated	35	7	8	2	1	2	17	13	1
Organic versus conventional farm management	58	36	24	6	3	13	19	10	3
Organic versus conventional versus integrated farmmanagement	9	4	1	1	0	1	4	3	0
Integrated versus conventional farm management	18	0	3	0	0	1	8	7	1
Organic versus integrated farm management	6	2	1	0	1	0	4	1	0

Table 4 Number articles for tillage

	Invertebrates	*Natural enemies*	*Soil organisms*	*Plants*	*Farmland birds*	*Mammals*	*Plants of cons concern*
Total tillage publications	26	7	19	6	10	1	2
Ploughing	23	7	17	4	7	0	1
Non inversion tillage	12	5	8	3	6	0	0
Minimal tillage	17	2	14	2	3	1	2

More than one organism/functional group may have been investigated in a study. The number of articles for each taxonomic group will reflect (at least in part) the search terms used.

farming system, (for example in comparisons of collembola numbers in conventional, organic or integrated farm management winter wheat [40]).

Field margins were the most commonly studied of the remaining biotopes (*n = 215 articles*). Studies on field margins, hedgerows, set-aside, meadows and semi-natural grassland were particularly varied, from investigating them under different farming systems and testing different management or restoration options to comparing their use with other biotopes by a number of different organisms.

Studies of interventions that were specifically designed as agri-environmental options were generally manipulative. Examples of this included, wild bird cover, conservation headlands, skylark plots, beetle banks and the Lapwing option. These type of options tend to be simple to implement on small scales, and are often targeted towards particular organisms so may be more conducive to manipulative studies. However, some named agri-environmental scheme options were more studied than others and, where named, often tended to be UK schemes. For example, conservation headlands (*n = 24 articles*), wild bird cover (*n* = 23) and beetle banks (*n* = 16) are all UK current or historic agri-environment schemes.

Many studies on hedgerows were correlative as were all those on green lanes. The numbers of correlative and manipulative studies were similar for semi-natural grassland, set-aside, field margins, lowland wet grassland, meadows, ditches and overwintered stubbles. The numbers of articles reporting the effect of specific biotope or agri-environment interventions are detailed in Figure 10.

Table 8 shows cross-tablulations of the number of articles found for biotope intervention and organism type. As above, most research investigated invertebrates, plants and farmland birds. Publications on mammals were much less common, being mainly studied in field margins, hedgerows and set-aside. No studies were found for amphibians and reptiles.

The subject organism often reflects the intent of the intervention. For example, wild bird cover tries to provide winter seed for farmland birds and most studies (17 of 24) were on birds but some investigated affects on other organisms either in addition or instead. Some options designed for birds (*e.g.* overwinter stubbles and skylark plots) often examine plants and invertebrates as these provide food sources for birds. Field margins on the other hand provide a habitat for a wide range of organisms but there has been a disproportionate amount of research on invertebrates.

Plants of conservation concern are often found within arable crops and grass fields. Five articles were found for plants of conservation concern in arable crops, and 2 for grass field.

Studies of the effect of farming system on the biodiversity of biotope

As before, fallow studies were included within set-aside, riparian buffer strips in field margins and Ecological Compensation Area meadows in meadow management. In the database these are listed separately.

Table 5 Number of articles for grassland system

	Invertebrates	*Plants*	*Farmland birds*	*Plants of cons concern*	*Pollinators*	*Natural enemies*	*Soil organisms*	*Invert food for farmland birds*
Intensive grassland	28	18	12	0	5	6	3	7
Extensive grassland	22	13	8	1	5	6	3	6
Intensiveversus extensive grassland	19	7	7	0	4	5	3	6

More than one organism/functional group may have been investigated in a study. Where an article is categorised as 'intensive versus extensive grassland', it will also be included within the 'intensive grassland' and 'extensive grassland' categories. The number of articles for each taxonomic group will reflect (at least in part) the search terms used.

Table 6 Number of articles for grazing

	Invertebrates	*Plants*	*Farmland birds*	*Mammals*	*Plants of cons concern*	*Pollinators*	*Natural enemies*	*Soil organisms*	*Invert food for farmland birds*
Grazing intensity unspecified	17	13	10	2	2	7	3	1	1
Intensively grazed	14	5	5	1	0	2	5	1	3
Extensively grazed	15	6	8	2	0	4	4	1	3
Intensive versus extensively grazed	13	4	5	1	0	2	4	1	3

More than one organism/functional group may have been investigated in a study. Where an article is categorised as 'intensive versus extensively grazed', it will also be included within the 'intensively grazed' and 'extensively grazed' categories. The number of articles for each taxanomic group will reflect (at least in part) the search terms used.

Few studies investigated the effect of different farming systems on the biodiversity of biotope. Of the 23 biotopes/interventions listed in Table 8 only eight have been investigated for farming system. These are shown in Table 9.

Most studies were on the biodiversity of field margins in either organic or conventional farming systems, or systems that compared the two. The number of articles for each type of organism on field margins in different farming systems is shown in Table 10. The majority of these were correlative and investigated invertebrates. Pollinators, in particular butterflies, predominated in comparisons between conventional and organic systems. There were few studies of field margins on integrated farms.

There were about half as many studies of hedgerows as of field margins but their distribution between farming system and group of organism was similar. The number of articles for each type of organism for hedgerows in different farming systems is shown in Table 11. There were few studies of the remaining habitats.

Discussion

General trends

- The amount of literature found each year generally increased up to the present day. This is likely to reflect changes in policy on the farmed environment in temperate Europe, but could also be affected by known bias in the searches.
- Most studies were from Western Europe, in particular the UK where there has been a long history of such research. This will also reflect bias in the search strategy.
- The literature found was predominantly published in peer reviewed journals.
- Invertebrates were the most frequently studied type of organism, possibly because invertebrates are easily observed, provide food resource for other organisms, and may form important functional groups in agriculture (*e.g.*, natural enemies and pollinators).

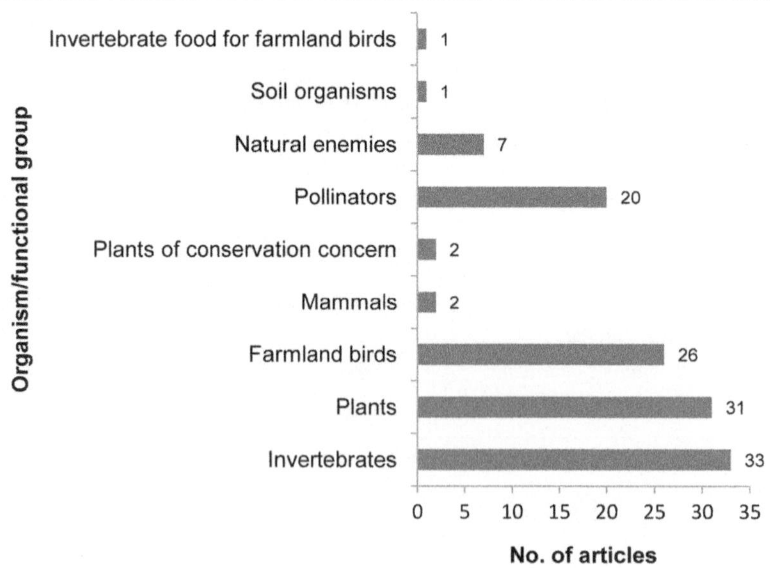

Figure 8 Organism or functional group for agri-environmental investigations. The differences will reflect, at least in part, the search terms used.

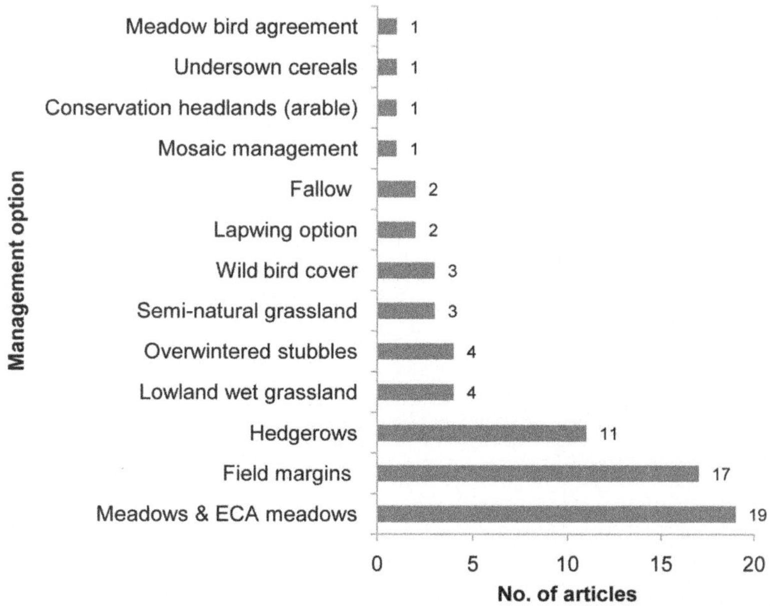

Figure 9 Management options specified in articles where studies investigated farms under agri-environmental schemes.

Some believe they are good indicators of overall biodiversity health, for example butterflies [41].

- Studies of plants and farmland birds were also found in high numbers, even though plants were not specifically searched for. Both are also believed to be good indicators of overall farmland biodiversity health [41]. Furthermore, farmland birds are valued by the public and policy makers.
- There were more studies of pollinators and natural enemies than for soil organisms and invertebrate food for farmland birds.
- Research into the use of field margins as a mitigation measure for biodiversity far outweighed the available research for other biotopes and agri-environment options.
- In terms of how farming systems impact on biodiversity, organic farming has been studied more often than integrated farming.

Gaps in the research

- Little research on amphibians and reptiles was found, particularly in studies of agri-environmental schemes. This will partially reflect biases in the search, however, there have been general concerns that these groups have been overlooked in agricultural research [42].

Table 7 Management options investigated on farms under agri-environment agreements and organism studied

	Invertebrates	*Plants*	*Farmland birds*
Meadows (including Swiss ECA meadows)	11	11	4
Field margins	15	8	1
Hedgerows	9	6	2
Lowland wet grassland	1	2	2
Overwintered stubbles	0	1	3
Semi-improved grassland	1	1	2
Wild bird cover (Option for birds UK, France)	0	1	2
Lapwing Option (UK option for birds)	0	0	2
Fallow	1	1	0
Mosaic management (Dutch option for birds)	0	0	1
Conservation headlands (unsprayed or unfertilised arable strips)	0	1	0
Under-sown cereals	0	1	0
Meadow bird agreement (Dutch option for birds)	0	0	1

Some options are specific to one region or country and some may no longer exist in the form stated.

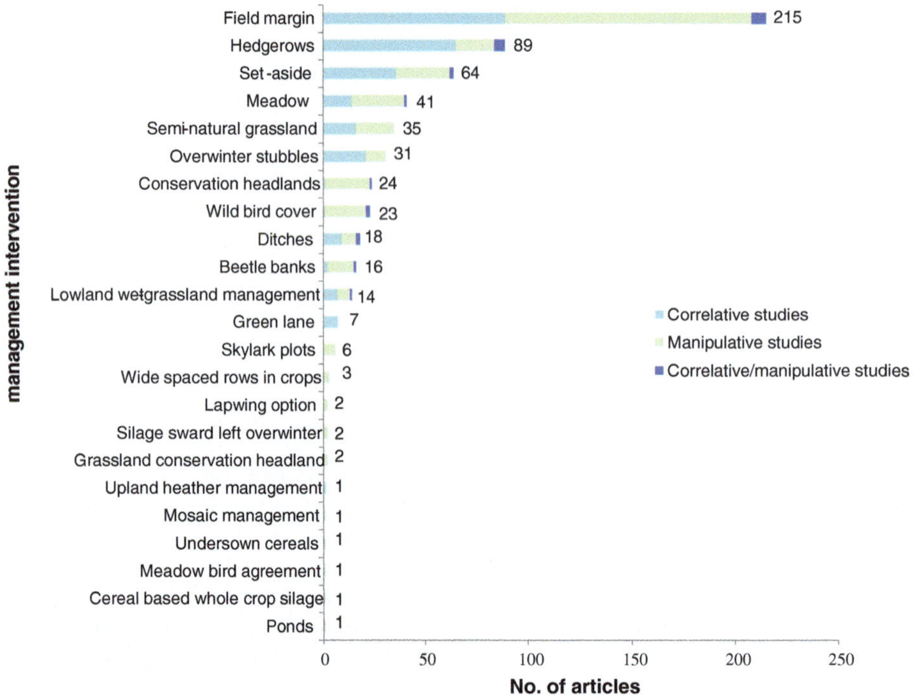

Figure 10 The number of articles reporting the effect of different management interventions on biodiversity. The total number of articles is at the end of each column. For this figure, arable and grass fields are excluded. Fallow studies are included in the set-aside category, riparian buffer strips as part of field margins and the Swiss Ecological Compensation Areas meadows as part of meadow management. Although ponds were excluded from the database, one article is included here as it relates to riparian buffer strips.

- There also appeared to be research gaps for mammals, plants of conservation concern, invertebrate food for farmland birds and soil organisms. Specific literature searches for these organisms would confirm this.
- Some agri-environment scheme options appeared to be more studied than others. For example, conservation headlands, wild bird cover and beetle banks compared with others such as undersown cereals and cereal based whole crop silage. Searching terminology may have biased the findings for some options – '*bird*' for example was included as a search term, so may have increased the number of references found for 'wild *bird* cover'. Variations in local terminology may have also introduced bias. Where more focused information is required, searches for named agri-environment options tailored to the region and time period of interest are recommended.
- Little research comparing the impact of farming system, biotope or agri-environment options on biodiversity was found for Eastern European countries. It may be that Eastern European countries are less likely to publish in English, although this was not tested for.
- There was little research available relating to the impact of integrated farming on biodiversity.
- Further evidence is needed to test the efficacy of agri-environment schemes under different types of farming system. The majority of the research found focused on field margins and organic farming.

Recommendations for further development of the systematic map

The systematic map is easily updatable. Any updating would provide the opportunity to improve the database.

Recommendations for updating include:

- Examination of the full text of included papers, where possible, in order to confirm coding and categorisation.
- Extension of the database to include additional coding for experimental design and data analysis within papers. This would help facilitate critical appraisal of the included research and also enable more efficient identification of topics suitable for future systematic reviews.

Potential systematic review topics

The systematic map highlights subtopics that appear to be populated with enough primary research to inform specific systematic reviews. Areas where studies were manipulative rather than correlative may be likely to

Table 8 Number of articles for each combination of biotope type and taxonomic group

	Invertebrates	*Plants*	*Farmland birds*	*Mammals*	*Reptiles*	*Amphibians*	*Plants of consconcern plants*	*Pollinators*	*Natural enemies*	*Soil organisms*	*Invertebrate food for farmland birds*
Field margins	153	68	29	18	1	2	4	65	46	8	12
Ponds/pools	1	0	1	0	0	1	0	0	0	0	0
Semi-natural grassland	15	20	9	2	0	0	1	9	2	0	1
Beetle bank	11	5	2	2	0	0	0	1	9	0	3
Hedgerows	41	32	27	10	0	0	0	8	9	3	3
Grassland conservation headland	2	2	1	0	0	0	0	0	0	0	1
Set-aside	32	18	26	7	0	0	3	11	6	2	7
Lowland wet grassland	4	6	8	0	0	0	0	1	1	1	0
Meadows	20	21	8	1	0	0	0	12	8	0	1
Ditches	6	4	8	1	0	1	0	2	1	1	1
Wild bird cover	9	6	17	1	0	0	0	7	0	0	2
Lapwing option	0	0	2	0	0	0	0	0	0	0	0
Overwinter stubbles	3	7	27	0	0	0	0	0	1	1	2
Mosaic management	0	0	1	0	0	0	0	0	0	0	0
Wide spaced rows in crops	2	2	3	0	0	0	0	0	0	0	1
Skylark plots	3	3	5	0	0	0	0	0	0	0	2
Cereal based whole based silage	1	0	0	0	0	0	0	0	0	0	1
Conservation headland (Arable)	15	12	5	1	0	0	3	8	6	0	3
Undersown cereals	0	1	0	0	0	0	0	0	0	0	0
Silage sward left overwinter	1	0	2	0	0	0	0	0	0	0	1
Meadow bird agreement	0	0	1	0	0	0	0	0	0	0	0
Upland heather	1	1	0	0	0	0	0	0	0	0	0
Green lanes	5	5	2	0	0	0	0	3	0	0	0

yield the most robust data for meta-analysis, although critical evaluation of the primary research papers would confirm this.

Invertebrates were the most commonly studied type of organism, with over 50% of studies manipulative. Pollinators (particularly Lepidoptera and Bombus spp.) and natural enemies (particularly arachnids and carabids) were the most frequently studied invertebrates, and abundance was the measure of outcome most frequently mentioned. Questions investigating which interventions affect the abundance of these invertebrates would be suitable for systematic review. Plants were also highly studied and tended to include data from manipulative experiments, although in some cases plants were used as a comparator for invertebrate abundance studies. There were large numbers of bird studies, but many were correlative and this would need to be taken into account in further analysis.

Table 9 Number of articles for each type of biotope in different farming systems

	Field margins	*Hedgerows*	*Semi-natural grassland*	*Meadows*	*Ditches*	*Stubbles*	*Conservation headlands (arable)*	*Set-aside*
Organic	**26** *(4 22 0)*	**9** *(2 6 1)*	**2** *(0 2 0)*	**2** *(1 1 0)*	**2** *(0 1 1)*	**3** *(1 2 0)*	**0**	**4** *(0 4 0)*
Conventional	**23** *(6 17 0)*	**10** *(2 7 1)*	**2** *(0 2 0)*	**2** *(2 0 0)*	**2** *(0 1 1)*	**2** *(0 2 0)*	**2** *(2 0 0)*	**4** *(1 3 0)*
Integrated	**6** *(3 3 0)*	**3** *(0 3 0)*	**0**	**1** *(0 1 0)*	**0**	**1** *(1 0 0)*	**1** *(1 0 0)*	**1** *(0 1 0)*
Organic vs conventional	**18** *(2 16 0)*	**7** *(2 4 1)*	**2** *(0 2 0)*	**1** *(1 0 0)*	**2** *(0 1 1)*	**2** *(0 2 0)*	**0**	**2** *(0 2 0)*
Organic vs conventional vs integrated	**1** *(0 1 0)*	**1** *(0 1 0)*	**0**	**0**	**0**	**0**	**0**	**0**
Integrated vs conventional	**2** *(2 0 0)*	**1** *(0 1 0)*	**0**	**0**	**0**	**0**	**1** *(1 0 0)*	**0**
Organic vs integrated	**2** *(0 2 0)*	**1** *(0 1 0)*	**0**	**1** *(0 1 0)*	**0**	**1** *(1 0 0)*	**0**	**0**

The number of manipulative, correlative and combined articles respectively is in brackets. Where a study compared farming systems for example 'organic versus conventional', the study was also listed as 'organic' and 'conventional'.

Field margins were the most commonly studied biotope, with 50% of the studies manipulative, and the majority of field margin experiments also investigating invertebrates.

Specific systematic review questions that are likely to yield data suitable for meta-analysis may include:

'What is the comparative effectiveness of field margin management interventions (X, Y and Z) for enhancing invertebrate abundance?'

Or more specifically

'What is the comparative effectiveness of field margin flower mixtures for enhancing pollinator abundance?'

Other areas that, although less highly populated with papers, may still enable systematic review analysis could include investigations into the impacts of ploughing and other tillage on invertebrates, and the impacts of grazing intensity on invertebrates. Comparisons of farming system were less frequent and appeared to be disparate – possibly making these questions less suitable for further analysis.

Limitations of the systematic map

The high volume of this literature found in this exercise demonstrates the size of this topic. Time, manpower and financial constraints precluded study of the full texts. The broad topic also complicated compilation of the database.

Limitations during searching

- Invertebrates and farmland birds were specified in the search terms as indicators of biodiversity at the request of the stakeholders. Amphibians, reptiles, mammals, and plants were not, but were still included in the systematic map database where found.
- The databases used in the searching phase, document publications starting after a particular date/year, so the results will not include a comprehensive assessment of earlier articles.
- The search was limited to literature published in English, and some of the databases were compiled for UK usage. This will have resulted in an international bias.

Limitations during the exclusion phase

- Exclusion of references was based upon the abstract only, due to the volume of literature captured, and

Table 10 Number of articles for each type of organism on field margins in different farming systems

	Invertebrates	*Plants*	*Farmland birds*	*Mammals*	*Plants of cons concern*	*Pollinators*	*Natural enemies*	*Soil organisms*	*Invertebrate food for farmland birds*
Organic	19	8	2	2	1	9	6	0	2
Conventional	15	5	1	3	1	7	5	1	3
Integrated	5	0	1	2	0	1	3	1	1
Organic vs conventional	12	5	1	1	1	6	2	0	2
Organic vs conventional vs integrated	0	0	0	1	0	0	0	0	0
Integrated vs conventional	2	0	0	0	0	1	2	0	1
Organic vs integrated	2	0	0	0	0	0	1	0	0

Where a study compared farming systems for example 'organic versus conventional', the article was also listed as 'organic' and 'conventional' separately.

Table 11 Number of articles for each type of organism for hedgerows in different farming systems

	Invertebrates	*Plants*	*Farmland birds*	*Mammals*	*Pollinators*	*Natural enemies*	*Soil organisms*	*Invertebrate food for farmland birds*
Organic	3	5	1	1	0	1`	1	0
Conventional	2	6	2	1	0	0	1	0
Integrated	2	0	1	0	0	1	1	0
Organic vs conventional	1	5	1	1	0	0	0	0
Organic vs conventional vs integrated	1	0	0	0	0	0	1	0
Integrated vs conventional	0	0	1	0	0	0	0	0
Organic vs integrated	1	0	0	0	0	1	0	0

Where a study compared farming systems for example 'organic versus conventional', the article was also listed as 'organic' and 'conventional'.

so depended upon the detail provided. Some important information will be missing, and other information may be misleading.

Limitations in interpretation

- Counting studies may imply either that there is an 'evidence gap', or that there is a large amount of research for a subtopic, but does not consider the quality and robustness of studies and so could be misleading (i.e. a small quality of high quality studies will provide more robust evidence than a large number of low quality studies.)
- As no quality appraisal of research has taken place in this systematic map, we have deliberately not indicated whether individual studies reported a positive or negative impact of interventions on the studied outcome (i.e. biodiversity). Users of the map are advised to read the papers relevant to the intervention or organism of interest in order to assess their quality and robustness.
- Ponds were excluded (the one reference to ponds was included because riparian buffer strips were also studied). This will have affected the number of wetland taxa studies included.
- There may appear to be research gaps for some topic areas due to incomplete reporting within the abstracts of included studies.
- The summary statistics report the number of articles rather than the number of studies for each topic area. This may bias the number of results found in favour of those with multiple outputs of one study.
- Coding of each study was carried out independently for each factor. For some factors, false relationships will be indicated. For example, where a study investigated more than one biotope, specific management options will show for all biotopes investigated (*e.g.* cutting of field margins may be specific to them, but if other biotopes such as hedgerows were included in the same study, the independent coding would imply that cutting was also studied for hedgerows.)
- Keywording should accurately reflect the range of topics in the field. However, names for interventions, such as agri-environment scheme options, have changed historically, and vary between regions and schemes. This should be taken into consideration when searching the database.

Conclusions

This systematic map provides a large database of research on the primary topic: 'The effectiveness of Integrated Farm Management, Organic farming and Agri-environment Schemes as interventions for conserving biodiversity in temperate Europe, with an emphasis on birds and invertebrates', subject to the limitations described. The map is up-to-date to June 2010 (with the exception of the Conservation Evidence search carried out in September 2010).

More studies were included in the final map than would have been in a systematic review, but full data extraction and critical evaluation of the included papers was not carried out. Papers are listed as either correlative or manipulative, and where available, basic experimental design is also shown in the database. Beyond this there is no indication of the quality of the papers.

Implications for policy and management

The map provides a research tool for users interested in this topic area. It can be used to gather information on a wide array of sub-topics that can be cross-tabulated. As the map does not include quality appraisal or data synthesis of studies, users should retrieve the full text copies of studies of interest in order to assess them fully.

Implications for future farmland biodiversity research

The systematic map highlights a variety of areas for which there appear to be evidence gaps. Some potential questions to be investigated by further primary research include:

'How effective are agri-environment options in providing for amphibians and reptiles?'

'How effective are agri-environment schemes under differing farming systems?'

It would also be useful for more research to investigate the impacts of farming interventions on biodiversity in Eastern European countries, as research appeared to be generally lacking for these regions.

Implications for future systematic maps

This systematic map was one of the first of its kind in environmental management and a number of implications for future systematic maps in the field have arisen. Systematic maps have the ability to provide a useful tool to investigate wider topic areas than systematic review, and potential to include non-experimental evidence. For example, in this case we have included correlative studies that may not be suitable for systematic review analysis. It may also be feasible to include in systematic maps, research that, for example, investigates the costs of interventions, where this may be of use to decision makers. In order to make systematic maps more robust for the future, it may be usful to include ways in which to critically evaluate studies. As this study has illustrated, questions that are too broad or disparate can cause issues of repeatability with systematic maps as well as with systematic reviews.

As systematic mapping for environmental evidence is new, different approaches are likely to be used by different authors. A Systematic Mapping Methods group hosted by the Collaboration for Environmental Evidence has been established to evaluate their use in the evidence-based framework for ecology and conservation, and will attempt to deal with some of the issues that have arisen.

Additional files

Additional file 1: Operational definitions and clarification of terms.

Additional file 2: Coding categories employed in the systematic map database.

Additional file 3: Database of farmland biodiversity research searchable by subtopic.

Additional file 4: How to search the database.

Additional file 5: Record of searches.

Additional file 6: References included in the systematic map database.

Additional file 7: Systematic reviews and meta-analyises of relevance.

Competing interests

The authors declare that they have no competing interests.

Authors' contributions

NPR designed the study, and carried out the 2008 searches and sorting. KLJ carried out the 2010 searches and sorting. Both authors worked on the systematic map design and coding and on the manuscript. Both authors read and approved the final manuscript.

Acknowledgements

KLJ was funded by the John Oldacre Foundation. Stakeholder meetings, training and some consumable costs were funded by a British Ecological Society-Ecology into Policy Grant.

References

1. Robinson RA, Sutherland WJ: **Post-war changes in arable farming and biodiversity in Great Britain.** *Journal of Applied Ecology* 2002, **39**:157–176.
2. The reform of the CAP towards 2020: *Consultation Document for Impact Assessment.* Brussells: European Commission; 2010.
3. Donald PF, Green RF, Heath MF: **Agricultural intensification and the collapse of Europe's farmland bird populations.** *Proceedings Royal Society London Ser. B* 2001, **268**:25–29.
4. Fuller RJ, Gregory RD, Gibbons DW, Marchant JH, Wilson JD, Baillie SR, Carter N: **Population declines and range contractions among lowland farmland birds in Britain.** *Conservation Biology* 1995, **9**:1425–1441.
5. Sotherton NW, Self MJ: **Changes in plant and arthropod diversity on lowland farmland: an overview.** In *The Ecology and Conservation of Lowland Farmland Birds.* Edited by Aebischer NJ, Evans AD, Grice PV, Vickery JA. Southampton: British Ornithologists' Union; 2000:26–35. 27-28 March 1999.
6. Flowerdew JR: **Mammal biodiversity in agricultural habitats.** In *Biodiversity and Conservation in Agriculture.* Edited by Kirkwood RC. UK: British Crop Protection Council; 1997:25–40.
7. Boatman ND, Parry HR, Bishop JD, Cuthbertson AGS: **Impacts of Agricultural Change on Farmland Biodiversity in the UK.** In *Issues in Environmental Science and Technology. No.25. Biodiversity under Threat.* Edited by Hester RE, Harrison RM. Cambridge UK: Royal Society of Chemistry; 2007:1–32.
8. Boatman ND: **How do farming practices influence biodiversity?** In *Farming, Forestry and the Natural Heritage: Towards a More Integrated Future.* Edited by Davidson R, Galbraith C. Edinburgh: Scottish Natural Heritage; 2006:39–57.
9. Benton TG, Vickery JA, Wilson JD: **Farmland biodiversity: is habitat heterogeneity the key?** *Trends in Ecology and Evolution* 2003, **18**:182–188.
10. Pretty J, Sutherland WJ, Ashby J, Auburn J, Baulcombe D, Bell M, Bentley J, Bickersteth S, Brown K, Burke J, Campbell H, Chen K, Crowley E, Crute I, Dobbelaere D, Edwards-Jones G, Funes-Monzote F, Godfray HCJ, Griffon M, Gypmantisiri P, Haddad L, Halavatau S, Herren H, Holderness M, Izac A-M, Jones M, Koohafkan P, Lal R, Lang T, McNeely J, *et al*: **The top 100 questions of importance to the future of global agriculture.** *International Journal of Agricultural Sustainability* 2010, **8**(4):219–236.
11. Commission of the European Communities: *Perspectives for the Common Agricultural Policy.* Brussels: COM, 85; 1985:333.
12. Kleijn D, Sutherland WJ: **How effective are European agri-environment schemes in conserving and promoting biodiversity?** *Journal of Applied Ecology* 2003, **40**:947–969.
13. European Court of Auditors: *Is Agri-environment support well designed and managed? Special report no.7//2011.* Luxembourg: European Court of Auditors; 2011.
14. Kleijn D, Baquero RA, Clough Y, Diaz M, Esteban J, Fernadez F, Gabriel D, Herzog F, Holzshuh A, Johl R, Knop E, Kruess A, Marshall EJP, Steffan-Dewenter I, Tscharntke T, Verhulst J, West TM, Yela JL: **Mixed biodiversity benefits of agri-environment schemes in five European countries.** *Ecology Letters* 2006, **9**(3):243–254.
15. Wittingham MJ: **Will agri-environment schemes deliver substantial biodiversity gain, and if not why not?** *Journal of Applied Ecology* 2007, **44**(1):1–5.
16. Marshall EPJ, West TM, Kleijn D: **Impacts of agri-environment field margin prescriptions on the flora and fauna of arable farmland in different landscapes.** *Agriculture, Ecosystems and Environment* 2005, **113**:36–44.
17. Berry P, Ogilvy S, Gardner S: *Integrated Farming and Biodiversity – English Nature Research Reports no. 634.* Adas: English Nature; 2005.

18. Wickramasinghe LP, Harris S, Jones G, Vaughan N: **Bat activity and species richness on organic and conventional farms: impact of agricultural intensification.** *Journal of Applied Ecology* 2003, **40**(6):984–993.
19. Chapple DG, Wade DR, Laverick RM, Eldridge PJ: **Whole farm integrated management and farmland birds.** *Aspect of Applied Biology* 2002, **67**:129–134.
20. Feber RE, Smith H, Macdonald DW: **The effects on butterfly abundance of the management of uncropped edges of arable fields.** *Journal of Applied Ecology* 1996, **33**(5):1191–1205.
21. Hole DG, Perkins AJ, Wilson JD, Alexander IH, Grice PV, Evan AD: **Does organic farming benefit biodiversity?** *Biological Conservation* 2005, **122**:113–130.
22. Bengtsson J, Ahnstrom J, Weibull AC: **The effects of organic agriculture on biodiversity and abundance: a meta-analysis.** *Journal of Applied Ecology* 2005, **42**(2):261–269.
23. Fuller RJ, Norton LR, Feber RE, Johnston PJ, Chamberlain DE, Joys AC, Mathews F, Stuart R, Townsend MC, Manley WJ, Wolfe MS, MacDonald DW, Firbank LG: **Benefits of organic farming to biodiversity vary among taxa.** *Biology Letters* 2005, **1**(4):431–434.
24. Gabriel D, Sait SM, Hodgson JA, Schmutz U, Kunin WE, Benton TG: **Scale matters: the impact of organic farming on biodiversity at different spatial scales.** *Ecology Letters* 2010, **13**(7):858–869.
25. Coren E, Fisher M: *The conduct of systematic research reviews for SCIE knowledge reviews.* UK: Social Care Institute for Excellence; 2006. www.scie.org.uk.
26. Grant MJ, Booth A: **A typology of reviews: an analysis of 14 review types and associated methodologies.** *Health Information and Libraries Journal* 2009, **26**:91–108.
27. Clapton J, Rutter D, Sharif N: *SCIE Systematic mapping guidance. April 2009.* UK: Social Care Institute for Excellence; 2009. www.scie.org.uk/publications/researchresources/rr03.pdf.
28. Bates S, Clapton J, Coren E: **Systematic maps to support the evidence base in social care.** *Evidence and Policy* 2007, **3**(4):539–551.
29. Sutherland WJ, Armstrong-Brown S, Armsworth PR, Brereton T, Brickland J, Campbell CD, Chamberlain DE, Cooke AI, Dulvy NK, Dusic NR, Fitton M, Freckleton RP, Godfrey CJ, Grout N, Harvey HJ, Heldley C, Hopkins JJ, Kift NB, Kirby J, Kunin WE, MacDonald DW, Marker B, Naura M, Neale AR, Oliver T, Osborn D, Pullin AS, Shardlow MEA, Showler DA, Smith PL, *et al*: **The identification of 100 ecological questions of high policy relevance in the UK.** *Journal of Applied Ecology* 2006, **43**:617–627.
30. Bradbury RB, Browne SJ, Stevens DK, Aebischer NJ: **Five-year evaluation of the impact of the Arable Stewardship Pilot Scheme on birds.** *Ibis* 2004, **146**(2):171–180.
31. Collaboration for Environmental Evidence: *Guidelines for Systematic Review in Environmental Management. Version 4.0. Environmental Evidence 2010.*: ; www.environmental evidence.org/Authors.htm/Guidelines.pdf.
32. Karla N, Newman M, Karla N, Newman M: *A systematic map of the research on the relationship between Obeisity and Sedentary Behaviour in young people. Technical report. In Research Evidence in Education Library.* London: EPPI-Centre, Social Science Research Unit, Institute of Education, University of London; 2009.
33. Sharif N, Brown W, Rutter D: *Systematic Map Report 03. The extent and impact of depression on BME older people and the acceptability, accessibility and effectiveness of social care provision.* University of London: Social Care Institute for Excellence. Social Science Research Unit, Institute of Education; 2008.
34. Feber RE, Johnson PJ, Firbank LG, Hopkins A, Macdonald DW: **A comparison of butterfly populations on organically and conventionally managed farmland.** *Journal of Zoology* 2007, **273**(1):30–39.
35. Irmler U: **Changes in earthworm populations during conversion from conventional to organic farming.** *Agriculture, Ecosystems & Environment* 2010, **135**(3):194–198.
36. Feber RE, Hopkins A: *Diversity of plant and butterfly species on organic farmland field margins in relation to management. In British Grassland Society Fifth Research Conference: 8-10 September 1997.* UK: University of Plymouth; 1997:63–64.
37. Pfiffner L, Niggli U: **Effects of bio-dynamic, organic and conventional farming on ground beetles (Col Carabidae) and other epigaeic arthropods in winter wheat.** *Biological Agriculture & Horticulture* 1996, **12**(4):353–364.
38. Hutcheon JA, Iles DR, Kendall DA: **Earthworm populations in conventional and integrated farming systems in the LIFE Project (SW England) in 1990-2000.** *Annals of Applied Biology* 2001, **139**(3):361–372.
39. Pywell RF, Meek WM, Carvell C, Hulmes L: **The SAFFIE project: enhancing the value of arable field margins for pollinating insects.** *Association of Applied Biologists* 2007, **81**:239–245.
40. Alvarez T, Frampton GK, Goulson D: **Epigeic Collembola in winter wheat under organic, integrated and conventional farm management regimes.** *Agriculture, Ecosystems & Environment* 2001, **83**(1/2):95–110.
41. Department for Food and Rural Affairs: *UK Biodiversity Indicators in your Pocket 2010. Measuring progress towards halting biodiversity loss PB13400.* UK: Defra; 2010.
42. ADAS: *Hedgerow Management and Wildlife Review: further update, edit and publication. Final Report for UK Department of Food and Rural Affairs (BD2108).* UK: Defra; 2004.

2

Assessing community-based conservation projects: A systematic review and multilevel analysis of attitudinal, behavioral, ecological, and economic outcomes

Jeremy Brooks[1], Kerry Ann Waylen[2] and Monique Borgerhoff Mulder[3*]

Abstract

Background: Community-based conservation (CBC) promotes the idea that long-term conservation success requires engaging with, and providing benefits for local communities. Though widespread, CBC projects are not always successful or free of controversy. With criticisms on all sides of the conservation debates, it is critical to have a better understanding of (1) whether CBC is an effective conservation tool, and (2) of the factors associated with the success or failure of CBC projects, and the scale at which these factors operate. Recent CBC reviews have typically examined only a single resource domain, have limited geographic scope, consider only one outcome, or ignore the nested nature of socioecological systems. To remedy these issues, we use a newly coded global comparative database of CBC projects identified by systematic review to evaluate success in four outcome domains (attitudes, behaviors, ecological, economic) and explore synergies and tradeoffs among these outcomes. We test hypotheses about how features of the national context (H-NC), project design (H-PD), and local community characteristics (H-CC) affect these four measures of success.

Methods: To add to a sample of 62 projects that we used from previous systematic reviews, we systematically searched the conservation literature using six terms in four online databases. To increase the number of projects for each country in order to conduct a multilevel analysis, we also conducted a secondary search using the Advancing Conservation in a Social Context online library. We coded projects for 65 pieces of information. We conducted bivariate analyses using two-dimensional contingency tables and proportional odds logistic regression and conducted multivariate analyses by fitting reduced form proportional odds logistic regression models that were selected using a forward stepwise AIC approach.

Results: The primary and secondary searches produced 74 new projects to go along with the 62 projects from previous reviews for a total of 136 projects. The analyses suggest that project design, particularly capacity building in local communities, is critical in generating success across all outcomes. In addition, some community characteristics, such as tenure regimes and supportive cultural beliefs and institutions, are important for some aspects of project success. Surprisingly, there is less evidence that national context systematically influences project outcomes.

(Continued on next page)

* Correspondence: mborgerhoffmulder@ucdavis.edu
[3]Anthropology Department, University of California, Davis, 210 Young Hall, One Shields Ave., Davis, CA 95616, USA
Full list of author information is available at the end of the article

(Continued from previous page)

Conclusions: Our study supports the idea that conservation projects should be carefully designed to be effective and that some characteristics of local communities can facilitate success. That well-designed projects can prevail over disadvantages relating to the pre-existing national and local context is encouraging. As the evidence base on CBC grows, it will be useful to repeat this analysis with additional search terms, and consider additional variables related to national context to further evaluate the role of broader socio-political and economic contexts.

Keywords: Community-based conservation, Conservation and development, Conservation evaluation, Conservation interventions, Socio-ecological systems, Community institutions, Evidence based conservation, Multi-level analysis

Background

Conservation practitioners continue to seek viable alternatives to strict protectionism, and it is increasingly argued that projects must achieve not only ecological but also economic, and social goals [1]. Since the 1980s, conservation efforts in developing countries have generally tried to incorporate the interests and views of local people, an approach often referred to as community-based conservation (CBC) [2]. A variety of approaches fall under the umbrella of CBC; each diverse in their details [3]. What unites these approaches, however, is that they typically aim to combine elements that link conservation with development, engage local communities as active stakeholders, and/or devolve control over natural resources. The shared rationale is that promoting socio-economic benefits, either directly or by compensating the costs associated with conservation, is important in both its own right, and as a key strategy for slowing deforestation and protecting endangered habitats and species.

Contemporary CBC nevertheless faces criticism on several fronts. Communities are often idealized as harmonious units [4], decentralization initiatives stall in their implementation because centralized governments are unwilling to cede power [5,6], and market-based approaches to community-based natural resource management [7] are challenged for assuming that resource commercialization is compatible with conservation goals [8]. Some conservationists anticipate sharp tradeoffs between conservation and economic development and fear that delegitimizing conservation as a priority will further water down already limited funds [9].

With such controversies unresolved, and a strong need to make more effective use of the billions of dollars devoted to CBC efforts globally [10], we need a better understanding of the factors associated with the success and failure of conservation projects, and the scale at which these factors operate. There are many arguments in the literature about how to improve the practice of conservation [11]. Here, we directly test some of these arguments using a large comparative database of CBC projects identified with a systematic literature review. We use a multi-level design and model-fitting approach to evaluate CBC success in four outcomes (attitudinal, behavioral, ecological, economic) by testing a series of hypotheses about how features of the national context (H-NC), project design (H-PD), and community-level characteristics (H-CC) affect measures of success. We also explore the evidence that synergies between pairs of outcomes may be more common than anticipated.

Insights from previous reviews

Despite the prominence of CBC strategies, and strong arguments for and against their effectiveness [12], there have been few *quantitative, comparative* evaluations of CBC successes and failures. Two previous systematic reviews have studied the determinants of conservation success, focusing on the use of development as a conservation tool [13] and the effect of local cultural context and project engagement with local culture [14]. These reviews provide valuable insights into determinants of conservation success and the use of systematic reviews to understand CBC project outcomes. Although these studies focused on various aspects of project design and local context, they did not consider the influence of the broader socio-political institutional framework, such as quality of governance, economic inequality, or development progress and they had somewhat limited sample sizes.

Several other qualitative [15-17] and quantitative [18-24] studies have been conducted on CBC or closely related topics. These studies suggest that a number of factors can be associated with project success and failure, including leadership, strong local institutions, local participation, capacity building, secure rights to land and resources, and provision and equitable distribution of economic benefits. Nevertheless most of these studies (a) examine only a single resource domain (e.g. forestry), (b) have limited geographic scope, (c) consider only one outcome (e.g. ecological success *or* economic success), and/or (d) ignore the nested nature of socio-ecological systems [25]. Nested analyses are particularly important because evidence from recent studies [23,26,27] suggests that national governance institutions, corruption, and standards of living can influence project outcomes. In fact, Tallis et al. [22] explicitly call for such a review to examine the trend that projects are successful in countries with effective government [28]. Therefore, this study was motivated by a desire to assess the effect of non-local factors on conservation projects in addition to the goal of increasing the sample size from previous reviews for a more robust statistical analysis.

Our work is influenced by the systematic comparative work undertaken by the Workshop in Political Theory and Policy Analysis at Indiana University led by Elinor Ostrom as well as by efforts to systematically examine the effectiveness of different environmental management and policy initiatives through the Centre for Evidence Based Conservation [29]. Ostrom and her colleagues argue for the need to collect standardized data with which to examine the success of common property institutions (CPIs) and linked socioecological systems (SESs) [25,30]. This work has produced some remarkable comparative articles [31-33], and influenced much other work, including some of the hypotheses introduced in Brooks et al. [13].

Of particular relevance to this study is Ostrom's introduction of the framework of decomposable systems and nested structures as a tool for analyzing CPIs and SESs [25,34]. Our goal here is to employ this framework to understand the interaction between multiple factors and across multiple scales for existing programs devoted to linking conservation and development. Useful initial advances in this area (see [21,35,36]) suggest the utility of this approach, highlighting various sets of biophysical, social, economic and institutional considerations to be important in different contexts. We believe that it is now important to analyze more precisely the context itself, in particular the external socio-political context – such as measures of development progress, income inequality, and the effectiveness of state-level institutions - within which a natural resource management scheme exists.

The 136 CBC projects in our sample focus on conservation challenges in managing forests, grasslands, wildlife, and fisheries in 40 countries, and are nested within national socio-economic and political contexts. As in the previous systematic reviews [13,14], we used four measures of success in the analyses (attitudinal, behavioral, ecological, economic). We use these four outcomes measures because many of the debates in conservation are a result of an interest in different outcomes [37-39] reflecting the multiple goals inherent to CBC: for more explanation of these domains and the relationships between them see the description of outcome variables below and [18].

We also want to note that this study contains several features that may not be typical for systematic reviews including the objective of testing multiple hypotheses, the use of secondary, national-level data, and a targeted search in a digital library. We describe each of these features below and justify their inclusion in the study.

Objectives and hypotheses

Objectives

We have four objectives for this review. The primary objective is to examine the overall effectiveness of community-based conservation. The second and third objectives build on the primary objective by exploring the factors that might affect CBC success. For these two objectives, we test a number of hypotheses about how aspects of national context, project design, and community characteristics relate to multiple project outcomes to understand *when and in what contexts* CBC is likely to succeed. As such, we also present nine hypotheses for objectives two and three in the next section. Because of the scope of this study, the number of hypotheses tested, and corresponding large number of variables used in the analysis, this section of the study should be viewed as a test of broad hypotheses that can provide insights to generate additional research questions and hypotheses and point to avenues for future research. Our final objective is to determine how well our coding matches the intent of the authors of the articles from which we extracted our data.

Objective 1. Is community-based conservation an effective conservation tool?
The overarching question for this review is whether and to what degree CBC is effective as a strategy for solving diverse and multi-dimensional conservation problems. With this objective, we are building on previous systematic reviews [13,14] by using a larger sample size.

Objective 2. Which aspects, if any, of national socio-economic and political contexts affect the outcomes of local-level CBC projects?
Using studies that provide information on at least two of the four outcome variables, we will assess the role of national socio-political context on the outcomes of community-based conservation projects. We are explicitly operationalizing the insights of Ostrom [25], and others [40], by examining how the success of different natural resource management strategies is both facilitated and constrained by aspects of the national socio-economic and political

contexts in which these strategies are implemented.

Objective 3. Which aspects, if any, of project design and community characteristics affect the outcome of CBC projects?
Using studies that provide information on at least two of the four outcome variables, we will expand the dataset built through the previous reviews [13,14] to note the growth in the evidence base and determine whether the results of these reviews (pertaining to local cultural context, market incursion, local participation, and access to resources) are refuted or more strongly supported. In doing so, we will examine which aspects of project design and community characteristics are associated with successful outcomes

Objective 4: Coding validity
The process of conducting an analytical systematic review by coding for variables in the literature is assumed to be valid. It is possible, however, that the way researchers code information about projects differs from the way the authors of project reports and articles intend their information to be interpreted. In this study, we asked the corresponding author of each article to respond to a questionnaire composed of subset of our coding sheet. Although we distributed the questionnaire primarily for the purpose of filling in missing values in the dataset, we used the opportunity to examine how well our coding matched the authors' responses to the questionnaire as a way of testing the validity of our coding. This test is an initial step at examining the ability of coders to accurately extract information from project reports and articles for the purposes of a systematic review.

Hypotheses for objectives 2 and 3

Conservation typically requires restraint in current resource use to obtain long-term benefits from the resource base. Therefore, conservation behavior requires individuals to absorb short-term costs and potentially share benefits with a larger number of people. As such, conservation creates a collective action problem [41] whereby individuals are reluctant to cooperate by forgoing resource use without assurances that others will do the same. Deductive theory [17,42] and inductive observations [43], provide insights into the conditions that both favor and disfavor the adoption of conservation behavior. From these insights, we predict that projects that contain the sets of factors described below will be more likely to succeed in all outcomes. The hypotheses that we are testing are derived from the theoretical perspectives on conservation behavior and common pool resource management noted above as well as from a careful reading of the CBC literature and the outcomes of previous CBC reviews and summaries [14,15,18,19,22,44]. The variables listed in brackets after each hypothesis (as italicized text within parenthases) are described more fully in Table 1 and summary statistics are provided in Table 2. Note that conceptually similar variables were combined for the multivariate analysis. For more discussion of why variables were combined prior to the multivariate analysis see section 3.9.2 - Multivariate analysis. In the text below, where a combined variable was used for the multivariate analysis, its name is presented within square brackets preceding the relevant separate variables.

National context (H-NC)

As noted above, conservation is akin to other forms of cooperation and faces similar obstacles. As such, conservation attitudes and behaviors require conditions of trust [27,45]. The degree of stability, transparency, and accountability of national-level governance institutions might influence project outcomes by affecting confidence in local-level institutions and governance [26]. Despite the push for decentralization, well-functioning central governments may also be necessary to counteract the patronage that may exist, or can potentially arise, in rural communities [46].

Hypothesis 1: *NC1 - National political context*: Success is more likely when projects are implemented in countries where there is greater transparency and stability in governance and where the populace has a voice in politics and enjoys common civil liberties. [NC-Governance, NC-Rights].

Conservation efforts often restrict access to, or use of, natural resources. In societies with low or highly unequal standards of living these restrictions may be unpopular and full conformity may not be possible. In addition, low standards of living and low access to education and healthcare may also limit conservation efforts because individuals in such conditions are more likely to value immediate benefits from resources more than anticipated future benefits [47].

Hypothesis 2: *NC2 - National socioeconomic context:* Success is more likely when projects are implemented in countries with a higher

Table 1 Description, measurement and coding of predictors and outcomes

National context	
H-NC1: National Political Context	
NC-Governance[i,ii]	World Bank scores for: voice in politics, government stability, government effectiveness, regulatory quality, rule of law, and corruption.
NC-Rights[i,ii]	Values for political rights and civil liberties from the FreedomHouse database.
H-NC2: National Socio-economic context	
NC-HDI[i]	Human Development Index score from the UNDP trends dataset
NC-Gini	Gini inequality coefficient. One value for each country from World Resources Institute country ranking of values from 2000-2007
Project Design	
H-PD1: Participation/ Engagement	
PD-Participation	Combines *PD-Impetus* (whether impetus for project came from the community), *PD-Establishment* (level of community involvement in project establishment, and *PD-Decision-making* (level of community involvement in daily project decision-making. All coded as: no community=1, some community =2, joint or complete community involvement=3. *Participation* is sum of the three, three categories: low (3,4), moderate (5,6), high (7–9).
PD-Engagement	Combines *PD-Approach local culture* (whether the project engaged with local cultural traditions and beliefs), and *PD-Approach local institutions* (whether the project engaged with local institutions and/or leaders). Variables coded: conflicted=1, mixed=2, engaged=3. *Engagement* is sum of both measures, three categories: low (2,3), moderate (4), high (5,6)
H-PD2: Access to and utilization of resources	
PD-Protectionism[iii]	IUCN ranking for protected area associated with the project two categories: Strict Nature Reserve/National Park, Other (national monument, habitat/species management area, protected landscape, managed resource area, no protected area)
PD-Resource use	Constraints on resource use, three categories: protected, regulated, unregulated
H-PD3: Project benefits	
PD-Economic benefits	Economic/development benefits provided by project and type of resource use[iv]. Four categories: ecotourism (indirect use of targeted species/habitat), CBC (community efforts to minimize resource use), compensation/ substitution (prohibition or minimized use of targeted resource but other benefits provided), enhancement (increasing marketable use of the targeted resource).
PD-Equity	Combines *PD-Equitable distribution* (are benefits produced by the project equitably distributed), and *PD-Elite capture* (are benefits produced by the project captured by local elites). Coded as yes, no. *Equity* coded as good (no elite capture and equitable distribution), poor (elite capture, inequitable distribution, or both)
H-PD4: Human/social capital	
PD-Capacity	Combines *PD-Capacity skills* (did project build skills to aid development or conservation efforts), and *PD-Capacity institutions* (did project build or reinforce local institutions). Coded as yes, no. *Capacity* is: no (neither), yes (one or both).
PD-Social capital	Whether the project produced non-economic benefits for the community or eroded community interactions or cultural beliefs, two categories: yes (benefits provided), no (no benefits or social capital eroded)
PD-Envt. education	Whether environmental education was a component of the project: yes, no
Community characteristics	
H-CC1: Market integration	
CC-Market access	The degree communities are market integrated including wage labor, selling and purchasing goods, and remoteness, three categories: low, moderate, high
CC-Threat	Threats to local natural resources and/or the protected area (e.g. logging, hunting, land clearance, commercial development). Up to three noted for each project and coded by motivation (subsistence = 1, mixed = 2, commercial = 3). Sum of the three threats divided into three categories: low (1,2), moderate (3–5), high (6–9)
H-CC2: Supportive local context	
CC-Local Institutions	Combines *CC-Supportive local culture* (do local traditions and beliefs support conservation: unsupportive, mixed, supportive) and *CC-Effective local government* (quality of pre-existing local governance institutions: ineffective, mixed, effective). *Local institutions* is: low (both ineffect./ unsupp. or one ineffet./unsupp. and one mixed), moderate (one is ineffect./unsupp. and one is effect./ sup.), high (both effect./supp. or one effect./supp. and one mixed)
CC-Tenure	Control or ownership over primary resources targeted by project, three categories: no community, mixed, total community (entity within community including private or communal ownership).
CC-Charisma	Presence of charismatic individual/group of individuals facilitating project? yes, no
H-CC3: Local population	
CC-Population size	Size of human population, three categories: low (< 5000), moderate (5001 – 50000), high (>50001)
CC-Population heterogeneity	Is targeted community ethnically or culturally diverse, two categories: low (one ethnic/cultural group present, or < 33% and > 67% of one ethnic group), high (multiple ethnic groups, >33% and <67% of community of one ethnic group and/or the author notes disharmony in community based on caste, class or ethnic divisions)[v]

Table 1 Description, measurement and coding of predictors and outcomes *(Continued)*

Controls	
CTR-Ecoregion status	Status of ecoregion(s) in project area [107]. When multiple exist only lowest status value is coded, three categories: critically endangered, vulnerable, relatively stable
CTR- Author discipline	Affiliation of first author, four categories: biological sciences, social sciences, interdisciplinary science or department, employed by an NGO
CTR- Years project running	Number of years the project has been running. Year of project initiation was subtracted from the year research for the project was conducted[vi].
Monitoring and Outcomes	
Monitoring	Type of monitoring and measurement for each outcome variable, three categories: quantitative, qualitative, author's judgment (when author suggests outcomes without published data to support the claim).
Attitudinal	Project outcomes with regard to local attitudes towards the project or conservation
Behavioral	Project outcomes with regard to local resource use
Ecological	Project outcomes with regard to condition of the habitat and/or key species
Economic	Project outcomes with regard to economic or other development benefits All outcomes coded as: success (most indicators show improvement), limited success (some indicators show improvement), failure (majority of indicators show no change or decline).
Additional Variables[vii]	
Rainfall	Average annual rainfall for the community or communities involved in the project (mm)
Elevation	Average elevation for the community of communities involved in the project (m)
Habitat types	Habitat types in, or adjacent to, the community or communities involved in the project as listed in the article
Subsistence	Subsistence type(s) exhibited by community members, including hunter-gatherers, fishers, pastoralists, horticulturalists, agrarians, mixed)
Resource importance	Importance to the local people of the resource(s) that was targeted by the project, five categories: fundmental direct (positive benefits), fundamental negative (pest species), non-essential (incidental), value added income, value added other (cultural value
National policies	Presence or absence of national policy transferring (or reaffirming) land and/or resource rights to local communities: yes, no
Policy implementation	Whether supportive land/resource rights policies were actually implemented: yes, no
Government involvement	Whether the national government is involved in project decision-making: yes, no
Government support	Whether the national government financially supports the project: yes/no
External involvement	Whether an NGO is involved in project decision-making or supports project decision-making, three categories: involved formally, not involved but supportive, not involved/conflicting
Economic Heterogeneity	Is the targeted community economically heterogeneous (income, wealth, land ownership, or livestock ownership, three categories (low, moderate, high)

Predictor variables grouped by hypothesis and general category.
i. Scores taken from the year closest to the date of research.
ii. Created a single value using the first factor score from a principal components analysis of the scores.
iii. See Dudley [121] for a description of IUCN protected area rankings and criteria.
iv. adapted from Abbott et al. [59].
v. modified from Oldekop et al. [95].
vi. When studies did not report the year that their research was initiated, this value was approximated. We calculated the mean value for the number of years between the initiation of research and the publication year for all studies in the sample and then subtracted this value from the year that study was published.
vii these variables were not included in the analysis because there was a lack of information provided by the articles in our sample.
The right-most column describes those variables that are conceptually similar and were combined for the multivariate analysis to reduce the number of predictors.

level of development progress as measured by the Human Development Index and in which there is greater economic equality as measured with Gini inequality coefficients. [NC-HDI, NC-Gini]

Project design (H-PD)

The design of projects can affect incentives to participate in conservation as well as the payoffs associated with foregoing resource use by determining who drafts project rules, what and how many benefits are provided and to whom, and the degree to which access to and use of resources is permitted. We broadly categorize these characteristics along the four dimensions of decentralization, utilization, effective benefit provision, and investment in human/social capital. We derive hypotheses for each of these dimensions below.

A large body of literature suggests that devolving decision-making and control to local communities can be beneficial from a conservation perspective [48]. Local bodies are thought to be more responsive to local conditions, have more detailed knowledge of resource dynamics, and have more incentive to harvest resources sustainably because they tend to feel more secure in their access to future benefits and thus discount future returns less than outsiders [48]. Similarly, by engaging with local cultural traditions and leaders projects may encourage greater local participation and insight and decrease the likelihood of failure due to cultural insensitivity [49].

Table 2 Summary statistics for all independent and dependent variables for raw data before imputing missing values

National context					
	Min	Max	Mean (SD)		NA
Governance	−2.32	2.93	0.001 (1.01)	–	9
Rights	−1.01	2.68	0.01 (1.01)	–	0
HDI	0.26	0.87	0.61 (0.14)	–	7
Gini	29.80	70.70	45.93 (8.73)	–	9
Project Design (number of responses for each coding category)					
Impetus	No community 98	–	Joint/Complete community 29	–	NA 9
Establishment	No community 51	Some community 43	Joint/Complete community 32	–	10
Decision-making	No community 22	Some community 45	Joint/Complete community 65	–	4
Approach local culture	Conflict 16	Mixed 21	Engaged 48	–	51
Approach local institutions	Conflict 11	Mixed 15	Engaged 56	–	54
Protectionism	SNR/NP 43	Other/no park 57	–	–	36
Resource use	Protected 42	Regulated 75	Unregulated 6	–	13
Intervention benefits	Ecotourism 37	CBC 21	Compensation/ substitution 53	Enhancement 24	1
Elite capture	Yes 31	No 104	–	–	1
Benefit distribution	Yes 49	No 87	–	–	0
Social capital	No 92	Yes 44	–	–	0
Capacity skill	No 65	Yes 66	–	–	5
Capacity institutions	No 92	Yes 44	–	–	0
Environmental education	No 80	Yes 54	–	–	2
Community characteristics (number of responses for each coding category)					
Market integration	Low 24	Moderate 49	High 28	–	35
Threat	Low 13	Moderate 30	High 45	–	48
Supportive local culture	Unsupportive 19	Mixed 32	Supportive 29	–	56
Effective local government	Ineffective 18	Mixed 26	Effective 39	–	53
Tenure	No community 30	Mixed 49	Total community 41	–	16
Charisma	No 116	Yes 20	–	–	0
Ethnic heterogeneity	Low 46	High 43	–	–	47
Population	High 22	Mod 33	Low 47	–	4
Controls					
Ecoregion status	Critically endangered 60	Vulnerable 39	Relatively stable 14	–	23
Author discipline	Biological Sci. 17	Interdisciplinary 17	NGO 38	Social Sci. 35	29
Years project running	Min 0	Max 28	Mean (st. dev.) 7.68 (5.10)	–	10
Outcomes (number of responses for each coding category)					
	Success	Limited Success	Failure		
Attitudinal	39	33	23	–	41
Behavioral	44	26	31	–	36
Ecological	45	13	19	–	59
Economic	29	36	53	–	18

Hypothesis 3: *PD1 - Participation/Engagement:* Success is more likely when there is more emphasis on participation in initiation, establishment, and day-to-day running of the project, and when the project engages positively with local governance institutions and with local cultural beliefs, practices and traditions. [PD-Participation *(combined PD-Impetus, PD-Establishment, and PD-Decision-making),* PD-

Engagement *(combined PD-Approach local culture and PD-Approach local institutions)*]

The evidence and arguments that resource utilization can be an effective conservation tool are less straightforward. While protectionism can sometimes result in ecological success [50], access to resources may produce economic and other benefits, which provides communities with an incentive to extract resources sustainably [51]. In addition, restricting or prohibiting access to resources, or providing insufficient compensation for the costs associated with the loss of access, may engender resentment [52] and subsequent biodiversity loss [53]. Our hypothesis is based on the latter logic.

Hypothesis 4: *PD2 - Access to and utilization of resources:* Success is more likely when projects emphasize greater levels of utilization and subsequently generate support for, and compliance with, conservation initiatives. [PD-Protectionism, PD-Resource use]

Linking resource use directly to performance payments [54], or providing alternative livelihoods [47], are examples of economic benefits that can lead to reduced harvests. The success of CBC projects may, therefore, depend on benefits provided through income from sales and wage labor opportunities, development infrastructure, or direct compensation. Because elite capture of benefits has been found to be a core obstacle to effective decentralization of resource management [55-57] economic benefits must also be equitably distributed among community members, with no capture of benefits by community elites.

Hypothesis 5: *PD3 - Project benefits:* Success is more likely when projects provide clear and well-directed economic benefits, and ensure that such benefits are shared equitably and without elite capture. [PD-Economic benefits, PD-Equity *(combines PD-Equitable distribution and PD-Elite capture)*]

Investing in human and social capital (including environmental education) in local communities may facilitate positive outcomes by lowering the costs associated with developing and enforcing local rules about resources use - referred to as transaction costs - and strengthening the ability of community members to coordinate [58]. Communities with greater capacity, knowledge, and social cohesion, are expected to have an easier time cooperating to manage resources because greater trust has developed [43,59]. For example, one study of enterprise-based projects found that training locals as managers and using community policing was a better predictor of project success than economic returns [60].

Hypothesis 6: *PD4 - Human/Social capital:* Success is more likely when projects invest in human and social capital. [PD-Capacity *(combines PD-Capacity skills and PD-Capacity institutions)*, PD-Social capital, PD-Environmental education]

Community characteristics (H-CC)

Finally, characteristics of local communities might affect the likelihood of achieving collective action for conservation. Integration with local and global markets, threats from outside markets, community institutions, and population size and heterogeneity may be particularly important [43,61].

Logic from neoliberal economics suggests that market integration enables rural communities to benefit from sustainably utilizing, protecting, and conserving their resources [7]. Market integration can provide substitutes for locally harvested resources [47], add value to local products [62], and/or provide external wage labor opportunities that result in decreased pressure on local resources for subsistence needs [47]. On the other hand, market integration can potentially increase pressure on vulnerable resources and habitats through opportunities for market sales and rising prices, which can create incentives for higher rates of extraction [63]. An additional problem is that new roads that often accompany or facilitate market integration can attract migrants and put even more pressure on local resources [64]. Our hypothesis is based on the view that success is most likely in communities that are market integrated.

Hypothesis 7: *CC1 - Market integration:* Success is more likely when projects are in communities that are more integrated into local and global markets and their species/habitats of interest face fewer preexisting threats. [CC-Market access, CC-Threat]

Strong community institutions can either incentivize or constrain the behavior of resource users [48] and effective local governance and charismatic leadership can inspire trust that these institutions will function as intended [23]. Clearly-defined rights for excluding outsiders and managing resources are also important because communities with these rules provide the security that rights over future harvests are protected [31,65]. Secure tenure over land and resources gives communities more buy-in, facilitates cooperation among users, enables greater flexibility in rules, and can result in communities valuing delayed returns to resources, all of which can contribute to good outcomes [15,17,31]. Finally, when community members are familiar with preexisting compatible institutions [66], one can expect more compliance and a higher likelihood of positive outcomes.

Hypothesis 8 *CC2 - Supportive local context:* Success is more likely when projects are in communities with effective local governance, institutions that align with project goals, well-defined land tenure, and strong leadership. [CC-Local Institutions *(combines CC-Effective local government and CC-Supportive local culture)*, CC-Tenure, CC-Charisma]

Finally, both community size and heterogeneity can affect project success [43]. There is no consensus regarding the relationship between population size and successful community resource management with theory and empirical studies suggesting that there may be a positive linear relationship [67,68], a negative linear relationship [32,69] or an inverted U-shaped relationship [70]. In the latter case, the expectation is that small populations are unable to absorb the costs of coordinating to develop and enforce management institutions whereas large populations suffer prohibitively high costs of such coordination [71]. We use as our hypothesis the prediction of a U-shaped relationship, which is supported by empirical research [70]. CBC outcomes also depend on the degree of community heterogeneity and whether that heterogeneity is economic, socioeconomic, or political [72]. Similar to the uncertainty about the effects of group size on collective action, there is conflicting evidence about the effects of heterogeneity on community resource management outcomes [73,74]. Here we make the simplest and most general prediction that socioeconomic heterogeneity has negative effects on outcome success [72,75].

Hypothesis 9: *CC3 - Local context:* Success is more likely when projects are in communities with moderate sized human populations, and in communities that are more culturally/ethnically homogenous. [CC-Population size, CC-Population heterogeneity]

We also controlled for the length of time the project has been running (CTR- Yrs. running), the first author's disciplinary background and affiliation (CTR-Auth discipline), and the status of the ecoregion in which the project was conducted (CTR-Ecoregion status).

Methods

Search strategy

The protocol for this systematic review (*CEE-09-021*) can be found at the following URL: http://www.environmentalevidence.org/SR82.html and the final coding protocol can be found in Additional file 1: Appendix A. The initial set of 62 projects in this study was taken from previous systematic reviews conducted by Brooks et al. [13] and Waylen et al. [14]. Six projects from Waylen et al. [22] were not included in this review because they were covered by more recent studies identified by the current search. JSB and KAW collected all articles used in the previous systematic reviews.

We expanded this initial sample by searching the publication databases ISI Web of Science, Anthropology Plus, and JSTOR, and by subsequently examining the first 500 hits for each search term on Google Scholar. We searched these databases for the years 2007 (when Waylen et al. [14] ended their search) to August 2009 (when the current review began). JSB conducted all new searches for the study, scanned all titles and abstracts and, where appropriate, completed the full text review. In cases of uncertainty, JSB sent a copy of the article to KAW for review and consultation.

All searches were conducted in English using the search terms listed below. Each search term phrase was used individually:

- community based conservation (including variants community-based conservation, "community based conservation", "community-based conservation")
- integrated conservation and development (including variant "integrated conservation and development")
- ICDP
- CBC
- community conservation (including variant "community conservation")
- community based natural resource management (including variants community-based natural resource management, "community based natural resource management", "community-based natural resource management)

Specialist search

A secondary search to find additional projects from each country already included in our sample was also conducted. This secondary search was conducted because we required as many cases as possible per country to investigate the effect of national predictors. The search was conducted using the digital library compiled by Advancing Conservation in a Social Context (ACSC) [76]. Because the ACSC library contains only sources that address conservation issues and projects, we searched the library by country name rather than the search terms listed above. JSB and a research assistant conducted this search. The research assistant was trained to identify projects in articles that met the search criteria. Following this training, the research assistant provided a summary of each of the projects in articles that were selected for a full review along with her inclusion/exclusion decision. The research assistant noted articles in which she was uncertain of the final decision and referred these to JSB.

Comprehensiveness of the search

We cannot be certain that this study includes every publication on relevant CBC and CBC-related projects, particularly since the search was conducted only in English. However, we believe our search was thorough, given that we considered 4,290 articles and our current sample of 136 projects is 5 times larger than our sample in 2006 [13] and over twice as large as the sample from 2010 [14]. In fact, our sample is larger than most, if not all, other reviews of community-based resource management (e.g. Gutierrez et al. [23] use 130 cases, Persha et al. [77] 84 cases, Oldekop et al. [78] 116 cases, Padgee et al. [19] 69 cases, and Cox et al. [44] 91 cases). While this does not provide evidence that our search captured all relevant projects, the systematic use of search terms means that we have a representative sample. Furthermore, we are certain that our search is replicable and we wanted the search to match, as much as possible, those from the previous reviews upon which we were building.

That said, the concepts and terminology related to CBC are constantly changing, so we would welcome readers to add to our search terms and search in other languages in subsequent studies. Search terms that future researchers should consider are "payment for ecosystem services", "co-management", "indigenous protected areas", and resource specific terms such as "community forestry", "community wildlife management", "community fisheries", and relevant variants of each. Although we captured projects that used these approaches in our search, more may exist. Search terms that are specific to natural resource domains are likely to have captured projects in our sample, (e.g. a study on community forest management is likely to include keywords or text matching our search terms), but this may not always be the case.

Study inclusion criteria

Figure 1 provides an overview of the process for identifying which projects were accepted in this review. Projects were accepted if they met six criteria:

1. The study was published in the primary or grey literature but not secondary sources. Where more than one acceptable article referred to the same project, we used the most recent article and used the older article to fill in any missing information.
2. The study had to provide information on a community-based conservation project, defined broadly as any conservation and development project or any community-based project in which conservation was the primary aim. We define projects as purposeful interventions (either externally initiated or internally initiated) that operate under unique institutional guidelines and have conservation goals. For instance, many interventions in Zimbabwe fall under the CAMPFIRE initiative and are thus considered to be the same general "project" (see criteria 1). The projects in this study vary in size and geographic scope, however, they all aim to directly produce changes in natural resource management by engaging with one or more communities. Each project in this study has separate project management. Thus, our pool may consist of Integrated Conservation and Development (ICD) projects, eco-tourism projects, National Park outreach projects that aim to incorporate communities in conservation planning and practice, extractive reserves, and general community-based natural resource management projects. We did not include interventions designed without conservation goals, such as ecotourism operations set up without explicit conservation aims, or projects reviewing the impact of a protected area on local communities in the absence of a specific project.
3. The subject of the CBC intervention could be any community or group of communities in a developing country (see #5 below). We included any study in which a CBC project was initiated in communities regardless of whether or not there was a protected area adjacent to the community or associated with the project.
4. Different community-based conservation projects aim to deliver multiple and variable goals, but generally are expected to deliver both societal and ecological improvements. Accordingly, for this study we felt it was important for authors to be cognizant of the multiple goals inherent to CBC, and that without multiple measures of success we would lack a more comprehensive understanding of the effectiveness of the project. Therefore, the study of the specific project had to measure at least two of

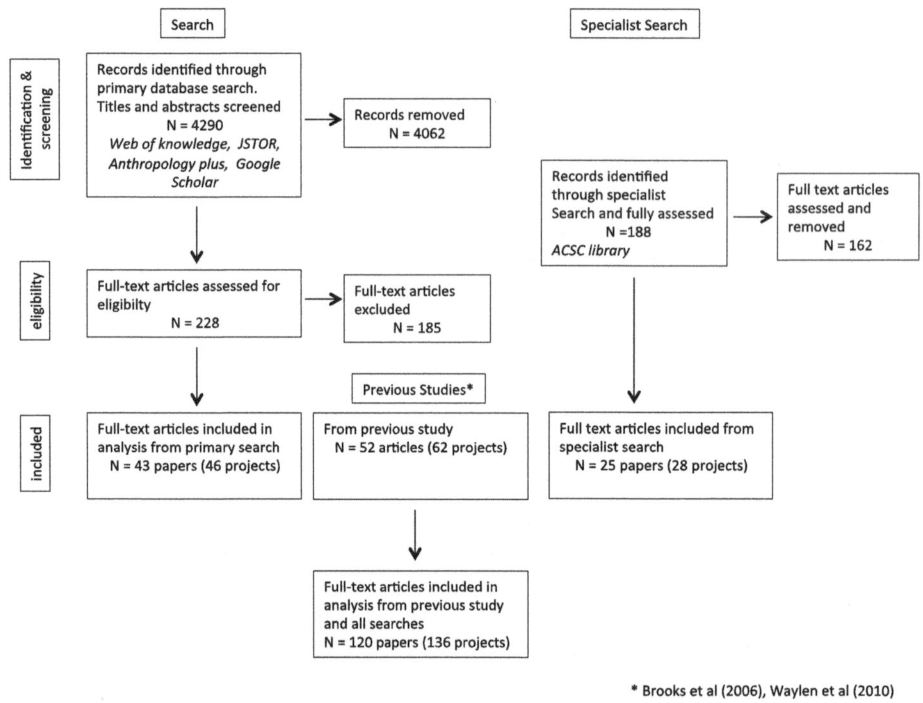

Figure 1 Systematic map of the search and inclusion process.

the four outcomes of interest (attitudinal, behavioral, ecological, economic). See section 3.8.2 Outcome variables for a more detailed description of the four outcome measures).

5. The project could not be exclusively located within a most-developed country as determined by 2009 gross national income (GNI) per capita data provided by the World Bank.
6. Finally, sufficient information had to be provided about the project. Projects for which there was missing information for more than 33% of important variables were not included. Projects for which an article provided only an overview or a descriptive explanation of implementation and outcomes were not used.

Decisions about whether to include projects were made jointly by JSB and KAW. Both researchers read articles for projects that qualified for a full review and made a decision about inclusion. If the researchers differed in their decisions, they discussed the case until reaching agreement. A research assistant assisted in evaluating the suitability of projects in the secondary search, but consulted with JSB who made the final decision on inclusion/exclusion.

Unit of analysis

There are three terms that are important to keep in mind in regards to the unit of analysis.

- "project" refers to the distinct CBC intervention that was analyzed and reported on
- "article" refers to the publication from which data on the project was extracted
- "study" refers to the research described in the article that pertains to the specific project.

Potential effect modifiers and reasons for heterogeneity

To test the aforementioned hypotheses we created a new dataset out of information extracted from the articles in our sample as well as secondary data. We collected information for predictor variables (effect modifiers) in three domains (national context, project design, and community characteristics) and attempted to include all theoretically relevant predictor variables that were feasible to measure given the data available in the articles. The variables in each domain listed below are the same as those listed within the previously stated hypothesis and are described more fully in Table 1.

National Context

- Governance
- Rights
- HDI
- Gini

Project Design

- Participation
- Engagement
- Protectionism
- Resource use
- Economic benefits
- Equity
- Capacity
- Social capital
- Environmental education

Community Characteristics

- Market integration
- Threat
- Local institutions
- Tenure
- Charisma
- Population size
- Population heterogeneity

Study quality assessment

One of our criteria for study inclusion was the quality of the study. Sources for which more than 33% pieces of data were missing were discarded. This metric generally corresponded with articles that the reviewers felt were poorly conducted or presented minimal detail and tenuous conclusions, and therefore served as a filter for project descriptions of particularly low quality. In addition, we assessed the quality of each study with a coding protocol that was based on work by the Cochrane Collaboration (chapters 8 and 13) [79], but modified to account for the realities of conducting research on CBC. We coded for potential conflicts of interest, how well the methods were conveyed, the general study design category, whether the study used appropriate control cases and accounted for confounding variables, and the type of analysis employed. See Table 3 for a full description of each quality assessment variable.

JSB and a research assistant conducted the study quality assessment. Both researchers coded 13 randomly selected projects to test the coding protocol and intercoder reliability. Cohen's kappa values for intercoder reliability ranged from 0.28 to 1. Variables with a kappa value < 0.50 (conflict of interest, study design category, confounds) were discussed by the researchers and changes were made to the coding protocol to increase clarity. Conducting the reliability test after variables were recoded to reduce the number of categories would also have produced higher Cohen's Kappa values indicating higher agreement. JSB coded 51 of the remaining projects, while the research assistant coded 75 of the remaining projects.

We combined measures of quality into an overall score, which was then used to rank projects into three categories. The study quality assessment had a possible range of 0 (highest quality) to 11 (lowest quality). Projects with a score of 0–3 were ranked as high quality, 4–7 as moderate quality, and 8–11 as low quality. We did not use the quality assessment ranking to weight the projects in the analysis because there was little variation in the quality of the projects in our sample. It is also important to reiterate the difficulty of conducting high-quality experimental studies with appropriate control cases when investigating CBC projects. We relied on generally low quality studies out of necessity for this study, but this does not mean we can assume they are reliable just because they are more feasible to conduct.

Data extraction

We refined the coding protocols used by Brooks et al. [13] and Waylen et al. [14] and extracted information on a number of variables related to national socio-economic

Table 3 Description of variables used in the quality assessment

Variable name	Description
Conflict of interest	Were any of the authors affiliated with the project described in the study in any way that could result in a conflict of interest (e.g. currently or previously employed by the project, served as an advisor on the project, or funded by the same source)? Coded as 1 (yes or unsure), 0 (no)
Methods description	How well were the methods in the paper reported. Coded as 1 (not very clear (missing information) or not reported) and 0 (very clear)
Study design category	How could the design of the study be characterized? Coded as 4 (case series or cross sectional study), 3 (case control study, cohort study, or historically controlled study), 2 (interrupted time series study), 1 (controlled before-and-after study or non-randomized control trial) and 0 (randomized control trial). Note only categories 3 and 4 were present in our sample
Controls	Were appropriate control cases included in the study? Coded as 1 (no or partial), and 0 (yes)
Confounds	Did the study account for and minimize effects associated with potential confounding variables? Coded as 1 (no or partial), 0 (yes)
Analysis type	What type of analysis did the author use? Coded as 2 (qualitative analysis or quantitative analysis with descriptive or observational statistics), 1 (analysis of variance, t-test, statistical correlation or other bivariate analyses), 0 (multivariate regression or other multivariate analyses).

and political context, project design, local community characteristics, type of monitoring of outcomes and outcome types. We collected 65 pieces of information for each project. The full coding protocol can be found in Additional file 1: Appendix A. For this review, we present only those variables directly relevant to the major hypotheses.

The final coding protocol was pre-tested on a sample of projects from Waylen et al. [14] that were discarded from this sample because our search uncovered more recent articles that addressed the same project. The coding occurred in two steps. First, JSB and KAW coded each of the 46 projects identified in the primary search and discussed disagreements to choose the appropriate coding. Then, JSB re-coded all projects from the prior reviews and coded all projects from the secondary (ACSC) search. K.A.W separately coded 47 (52%) of these remaining projects that required a second opinion. In all cases, coders based their decisions only on the information presented in the article.

Inter-coder reliability was assessed for the 47 projects that JSB and KAW coded separately by calculating Cohen's Kappa with the *irr* package [80] in R version 2.13.0 [81]. Cohen's kappa represents the proportion of agreement after accounting for the level of agreement expected by chance when coding categorical data (Cohen 1960) and Cohen's weighted kappa for ordinal data [82,83].

Predictors (Effect Modifiers) and outcome variables

We used multiple predictor variables, or effect modifiers, for each of the nine hypotheses. We provide a full description of national- and project-level variables as well as all four outcomes in Table 1. Where possible, coding of variables matched those of Brooks et al. [13] and Waylen et al. [14] for comparison. Many of the recorded predictor variables relate to social phenomena, which it is difficult to quantify on a numeric scale. Therefore, variables were coded as categorical or ordered categorical variables, with the exception of national-level variables and the number of years the project had been running, for which continuous data were used.

For each country represented in our sample, external sources were used to collect data on national-level characteristics from a variety of external sources. This was necessary because the authors of the studies/ publications included in this review rarely provide information on national governance, Human Development Index, political stability, and income inequality for the countries in which projects were conducted. Data for national level variables were collected from the following sources:

Governance: *World Bank* (http://info.worldbank.org/governance/wgi/sc_country.asp#). Scores were taken from the year closest to the date of research.

Rights: *Freedom House*: (http://www.freedomhouse.org/report-types/freedom-world). Scores were taken from the year closest to the date of research.

HDI: *The United Nations Development Programme* (http://hdr.undp.org/en/statistics/). Scores were taken from the year closest to the date of research.

Gini: *World Resources Institute*: Inequality coefficients ranked by country from data compiled from 2000–2007. One score was used for each country. Gini inequality coefficient data for the years 2000–2007 are no longer available from the World Resources Institute. However, the source of that compiled can be found at the World Bank (http://data.worldbank.org/indicator/SI.POV.GINI?page=2).

Outcome variables

Debates about conservation strategies often result from a focus on different outcomes. Some conservationists might emphasize a forest's ability to sequester carbon and provide a home for forest-dwelling peoples [84], whereas others might have an interest in the ecologically functional populations of species within the ecosystem [85]. As such, authors will disagree over conservation strategies [37], which, in many cases, illustrates the different perspectives of social versus natural scientists [39].

Because of these disagreements, and because CBC is based on the idea that multiple interrelated goals must be met to produce long-term conservation success, we included four outcome measures, attitudinal, behavioral, ecological, and economic. We coded each outcome variable according to the criteria below. In each case, coding decisions were based solely on information provided by the authors. Summary statistics for dependent variables can be found in Table 2.

Attitudinal success: Whether the attitudes of the community-members have changed towards conservation in general, the specific CBC project, and/or the protected area associated with the project.

Behavioral success: Whether levels of resource use, or other behaviors antithetical to conservation that were addressed by the project (e.g. killing nuisance wildlife), have decreased as a result of the project.

Ecological success: Whether the habitat and/or species of interest is in better condition as a result of the project. (e.g. the population size of a species has increased, or a given resource is more abundant).

Economic success: Whether the community has received economic (e.g. income, direct payments) or other development benefits (e.g. roads, schools, hospitals) as a result of the project.

Each dependent variable was coded on a three-level scale (success, limited success, failure). Project evaluations sometimes included multiple indicators for a given outcome variable. As such, outcomes were coded "success" when most indicators showed improvement, and "failure" when most indicators showed no change or decline. We considered "no change" to be a failure because CBC interventions are generally intended to correct a perceived problem. If, after the implementation of the CBC project, there is no change in circumstances for a given outcome, this suggests that the intervention was ineffective. Thus, we feel justified in coding the project as failing for that outcome. Outcomes were coded "limited success" if (a) the author reported conflicting results for multiple measures of an outcome (e.g. community members had positive attitudes towards conservation in general, but negative attitudes towards the nearby National Park), (b) the project benefited some community members but not all (e.g. income increased for 40% of households but did not change for the rest), and (c) success was produced for some communities that were part of the project, but not all (e.g. forest clearance declined in 3 of 5 communities that were a part of the study). Not all projects collected data for each of the four outcomes. As such, the sample size for each outcome ranges from 77 (ecological) to 118 (economic) (see Figure 2).

Author questionnaire

Missing information can be problematic for reviews like this because few articles address all of the variables for a given project that are of interest in this study. We attempted to reduce the number of missing values by systematically collecting information from the original authors of each source in our sample. A questionnaire was constructed that was composed of the questions in our coding sheet that we judged as most pertinent and which contained the greatest number of missing values. The corresponding author of each article was contacted via email with a request to complete the questionnaire.

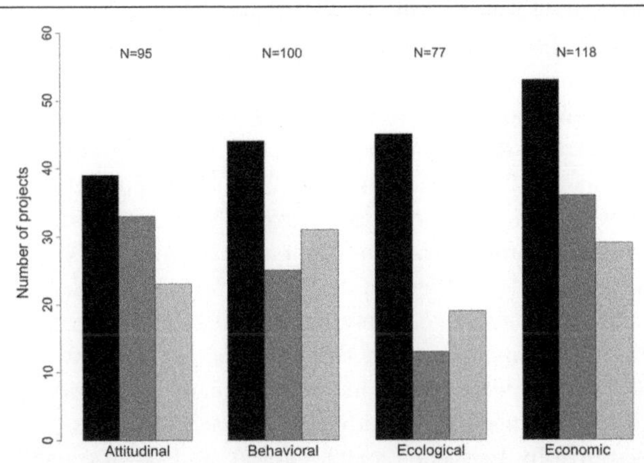

Figure 2 Outcomes of CBC projects from studies that measured at least two outcome measures (success =black, limited success = dark grey, failure = light grey).

Data synthesis and presentation

Bivariate analysis

All analyses were conducted using R statistical computing [81]. Please contact the corresponding author for scripts for each step of the analysis. Bivariate analyses were conducted using two-dimensional contingency tables for categorical predictors and proportional odds logistic regression for continuous predictors. The Goodman-Kruskal gamma statistic was used to summarize the association between predictors and outcomes and as a test statistic for Monte Carlo significance tests (see [13]). Because of the way the variables were ordered, a gamma value relatively close to 1 was interpreted as evidence in support of our hypotheses. A Monte Carlo p-value was obtained for each test of ordinal association as follows. For an observed table, 10,000 random tables (having the same row and column sums as the observed table) were generated under the null hypothesis of independence of predictor and outcome. These tables were generated with the function *r2dtable* in R. For each random table, the Goodman-Kruskal gamma statistic was calculated and stored. A one-sided p-value was calculated as the number of random gamma statistics greater than or equal to the observed gamma, divided by 10,000.

Models for each of the five continuous predictor variables and each outcome variable were fit using the proportional odds linear regression *polr* function in R. P-values were recorded from those models and added to the list of p-values obtained from the contingency tables.

Multiple testing was controlled for by adjusting significance levels for our (29 predictors * 4 outcomes) 116 significance tests using q values [86,87] to obtain approximate control of the False Discovery Rate (FDR). FDR is defined as the expected proportion of false-positive tests among tests called significant. The p-values obtained from the contingency tables and the regression models were supplied to the q value software (available at http://www.bioconductor.org/packages/release/bioc/html/qvalue.html), which calculates q values for the Monte Carlo significance tests. The "smoother" option was used in the q value software and the tuning parameter lambda was allowed to range in [0, 0.90]. Algorithms that take ordered significance levels (p values) from multiple hypothesis tests return a corresponding sequence of q values connected to the FDR for the tests [86]. Approximate control of the FDR is achieved by setting a threshold for the q values; for example, calling the tests having q values ≤ 0.05 significant implies that, of those tests, only about 5% are expected to be truly null-hypothesis cases.

Multivariate analysis

Proportional odds logistic regression models were fit to determine the best predictors of the four outcomes. The first step, however, was to modify the dataset in two ways. First, missing values were imputed to avoid possible sources of bias [88], and because the model-fitting procedure requires the removal of observations in which any of the independent variables is missing, which would have greatly reduced our sample size and constrained our ability to conduct this analysis. Five unique datasets were created using the Multivariate Imputation by Chained Equations *MICE* function in R. All predictor and outcome variables were used to impute missing values, though missing values in outcomes were not imputed. Second, after imputing missing values we combined variables that were conceptually similar to reduce the number of predictors in the model. The combined variables are depicted in Table 1. We examined Spearman's nonparametric correlations between predictors for each of the five imputed datasets. Because the Spearman's *r*-values were relatively low (see Additional file 2: Appendix B), we decided not to drop any variables from the analysis.

To minimize the opposing problems of omitting important variables and over-fitting the model [89,90], a reduced model was selected using a forward, stepwise AIC procedure [91]. In forward, stepwise AIC selection, variables are added, one at a time, to a model with no coefficients. The variable that produces the model with the lowest Akaike Information Criterion (AIC) score is selected. In subsequent steps, the remaining variables are added to the model one at a time and the variables already within the model are dropped until the model with the lowest AIC score is found. This process continues until no additions or deletions from the model can further reduce the AIC score [92].

Some authors have suggested that stepwise regression can lead to biased estimates [93] and have suggested other approaches to model selection [94]. However, we believe stepwise AIC is the best approach because (a) problems associated with model over-fit [90] compel us to select a reduced model with fewer predictors, (b) alternative approaches for model reduction, such as those that penalize coefficients (e.g. Lasso [95]), are difficult to employ with ordered outcome variables, and (c) it is essential for us to calculate robust standard errors to account for clustering at the country level, which is most straightforward with our approach.

Prior to fitting the models, the parallel slopes assumption that the relationship between all pairs of categories for a variable is the same was tested. The parallel slopes assumption was relaxed and a series of binary logistic regressions was run that included each predictor along with a dummy variable that was created by splitting the dependent variable into two levels (see [96]; pg 335). The linear predicted values we obtained from these regressions were plotted. If the differences in the values for each of the two levels in the binary regression are approximately equal for each category of the predictor variable, then the parallel slopes assumption has been met.

Most, but not all, of our predictors met the assumption of parallel slopes for the four outcomes. Therefore, partial proportional odds models were fit and forward, stepwise AIC model selection was used. However, the set of variables in the reduced partial proportional odds models was very similar to the set of variables in the reduced standard proportional odds models. The variables were the same for Attitudinal Outcomes, there was one additional variable in the partial proportional odds model for Behavioral Outcomes, and there was one variable that differed between the two models for Economic Outcomes. The first seven selected variables were the same for the two Ecological outcomes models, but diverge after that. Because standard proportional odds models are easier to interpret and because it is easier to calculate robust standard errors from standard models, we only present the results of the standard proportional odds models here.

After the stepAIC algorithm identified the model with the lowest AIC score, this reduced model was then fit using the *lrm* function, which produced the same coefficients as the *polr* function. The *robcov* function was then used to call the output from *lrm* to calculate robust standard errors for the coefficients. Robust standard errors were calculated to account for the clustering of projects at the country level.

The results from the five imputed datasets were pooled [97]. For cases in which a variable was not in the reduced model for all five imputed datasets, we entered a value of 0 for the estimate and standard error in the model for that imputed dataset. We then averaged the estimates for the three imputed datasets and calculated pooled standard errors [97,98]. See Additional file 3: Appendix A-D for the full results of the reduced-fit proportional odds logistic regression models and pooled values.

Separation problems [99] emerged for one imputed dataset for the Attitudinal analysis and for all imputed datasets in the Ecological analysis. Complete separation occurs when one level of a predictor is associated with only one outcome value. In our case, separation problems prevented the forward stepwise AIC model selection algorithm from converging on a reduced form model. For instance, none of the projects that were attitudinal failures were cases in which social capital was enhanced. This zero value indicates a nearly deterministic relationship between social capital and attitudinal outcomes. As such, for the fourth imputed dataset, social capital was removed from the set of predictors to allow the model selection algorithm to converge. To account for its removal in this imputed dataset, pooled estimates and standard errors were only calculated for social capital using the model outputs from the remaining four imputed datasets.

The separation problem was more prominent for the analysis of Ecological outcomes because of the smaller sample size. Market access and local institutions were removed from the set of predictors for all imputed datasets. In addition, combinations of resource use, charisma, and economic benefits were removed from the set of predictors for some of the imputed datasets. An investigation of the 2 × 2 tables of each predictor and Ecological outcomes suggests that the only variables for which there is a strong, and potentially deterministic relationship with Ecological outcomes is local institutions. However, because models would not converge with this variable included in the analysis for any of the imputed datasets, we cannot definitively say that local institutions are a significant predictor of ecological outcomes.

Comparison of researcher's coding with authors' responses

To compare our coding to the questionnaire responses of the corresponding authors of the articles in our sample we conduced an intercoder reliability test. We calculated Cohen's kappa for categorical data and Cohen's weighted kappa for ordinal data using the *irr* package in R [81].

Results

Review statistics

The primary search began in August 2009 and continued through February 2010. The primary search resulted in 4290 hits. After screening titles and abstracts, 228 articles were selected for full review from the initial

Table 4 Summary of search results

Database	Search term	Number of hits	Number selected for full review
Web of Science	Community based conservation	61	27
	Integrated conservation and development	19	4
	ICDP	29	1
	CBC	26	0
	Community conservation	34	9
	Community based natural resource management	163	7
Anthropology Plus	Community based conservation	1	1
	Integrated conservation and development	1	1
	ICDP	0	0
	CBC	0	0
	Community conservation	16	2
	Community-based natural resource management	0	0
JSTOR	Community based conservation	336	1
	Integrated conservation and development	176	1
	ICDP	12	0
	CBC	12	0
	Community conservation	415	1
	Community-based natural resource management	79	0
Google Scholar	Community based conservation	> 500 hits	82
Including grey literature	Integrated conservation and development	> 500 hits	25
	ICDP	> 500 hits	16
	CBC	> 500 hits	3
	Community conservation	> 500 hits	24
	Community based natural resource management	> 500 hits	23
	Total set aside for review from primary search		288
ACSC library	Search conducted using country names *		188
	TOTAL set aside for review:		416

* The ACSC search used country names because the library was composed of only conservation projects. The number of hits from the search for each country was not recorded.

search. Google Scholar produced the most hits and articles that were suitable for inclusion in the review based on our a priori criteria (see Table 4).

JSB compiled a list of countries represented in the sample and began a search of the ACSC online library for CBC articles in the relevant countries with the help of a research assistant. The secondary search of the ACSC digital library could not be conducted until a full list of countries represented in the sample derived from the primary search was created, so it was not completed until April 2010. An additional 188 articles were selected for full review from the specialist review of the ACSC library, which resulted in an additional 28 projects from 25 articles for the analysis.

In total, 74 projects from 68 articles were identified from both searches and these were added to the existing 62 projects carried over from the previous reviews. Overall, our sample includes 136 projects from 123 articles representing 40 countries (see Figure 1 for information about each aspect of the search). A full list of projects included in the sample is provided in Additional file 4: Appendix D. Because of the vast quantity of articles uncovered in our search, we did not keep records of the reasons that individual articles were discarded. However, projects were discarded when the article:

- did not describe a community-based conservation project
- described a project already covered by another more recent article (including interventions that fall under the same general policy, e.g. CAMPFIRE or ADMADE).
- provided an overview of CBC or of a project with no information on outcomes
- examined a project for which conservation was not one of the goals
- focused on the planning of a project before it had been carried out
- did not provide information on at least two outcomes
- was a short comment piece that lacked sufficient information on predictor variables
- used data to surmise about the potential effectiveness of an intervention, but did not address actual outcomes

Description of projects

The 136 CBC projects in our sample focus on conservation challenges in managing forests, grasslands, wildlife, and fisheries. This review also covered projects in communities with a variety of livelihood types and subsistence strategies including, among others, subsistence agriculture, hunting and gathering, fishing, pastoralism, wage labor, and various combinations of all of the above. We coded for this information, but did not include it in the analysis because of the frequency of missing values in the sample and because of the number of variables that we were exploring that are considered to be of greater importance.

Forty countries were represented in the sample. Forty percent of projects were in least developed countries, 33% were in lower middle income countries, and 27% were in upper middle income countries (see Table 5). The number of projects per country ranged from one (16 countries) to 19 (Tanzania) (Table 6). There were five or more projects from Brazil, India, Indonesia, Madagascar, Mexico, Nepal, Philippines, South Africa, and Tanzania. The majority of projects were located in Africa (N=65), followed by Asia/Oceania/Pacific Islands (N=44), and the Americas (N=30).

Study quality assessment

Over 80% [100] of the projects in our sample were considered to be low quality according our modified Cochrane Collaboration scale. The remaining 26 projects were in the moderate category (See Additional file 4: Appendix D for full list of results of quality assessment). The type of study design and analysis employed by the researchers largely determined the quality of the projects, which is to say that other biases and shortcomings contributed less to the low study quality rating. To demonstrate a causal link between a CBC intervention and the four outcomes of interest in this review, a study would have to use an experimental design and control for confounding factors like geographic location, ecosystem type and other environmental characteristics, the socio-economic make-up of the community and baseline conditions. Researchers would either have to conduct longitudinal studies after collecting sufficient baseline data, or replicate communities to have a treatment and control. Given the limited funding, scope and monitoring capacity associated with most CBC projects, taking these approaches is often infeasible.

Some recent studies have used quasi-experimental matching techniques to identify communities similar to the ones in their studies to act as controls. This approach has been used to explore the relationship between protected areas and poverty alleviation/exacerbation (e.g. [101,102]. In the context of CBC, this study design would allow one to ask how attitudinal, behavioral, ecological, and economic outcomes would have changed over time in the *absence* of an intervention. Such an approach would give scholars, policy makers, and conservation practitioners greater confidence that a CBC intervention had the effects they were presumed to

Table 5 Income groups correspond to 2009 gross national income (GNI) per capita (World Bank Atlas method)

Low income	Lower middle income		Upper middle income
Bangladesh	Belize	Papua New Guinea	Botswana
Cambodia	Bolivia	Philippines	Brazil
Ethiopia	Cameroon	Swaziland	Chile
Kenya	China	Thailand	Costa Rica
Madagascar	Ecuador	Vietnam	Malaysia
Malawi	El Salvador		Mexico
Mozambique	Guatemala		Namibia
Nepal	Guyana		Peru
Tanzania	Honduras		South Africa
Uganda	India		Suriname
Zambia	Indonesia		
Zimbabwe	Nigeria		

Source: http://data.worldbank.org/about/country-classifications/country-and-lending-groups. Data on Western Samoa and Taiwan was not provided.

have. Unfortunately, few if any such studies exist for CBC. We encourage scholars to employ such study designs where at all possible in the future.

We largely viewed the quality assessment as an exercise to illustrate the generally low quality of the studies in our sample *relative to the gold standard for experimental design*. This analysis was not intended to guide our main statistical analysis and was not used to more strongly weight studies in the moderate risk category. We feel it is important enough to highlight some of the deficiencies in CBC research and some of the more recent study design innovations.

Intecoder reliability tests between researchers

For the intercoder reliability test between coders, Cohen's kappa values ranged from a low of 0.35 (*charisma*) to a high of 0.91 (*protectionism*), with a mean value of 0.66 (Table 7). These results indicate that the level of agreement between coders was fair or better (>0.40) [103] for most of the key predictor variables in the analysis. However, seven of the predictors had poor agreement levels (kappa ≤ 0.40). Four of these predictors were excluded from the analysis. JSB and KAW reviewed the remaining three variables (tenure, local culture, charisma), modified and expanded the coding rules for each, and recoded them according to the new rules. The new level of agreement for each of the four outcome variables was substantial (kappa > 0.65) and averaged kappa = 0.78.

Table 6 Number of projects per country included in the sample

Country	Number of projects
Bangladesh	1
Belize	3
Botswana	3
Brazil	8
Cambodia	1
Cameroon	4
Chile	1
China	3
Costa Rica	4
Ecuador	2
El Salvador	1
Ethiopia	1
Guatemala	1
Guyana	1
Honduras	2
India	6
Indonesia	5
Kenya	4
Madagascar	10
Malawi	1
Malaysia	2
Mexico	5
Mozambique	2
Namibia	4
Nepal	9
Nigeria	1
Papua New Guinea	1
Peru	1
Philippines	7
South Africa	6
Suriname	1
Swaziland	1
Taiwan	4
Tanzania[a]	19
Thailand	1
Uganda	3
Vietnam	2
Western Samoa	1
Zambia	3
Zimbabwe	1

a. The number of projects is highest for Tanzania, in part, because Alcorn et al. (2002) (see Additional file 4: Appendix D) is a publication that reviewed multiple case studies in Tanzania.

Table 7 Results of the intercoder reliability test between researchers

Variable name	Cohen's Kappa
Project impetus	0.79
Establishment	0.56
Decision-making	0.62
Approach local culture	0.63
Approach local institutions	0.75
Resource use	0.86
Economic benefits	0.53
Equitable distribution	0.55
Elite capture	0.83
Capacity skills	0.54
Capacity institutions	0.84
Social capital	0.75
Environmental education	0.62
Market integration	0.52
Threat	0.59
Supportive local culture	0.37
Effective local government	0.56
Tenure	0.47
Charisma	0.35
Population size	0.90
Population heterogeneity	0.90
Attitudinal monitoring	0.64
Behavioral monitoring	0.86
Ecological monitoring	0.80
Economic monitoring	0.82
Attitudinal outcomes	0.66
Behavioral outcomes	0.86
Ecological outcomes	0.80
Economic outcomes	0.82

Kappa values greater than 0.40 indicate moderate agreement, greater than 0.60 indicate substantial agreement, and greater than 0.80 indicate almost perfect agreement [89].

Narrative synthesis

Readers can find citations for each project included in the study as well as its location, study quality rating, outcomes measured, type of monitoring of outcomes, and outcome type (success, limited success, failure) in Additional file 4: Appendix D. Overall, projects reported more successes than failures across all four domains (Figure 2). The proportion of projects that were considered successful was higher for ecological outcomes than the other three outcomes. The number of projects that provided information on the four outcomes ranged from 79 (ecological) to 121 (economic) with 97 and 102 projects providing information on attitudes and behaviors respectively.

Quantitative synthesis

This section provides a brief synthesis of the main review results. It is organized around the core question that our reviewed addressed as well as the nine hypotheses related to objectives 2 and 3.

Objective 1: Is community-based conservation an effective conservation tool?

There were more reported successes than failures of CBC interventions across all four outcomes. Forty-one percent of projects suggested attitudinal success whereas 24% suggested attitudinal failure. Forty-three percent of projects suggested behavioral success compared to 31% that suggested failure. Fifty-eight percent of projects suggested ecological success compared to 25% that suggested failure and 45% of projects suggested economic success compared to 25% that reported failure. See Figure 2 for a visual depiction and Table 2 for raw data.

An examination of two-way contingency tables using the Goodman-Kruskal gamma statistic and Monte Carlo p-values indicated a positive and significant association between all pairs of outcome variables (Table 8). This statistical relationship between all pairs of outcomes indicates a tendency for author's to report successful outcomes in more than one domain.

The amount and type of monitoring was similar to that found in our previous reviews [13,14] with the majority of projects failing to report on all four outcomes and most relying on qualitative monitoring or author's judgment as opposed to quantitative monitoring of outcomes (see Figure 3). Twenty-seven projects (19%) measured all four outcomes, 67 projects (48%) measured three outcomes, and 45 projects (32%) measured two outcomes. Eight (30%) of the projects measuring all four outcomes considered all outcomes to be successful and two (7%) considered all outcomes to be failures. Twelve

Table 8 Goodman-Kruskal gamma statistic and associated p-value for association between pairings of Outcome variables

Pairs of outcomes	Goodman-Kruskal gamma (*p-value*)
Attitudinal & Behavioral	0.67 (0.000)
Attitudinal & Ecological	0.53 (0.005)
Attitudinal &0 Economic	0.66 (0.000)
Behavioral & Ecological	0.87 (0.000)
Behavioral & Economic	0.52 (0.000)
Ecological & Economic	0.66 (0.000)

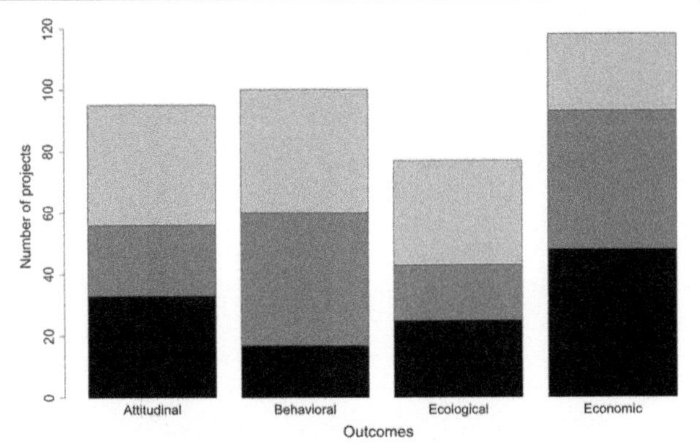

Figure 3 The frequency of type of monitoring for each outcome for studies that measured at least two outcomes: author's judgment (light grey), qualitative (dark grey), quantitative (black).

(18%) of the projects measuring three outcomes found each outcome to be successful and eight (12%) found each outcome to be a failure. Thirteen (29%) of the projects measuring two outcomes found them both to be successful, and seven (16%) found them both to be failures.

None of the projects quantitatively measured all four outcomes and 59 studies (42%) did not quantitatively monitor any of the four outcomes relying instead on the author's judgment. However, there was no significant relationship between the type of monitoring and success for any of the outcomes (*Attitudes*, $\gamma = 0.01, p>0.05$, *Behaviors*, $\gamma = 0.17, p>0.05$, *Ecological* $\gamma = -0.04, p>0.05$, *Economic* $\gamma = 0.18, p>0.05$), which indicates that qualitative studies and those that use the author's judgment were no more likely to report success or failure than quantitative studies.

Objective 2: Which aspects, if any, of national socioeconomic and political contexts affect the outcomes of local-level CBC projects?

Hypothesis 1: *NC1 - National political context*: Success when projects are implemented in countries where there is greater transparency and stability in governance and where the populace has a voice in politics and enjoys common civil liberties. [NC-Governance, NC-Rights]

Hypothesis 2: *NC2 - National socioeconomic context:* Success when projects are implemented in countries with a higher level of development progress and in which there is greater economic equality. [NC-HDI, NC-Gini]

The bivariate analysis shows that projects in countries with better national-level socioeconomic conditions have greater behavioral and attitudinal success, projects in countries with more equitable distributions of wealth have greater attitudinal success, and projects in countries with better overall governance have greater behavioral success (see Table 9). However, the full multivariate model with controls indicates that national context does not play an important predictive role in any domain of project success as the 95% confidence intervals cross zero in all cases (Figure 4a-d).

Objective 3: Which aspects, if any, of project design and community characteristics affect the outcome of CBC projects?

Project design

Hypothesis 3: *PD1 - Participation/Engagement:* Success when there is more emphasis on participation in initiation, establishment, and day-to-day running of the project, and when the project engages positively with local governance institutions and with local cultural beliefs, practices and traditions. [PD-Participation *(combined PD-Impetus, PD-Establishment, and PD-Decision-making),* PD-Engagement *(combined PD-Approach local culture and PD-Approach local institutions)*]

Hypothesis 4: *PD2 - Access to and utilization of resources:* Success when projects emphasize greater levels of utilization and subsequently generate support for, and compliance with, conservation initiatives. [PD-Protectionism, PD-Resource use]

Hypothesis 5: *PD3 - Project benefits:* Success when projects provide clear and well-directed economic benefits, and ensure that such benefits are shared equitably and without elite capture. [PD-Economic benefits, PD-Equity *(combines PD-Equitable provision and PD-Elite capture)*]

Hypothesis 6: *PD4 - Human/Social capital:* Success when projects invest in human and social capital. [PD-Capacity *(combines PD-Capacity skills and PD-Capacity institutions),* PD-Social capital, PD-Environmental education]

The bivariate results suggest that project design is important for success. Participation, engagement, project benefits, and human/social capital were particularly crucial, as variables from these clusters were significantly associated with each of the four outcomes (Table 9). The

Table 9 Coefficients from bivariate proportional odds logistic regression (continuous variables) and Goodman Kruskal gamma values from two-way contingency tables (categorical variables)

	Variable name	Attitudinal success	Behavioral success	Ecological success	Economic success
NC1	Governance		2.42		
NC1	Rights				
NC2	HDI	2.56	2.35		
NC2	Gini	1.94			
PD1	Impetus		0.36	0.68	
PD1	Establishment	0.41	0.38	0.47	
PD1	Decision-making	0.34	0.38	0.38	0.42
PD1	Approach local culture	0.55	0.56	0.70	0.56
PD1	Approach local institutions	0.46	0.63	0.81	0.45
PD2	Protectionism			0.45	
PD2	Resource use				
PD3	Economic benefits				
PD3	Equitable distribution	0.47	0.58	0.56	
PD3	Elite capture	0.62	0.60	0.44	0.45
PD4	Capacity skills	0.42	0.53	0.60	0.50
PD4	Capacity institutions	0.63	0.57	0.69	0.48
PD4	Social capital	0.78	0.63	0.50	0.53
PD4	Environmental education	0.34	0.33		
CC1	Market integration				
CC1	Threat				
CC2	Supportive local culture		0.34		
CC2	Effective local government		0.47	0.59	
CC2	Tenure	0.33		0.28	0.34
CC2	Charisma				
CC3	Population size		0.34		
CC3	Population heterogeneity				
CTR	Ecoregion status				
CTR	Author discipline				
CTR	Years project running				

Dark cells indicate a significant association between independent and dependent variables. Analyses were conducted on the dataset that included missing values (not imputed data).

results of the multivariate analysis (Figure 4a-d) support the bivariate results, in that several distinct aspects of project design are important predictors of success across all four outcomes (95% confidence intervals do not cross zero). Attitudinal success is most likely when the project produces or enhances social capital within communities (*PD-Social capital*), when the project emerges at the impetus of the community and the community is involved in project establishment and daily management (*PD-Participation*), and when benefits are equitably distributed without elite capture (*PD-Equity*). Behavioral success is positively affected by whether the project invests in the capacity of local individuals and institutions (*PD-Capacity*). Ecological success is promoted when the project engages positively with local cultural traditions and governance institutions (*PD-Engagement*) (see *note on interpretation* below), invests in the capacity of local individuals and institutions (*PD-Capacity*), and when the project emerges at the impetus of the community and the community is involved in project establishment and daily management (*PD-Participation*). Finally, economic success is most likely when the project builds capacity in the local community (*PD-Capacity*).

Community characteristics

Hypothesis 7: *CC1 - Market integration:* Success when projects are in communities that are

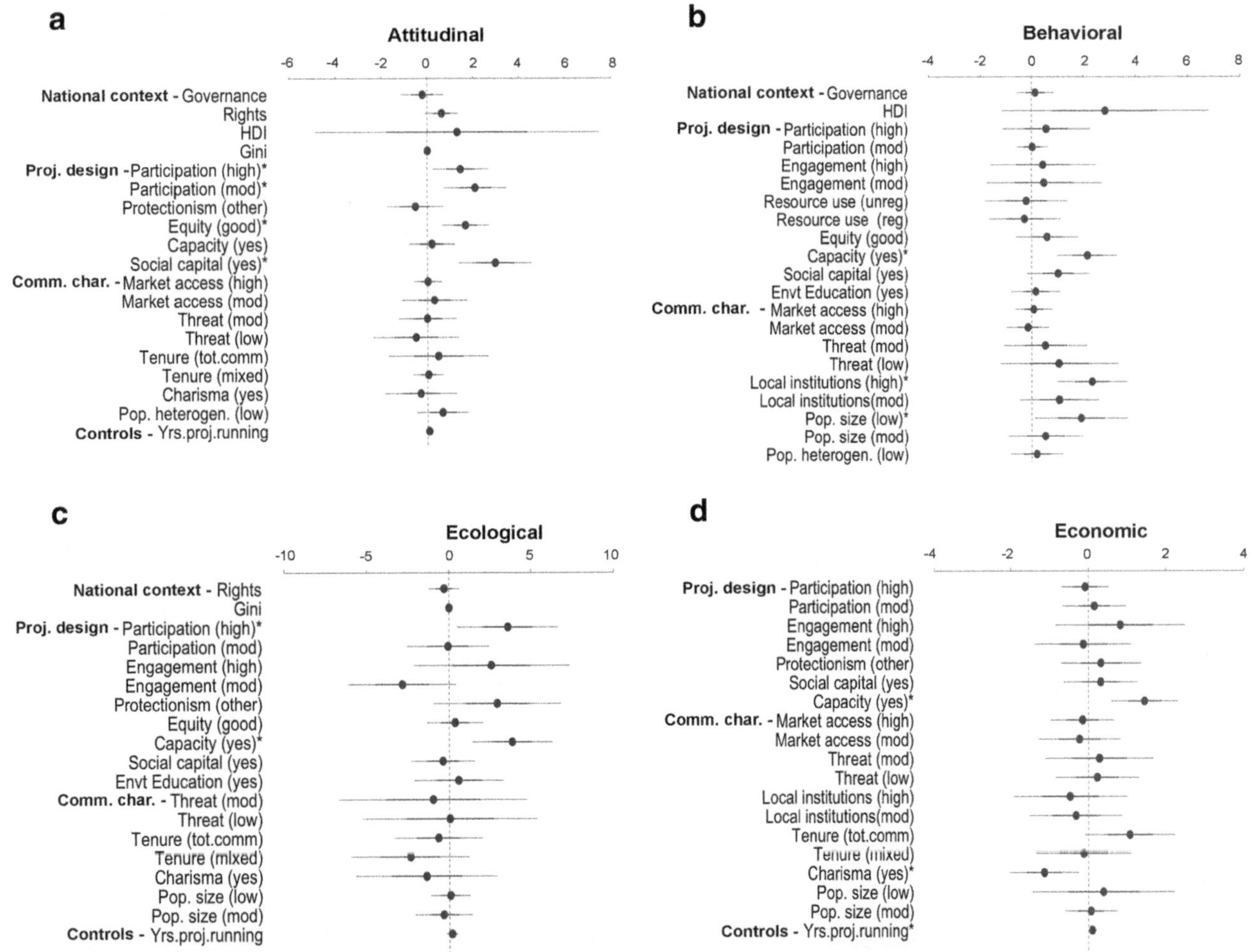

Figure 4 a – 4d^b. Plots of the pooled coefficients 50% confidence intervals and 95% confidence intervals (X-axis) for variables remaining in the reduced-fit model for each outcome variable as selected by forward, stepwise AIC. Asterisks indicate a significant association with an outcome.

more integrated into local and global markets and their species/habitats of interest face fewer preexisting threats. [CC-Market access, CC-Threat]

Hypothesis 8: *CC2 - Supportive local context:* Success when projects are in communities with effective local governance, institutions that align with project goals, well-defined land tenure, and strong leadership. [CC-Local Institutions *(combines (CC-Effective local government and CC-Supportive local culture)*, CC-Tenure, CC-Charisma]

Hypothesis 9: *CC3 - Local context:* Success when projects are in communities with moderate sized human populations, and in communities that are more culturally/ethnically homogenous. [CC-Population size, CC-Population heterogeneity].

Community characteristics were less likely to be associated with project outcomes than project design variables in the bivariate analysis (Table 9). Tenure showed the strongest effect and was associated with three outcomes. There was a similar trend in the multivariate analysis as only three variables were significant predictors and these were restricted to two domains - behavioral and economic (Figure 4a-d). Supportive cultural beliefs and effective institutions (*CC-Local institutions*) and smaller population size (*CC-Population size*) are associated with behavioral success, and strong tenure rights over the primary resources targeted by the project (*CC-Tenure*) (see *note on interpretation* below), are associated with economic success. Interestingly, the presence of charismatic leadership was negatively associated with the likelihood of economic success, although the apparent under-reporting of charismatic individuals may explain this result. Only 15% of projects in our sample reported a charismatic individual whereas, 69% of authors responding to our questionnaire reported that a charismatic individual was involved in the project.

Control variables

None of the control variables was significantly associated with the outcomes in the bivariate analysis. However, the length of time the project has been running (CTR-Yrs. running) was significantly and positively associated with economic success in the multivariate analysis[a]. There were no effects of author discipline or ecoregion status.

Additional effect modifiers

It is often difficult for project evaluations to include all information that may be relevant to this study because of the range of variables over space and time that could conceivably impact CBC projects. These factors might include historical data, ecological data, and socio-economic and political information from multiple scales and time periods. As a result, there is often missing information from project evaluations in articles. There were several variables that we intended to include in the analysis but were unable to because of a lack of information provided for the projects in our sample. These variables are described at the bottom of Table 1 and include:

- ecological variables including average rainfall, elevation, and habitat type(s)
- subsistence type(s) in the local community
- importance of targeted resource(s) to local peoples
- presence/absence of supportive national policies
- degree of implementation of national policies related to CBC
- type and quality of interaction between communities and government agencies/employees
- type of involvement of NGO or other external aid organization
- economic heterogeneity in communities

The implications of excluding these effect modifiers from the analysis are discussed below.

Note on interpretation

We presented our results in Figure 4a-d to simplify the process of interpreting proportional odds models, which can be difficult. Categories of variables whose 95% confidence intervals do not cross zero are considered to be significantly associated with outcomes relative to the reference category for that variable, which is not shown in the model output. In proportional odds models, coefficients can be used to compare the odds of a fixed outcome between different values of the predictor. For instance, Participation (high) is a significant predictor of Attitudinal outcomes (Figure 2a). The significant coefficient for Participation (high) suggests that projects with high levels of participation are more likely to have attitudinal success than projects with no or low levels of participation. In this model, we compare the level of the variable reported in the model output (high) with that of the omitted reference value, which in this case is Participation (no/low).

Readers may be aware that we present Engagement and Tenure as being important variables for Ecological and Economic outcomes, respectively, despite there being no indication of a significant relationship in Figure 4c or d. The reason for this is because of the details of interpreting the effect of categorical variables in proportional odds models.

Models can be set such that variables have different reference categories. In this study, we set the reference category so that the categories that match our hypotheses are visible in the model output (e.g. Participation (high) is expected to be associated with successful outcomes, so Participation (no/low) was set as the reference category). However, we could have set the reference categories differently. Models with different reference categories are algebraically equivalent but can provide different insights into key relationships between categories of a given variable. For instance, Figure 2c shows how the high and moderate levels of Engagement relate to the reference category (no/low) but not to each other. A different parameterization of this model in which Engagement (high) is the reference category allows one to consider the relationship between moderate and high (not shown) and indicates that projects with moderate levels of Engagement are significantly less likely to result in Ecological success than projects with high levels of Engagement. The same is true for our discussion of the relationship between Tenure and Economic success despite no indication of significance in Figure 2d. A different parameterization of this model with Tenure (community) as the reference category indicates that projects in communities with mixed tenure regimes are significantly less likely to result in economic success than projects in communities with community-held tenure rights.

Objective 4: Coding validity

Questionnaires for 54 projects (40%) were returned. Functioning email addresses for twenty authors could not be located and eight authors were unable to complete the survey due to time constraints or an inability to remember the details of a project.

Author's responses were used to fill in missing values for a number of variables including *tenure, market access, population size, population heterogeneity, effective local government, supportive local culture, capacity skill,*

Table 10 Results of inter-coder reliability test comparing researchers coding of a paper with the responses of corresponding authors to a questionnaire distributed by the researchers

Variable name	Cohen's Kappa
Project impetus	0.49
Establishment	0.47
Decision-making	0.61
Approach local culture	0.84
Approach local institutions	0.58
Resource use	−0.14
Economic benefits	0.10
Equitable distribution	0.32
Elite capture	0.25
Capacity skills	0.18
Capacity institutions	0.08
Social capital	0.24
Environmental education	0.26
Market integration	0.13
Supportive local culture	0.32
Effective local government	0.46
Tenure	0.35
Charisma	−0.09
Population size	0.78
Population heterogeneity	0.62
Attitudinal outcomes	0.66
Behavioral outcomes	0.56
Ecological outcomes	0.67
Economic outcomes	0.52

Kappa values greater than 0.40 indicate moderate agreement, greater than 0.60 indicate substantial agreement, and greater than 0.80 indicate almost perfect agreement [89]. Negative values close to zero indicate that the disagreement between researchers and authors was likely due to chance, while larger negative values suggest that researchers and authors had a different perception of how to code for that variable. Note that not all variables were included in the questionnaire for authors.

approach to local culture, approach to local government, project impetus, establishment, decision-making, and *resource use*. Author responses were used most often to replace missing values for *effective local government, supportive local culture, approach to local culture, approach to local government,* and *population heterogeneity*. For each of these variables, between 20 and 30 missing values were replaced with the author's response.

The results of the intercoder reliability test to score the level of agreement between our coding and the corresponding authors' responses to the questionnaires can be found in Table 10. The results suggest that there was a fair or better (>0.40) level of agreement with the corresponding author for half of the variables. Negative kappa values indicate disagreement between researchers and authors. It is important to note that there were high levels of agreement for all four of the outcome variables.

Discussion

Using a considerably larger sample size and suite of predictor variables than our previous reviews, our results both support findings of prior studies and provide new insights into the role of project design, national context, local community characteristics, and outcome synergies. Although we do not find a systematic effect of national level variables on CBC outcomes, we do find some support for the importance of community characteristics, and strong support for success depending on several aspects of project design. Before discussing the wider implications of our results, we start by reviewing potential effect modifiers in this study, and how they may affect interpretation of our results.

Reasons for heterogeneity

While there is evidence that CBC can be an effective conservation tool, it is clearly not always successful. Here we discuss the ways in which particular variables might affect CBC success.

The lack of evidence for significant effects of national level indicators on projects success fails to support our hypotheses (*NC1 & NC2),* which represent the view that transparent and effective national governance influences project success through various mechanisms. These mechanisms might include enhanced trust as well as conditions that facilitate valuing future returns from resources more highly among its citizens. This result also runs counter to the finding from the only other quantitative study of national indicators, that HDI is important for successful fisheries co-management [23].

Conversely, project design variables were important for successful CBC outcomes. In fact, three of the four project design hypotheses (*H-PD1, -PD-3, and -PD4*) were supported to some degree, consistent with previous reviews emphasizing the importance of (a) building local institutional capacity [16] and training and skills development [15,22], (b) equitable distribution of benefits including avoidance of elite capture [16,104], (c) engagement with local institutions and cultural beliefs and traditions [14,105,106], (d) the provision of social capital and other intangible social benefits [15,16,23], and (e) participation in rule making and day-to-day decision-making associated with the project [16,19,77]. In our study, capacity building was a significant predictor of all but attitudinal outcomes indicating that it may be a particularly important component of project design. In addition to the straight-forward skills that may be necessary to maintain the project, Baland and Abraham [46] argue that capacity building may also help combat elite capture by ensuring that project finances are not simply funneled to a handful of local leaders.

Several examples from projects included in our review illustrate the importance of these project design variables. For instance, in a project in Yunnan, China, community members were trained to monitor resources [107]. Through this process the community gained an awareness of broader forest management needs beyond timber and drafted their own rules for sustainable harvesting of forest products. A project in Tanzania had similar results in that participatory monitoring efforts led to reductions in wildlife traps found in local forests and general improvements in forest quality [108]. Additionally, evidence from a project in Costa Rica suggests that providing opportunities for community learning was an important component of successful and sustainable sea turtle egg-harvesting [109].

Finally, although not considered a feature of project design in our study, we found (unlike previous reviews [20,55]) that projects that have been established for a longer period of time are more likely to have economic success than more recently initiated projects. This result provides evidence that development opportunities and income generation may not emerge quickly and that CBC projects may require time before measureable economic success is achieved.

On the other hand, community characteristics were less important for CBC success than were project design variables. Only two hypotheses in this domain - *H-CC2* and, to a lesser extent *H-CC3* - were supported. Strong tenure rights were related to economic success and both population size and local culture and institutions affected behavioral success. This does not mean that community characteristics are unimportant. In fact, many reviews point to aspects of local context that are thought to be key to securing successful outcomes, such as a supportive local belief system [15,20] and well defined property rights and local tenure regimes [15,19,22]. In addition, we want to reiterate that the *local institutions* variable could not be included in the multivariate analysis for ecological outcomes (see the paragraph on separation problems in Section 3.9.2 Multivariate analysis). This issue leaves open the question of whether supportive local cultures and effective local institutions are important for ecological success.

Among the cases in our sample, there are several examples of the importance of local context. Scanlon and Kull [110] note that cultural pride and a shared sense of belonging (as well as the provision of economic benefits) were important for attitudinal and behavioral success in a community conservancy in Namibia. Further, Bajracharya et al. [111] note that a history and culture of cooperation in addition to strong traditional management institutions (*Ban Samiti*) were crucial for the behavioral and ecological success of the Annapurna Conservation Area in Nepal. Finally, secure sea tenure provided an important foundation for the success of a marine protected area in the Solomon Islands [112]. These anecdotes provide examples of some of the ways in which local cultural contexts can affect project success. Local contexts might relate to lower rates of discounting the value of future resource availability, make communities responsible for broader environmental impacts (externalities), motivate sustainable resource use, increase accountability, and/or increase the salience of trust, reciprocity, and social norms [48,58]. However, our findings again suggest that well-designed projects can, in many instances, overcome unfavorable features of the local cultural and institutional context.

Several variables that we expected to be significantly associated with project success were not. In the realm of community characteristics, we found no evidence that market access was positively or negatively associated with any measure of success. This is important because some individual studies and reviews show positive associations between project outcomes and some measures of market integration [13,21,23], while others find that market integration can lead to greater resource extraction and ecological degradation [44,77,100,113]. The lack of a significant association between market integration and any of the outcome variables may be attributable to two factors. First, we found coding for market integration to be problematic due to the limited information often presented in the project evaluations we coded. Authors of several previous reviews have also noted similar difficulties in adequately extracting measures of market integration [19,78]. Second, it is likely that market effects are contingent on any number of other variables including the nature of the resources in question, the size and make-up of the community, and the type of market. These aspects of market integration deserve closer quantitative examination.

Local ecological conditions are also mentioned frequently in the literature [15,16,21,77], but had no discernable impact in our study. Few articles in our sample provided detailed information about ecological characteristics so it is not entirely surprising that our coarse measure of ecoregion status was not associated with project outcomes. Future research could benefit from including measures of habitat types, elevation, rainfall and other ecological characteristics from outside sources.

In addition, there are several variables for which we attempted to collect information but that we omitted from the analysis because of insufficient information in the projects in our sample. These variables could have affected our results in a number of ways. For instance:

- Local ecological conditions like rainfall, elevation or predominant habitat type could affect the abundance and spatial heterogeneity of valuable resources in any given year or set of years, which may not be captured in short-term project evaluations. Such variation could influence the ecological outcomes that are reported at a given point in time.
- Historical inter-community and intra-community dynamics may affect levels of trust and cooperation as well as resource use patterns in ways that may not be visible to researchers.
- Projects related to resources that are fundamental for subsistence may resonate more strongly in communities and thus result in greater community participation in all stages of the project than those related to resources that create a fundamental hardship for local peoples (e.g. land for agriculture vs. crop damage from wildlife). That is, levels of local participation may be higher for projects targeting resources that have a particular level of importance to a community.
- The *implementation* of national policies related to CBC can differ greatly between nations [5,6] making it difficult to ascertain how important national policies actually are for CBC outcomes.
- The degree and type of involvement of local or international NGOs and/or multilateral aid organizations could conceivably affect project outcomes. Such organizations may provide funding, consultation, or any number of other services and the nature of the organization (local, international, conservation-focused, development-focused, etc.) could influence the way a project is designed and implemented and how long funding is provided and for what activities. Similarly, one reviewer suggested that consideration be given to overlap in funders in the event that major funding organizations may influence the ways in which projects are designed and implemented. Unfortunately, many projects have multiple sources of funding making it difficult to discern how much funding came from a given source, not to mention how much influence such funding would have allowed. That said, future research could consider the potential that projects funded by the same agency or organization are designed and implemented in similar ways.

In short, there is the potential that additional factors could directly affect outcomes or mediate the effect of particular predictors on outcomes.

Review limitations and potential biases

The review was somewhat limited by the omitted variables noted above, although this was unavoidable since it depended on the information presented in the articles we reviewed. A larger set of search terms (e.g. community forest management, community forestry, community fisheries, community wildlife management, payment for ecosystem services, co-management) might have also increased the sample size. Given (a) the number of variables identified through theory, modeling, and empirical studies that potentially influence CBC outcomes, and (b) the multiple scales at which those variables operate (household, community, region, nation), a larger sample could help further tease out key relationships between predictors and outcomes and aid in examining the role of national socio-economic and political contexts.

This shortcoming is partially offset by fact that the current sample is nearly five times larger than that from Brooks et al. [13], and over twice as a large as that use in the analysis by Waylen et al. [14]. Further, it is unlikely that there would be any systematic bias as a result of the terms that we did or did not use in the search.

The generally short time frame of many of the projects in our sample (an average of 7.7 years from the initiation of the project to the date research was conducted) may limit our ability to determine whether certain successful outcomes might presage other types of success in the future. For instance, while resource use may be considered to be economically beneficial at the time the data were collected, we have no evidence that the harvest of that resource is sustainable over the long term. Similarly, the relatively short time frame also precludes us from fully understanding the nature of synergies and tradeoffs.

As for any review there is the potential for publication bias, in which authors fail to publish or report on projects that have not produced successful outcomes. However, given (a) the controversy and debates that highlight multiple perspectives on CBC and suggest that there is as much interest in documenting failure as documenting success, (b) the fact that 26% [103] of all outcomes were reported as failures, and (c) the absence of a relationship between author discipline and any of the four outcomes that would indicate a disciplinary bias in reporting results, we feel that publication bias presents a minimal problem for this analysis.

It is also important to note that the 136 CBC projects in our analysis may not represent a random sample of CBC projects because information on all existing CBC projects is not available. It would be impossible to obtain a true random sample of projects because of the absence of documentation and reporting on all CBC projects. However, because we selected cases without knowledge of the outcomes, we believe it is appropriate to generalize the findings beyond this sample.

Additional biases may include, [1] bias arising due to poor study designs, [2] author bias/conflict of interest, and [3] bias in how interventions are described and which aspects are reported. It is impossible to eliminate or even identify all these sources of bias, though it is important to be cognizant of them.

Review conclusions

Implications for policy and management

Our primary objective was to examine whether and to what extent CBC interventions can be effective tools for conservation. Because CBC is inherently intended to address multiple, interrelated goals, we included four measures of success in our analysis. Our results indicate that CBC can be an effective tool as we found that there were more instances of success than failure for all four outcome measures and that this was especially true for ecological outcomes. However, we note that the number of failures is still large and the presence of a number of "limited success" results suggests that one should not be too optimistic about the likelihood of CBC success in all cases.

It is important to note that some caution is needed in fully interpreting our results. First, reported outcomes are sometimes based on a narrow set of measures and come from studies of varying quality. For instance, economic success could be a function of increased income through the sale of goods and/or access to wage labor, construction of a basic health center or school, improved access to subsistence resources, or other factors. Authors may only report on a subset of these benefits and benefits from some activities could accrue more slowly than others. The same problem could hold true for all of the outcome measures. Second, we determined that many of the studies that make up our sample are not of high quality relative to controlled laboratory studies. However, we recognize that a host of obstacles exist that often prevent researchers from conducting studies of higher quality including time and funding limitations, the absence of true control communities, and the difficult logistics of working in multiple communities often in remote places.

Finally, it is important note that outcomes can change over time. For instance, ecological and economic success may not persist if harvest rates become unsustainable or market fluctuations reduce the value of a resource. Practitioners must pay careful attention to the multiple measures of each outcome that are possible, how these multiple measures change over time, and how these measures might differ among households and communities involved in the project.

In regards to objectives two and three, there are a number of important trends that we would like to highlight from the results of this analysis:

- Project design variables appear to be critical and well-designed projects may be able to overcome national-level and community-level circumstances that might otherwise inhibit project success
- The aspects of project design that are most important are capacity building (skills and institutional capacity), equitable distribution of resources (including avoidance of elite capture), creating or enhancing social capital, engaging with local cultural traditions, institutions, and leaders, and ensuring local participation in project initiation, design, and or daily management
- Community characteristics were significantly associated with some outcomes, though this wasn't as critical a domain as project design. The key community characteristics were supportive and effective local cultural institutions, low population size, and locally held tenure.
- Our results suggest that working closely with local communities within the confines of their traditions and institutions, preparing and training communities with relevant skills and organization, emphasizing non-tangible, non-economic benefits, and ensuring equitable benefit distribution are all more important than the types of benefits communities receive and their degree of access to resources.

The key result for objective two is that national context variables were not important predictors of project success. The significant associations between national-level predictors and outcomes in the bivariate analysis disappeared when national context, project design, and community characteristics were considered together in the multivariate analysis.

Because countries were not equally represented by projects, and because there remains the potential that national-level indicators that were not included in this study may be important, we do not conclude that higher-level institutions do not matter. Rather, we suggest that well designed projects can be successful even in national contexts that are not typically viewed as conducive to success (such as low HDI, rampant corruption, unstable governments or poor regulatory quality). This result is encouraging because conservation practitioners typically cannot change national development progress, governance, or political rights and freedoms, and it suggests that conservation projects can succeed in challenging socio-economic and political contexts.

The key finding for objective three is that project design variables were important predictors of success for all outcomes. Our results show the importance of design features that include emphasis on community participation, capacity building, and equitable distribution of economic benefits. In short, well-designed projects with many of these facets can help overcome pre-existing obstacles of society and setting. In contrast to project design variables, there is considerably less evidence for consistent effects of local community characteristics on project outcomes, although local tenure was associated with economic success and small populations and supportive local culture and institutions were associated with behavioral success.

Our results largely support those of previous systematic reviews of CBC [13,14]. However, we recognize that groupings of low quality studies have a higher likelihood of systematic bias and, as such, that both our review and the two previous reviews may be affected by such bias. Because of this concern, we think it is important to note that our results also in many ways match those of additional reviews and empirical studies that have examined fisheries co-management [23], integrated conservation and development projects [15,22,114] community forest management [16,19,21,77,78] and payment for ecosystem services [115]. In addition, our results also support several of our theoretically grounded hypotheses. These factors give us some degree of confidence in the variables that we have identified as key factors associated with successful outcomes of CBC projects. Taken together, our review along with other reviews, empirical studies and theoretical papers suggest that projects that balance economic incentives, community empowerment, and secure rights can succeed [116].

We conclude that conservation practitioners and policy makers should not avoid implementing projects in places where national, or local, socio-economic and political conditions are not thought to be suitable for conservation success; in fact these may be places where conservation efforts are most needed. Rather, we suggest that CBC projects conducted in such 'inhospitable' contexts must pay particular attention to quality project design that emphasizes capacity building, participation, the importance of social capital, and engagement with local traditions and institutions.

Implications for research

CBC research

This review highlights six key issues pertinent to future research on CBC.

1. Integrating thorough and systematic monitoring into project design

First, we would like to emphasize the need for more thorough and systematic monitoring of CBC projects. This statement echoes the conclusions reached in previous systematic reviews [13,14] as well as by other scholars [29,117]. This systematic review was, to some degree, constrained by the lack of standardized reporting of key independent variables and outcomes. As noted above, only 19% of projects in this review monitored all four outcomes, and no study monitored all four outcomes quantitatively. Further, the rate of quantitative monitoring was below 50% for each of the outcomes. This is not to say that qualitative monitoring cannot be useful. Indeed, qualitative data collected in well-designed projects can provide critical insights, particularly into complex social issues. For instance, our results indicate the importance of project design features like participation, capacity building, and strengthening social capital, but do not tell us how precisely to implement such features. That said, we view quantitative approaches as critical for testing hypotheses derived from nuanced qualitative work (e.g. 6, 8, 24), and conducting the systematic comparisons that are indispensable for guiding a broader understanding of the challenges and opportunities of CBC.

Of course, there are financial and temporal constraints on projects that preclude more thorough monitoring of outcomes. That said, given the tremendous amount of money directed to CBC projects, and the repeated calls for more, and better, monitoring of outcomes, it is surprising that the quality and amount of monitoring has improved little since the reviews we conducted in 2006 and 2010.

2. Exploring multiple measures of success

We also emphasize the importance of including *multiple* measures of success. Rigorously collected data that address only one or two measures of success have limited analytical value because CBC projects almost inevitably have ecological, economic, and social consequences. Without measures of success that span these distinct dimensions, the overall effectiveness of CBC will be difficult to determine. Without a more statistically powerful study that includes ecological, economic, attitudinal, behavioral, and/or additional outcomes of interest, conservationists face a situation in which a given project is judged a success by an economist based on increased income for local inhabitants and a failure by an ecologist and an anthropologist based on, respectively, a critical population decline for an important species and negative community attitudes towards conservation efforts. Insofar as the CBC paradigm is based on the assumption that human and ecological well-being are inextricably linked, proper support for this paradigm will need to demonstrate empirically such an interdependence of measures of success. In fact, a larger sample of projects reporting rigorously collected and analyzed data from well-designed studies would help immensely with the logical extension of the current study. While we have investigated which factors are associated with each of four outcomes, we have, thus far, been unable to answer the question of which factors are associated with particular patterns of synergies and tradeoffs among the four outcomes. We strongly encourage researchers to pursue these questions in future studies.

3. Addressing the multiple indicators of success or failure for each outcome

The third issue, which relates to the limited amount of outcome monitoring and quality of reporting, is the issue of *what* is monitored and reported for outcome variables. For each outcome, a multitude of indicators could be monitored that might provide conflicting information. For instance, in the context of attitudinal outcomes, community members could express positive attitudes towards conservation in general, but negative attitudes towards a national park or CBC project. For behavioral outcomes, community members might reduce their harvest of fuelwood because of the intervention but maintain or increase their encroachment into a protected area to collect other non-timber forest products. For ecological outcomes, the population of one species could have increased as the result of a project, but habitat degradation due to a secondary threat could continue to negatively affect other species. In the context of economic outcomes, a project might build a community school, but also reduce access to critical subsistence resources, which can negatively affect household income. In each case, measuring one aspect of an outcome domain would give only partial insight into the effects of the project. The number of, and variability in, potential indicators highlights the importance of the project having clear goals and of the article's authors clearly stating, to the best of their knowledge, those goals in the text. Primary researchers and evaluation teams need to be clear about the specific problem that the project is designed to address and the goals of the project in each of the four outcome areas.

4. Greater attention to standardizing relevant predictors

The fourth issue relates to which predictor variables are measured. The CBC literature is dominated by case studies that do not follow standardized guidelines for the type, quality, and amount of data that are collected. These problems are, in part, due to the interdisciplinary nature of CBC. Researchers from disciplinary perspectives ranging from ecology to political science to anthropology to human geography have an interest in CBC. With each disciplinary perspective comes a unique set of questions, theoretical backgrounds, and methodological toolkits that inevitably produce different kinds of data for different predictors. This diversity makes it difficult to compare interventions in diverse contexts because there are inevitably gaps in the information that is presented in any given publication.

There are efforts to remedy the lack of standardized data collection and reporting [118,119], although even these attempts have produced disagreements about which variables are most critical and how best to collect and organize data related to the inputs, design, and contexts that affect CBC outcomes. Further, while standardization will help with problems of what is reported, it will be of less help in improving study designs.

5. Exploring non-local influences of CBC

This study found little evidence for systematic effects of national level variables on project outcomes when other factors are also considered. However, this finding may be influenced by the variables chosen to represent effects at this level, or the possibility that national level variables interact with, or mediate, other influences with less direct effect. Furthermore, an institutional perspective suggests that there is no easy dichotomy between 'local' and 'national'. Depending on context, it is entirely likely that a range of other factors (e.g. at the regional level) may also play an effect. We suggest a fruitful and useful area for future research is disentangling whether and how influences at different scales affect outcomes of CBC projects.

6. Improving the quality of reporting

The final and most general issue is the quality of the studies conducted on CBC projects. As noted above, many studies of CBC projects fail to use appropriate control cases or to consider potentially important confounding variables, such as baseline socio-economic conditions, geographic location, or ecological conditions and variation. Although in-depth, qualitative case studies can be valuable, many such studies do not account for the variety of variables that might otherwise make them ideal examples of robust qualitative research. We, therefore, believe that high priority should be given to longitudinal rather than cross-sectional studies of CBC projects wherever funding and manpower allows as well as to studies that employ quasi-experimental designs in cases where "control" communities can be identified. Such approaches address the problems of endogeneity (such as CBC projects being set up in already troubled areas) and the lack of counterfactuals (assuming that changes in outcomes would not have happened in the absence of the intervention).

In addition to these broader implications for conducting CBC research we also have recommendations for future systematic reviews of CBC or related literatures that have emerged from this project.

1. Practical constraints and feasibility of extensive coding protocols

Both coders recognized the cognitive limitations of trying to code for 65 pieces of information in project reports and articles. While it is important to collect as much information from articles as possible, particularly for a topic as complex and interdisciplinary as CBC, there are diminishing returns to adding extra variables to a coding protocol. It is cognitively taxing to keep 65 pieces of information in mind while reading and coding for project information. We recommend that future reviews note the variables for which we were unable to collect adequate information as well as variables that were not significantly associated with any of the four outcomes in the bivariate or multivariate analyses. By excluding some of these variables, coders can focus their attention on more important factors.

2. Supplementing data from the sources' authors can be challenging but may be beneficial

The author's survey was beneficial for reducing the amount of missing information in our dataset, thus limiting the amount of imputed data. There were, however, some difficulties in the process of collecting author responses. We sought to minimize the survey length to avoid respondent fatigue and the potential for non-response. In doing so, we kept the directions and explanation for each question (which were distilled descriptions of our coding protocol) compact. We pretested the questionnaire with scholars who were both familiar and unfamiliar with CBC to make sure our instructions were clear. However, some respondents requested more detailed explanations of the variables about which we were inquiring as well as explanations of the distinctions between the response levels offered. While balancing survey length with survey clarity is always difficult, in hindsight it may have been better to reduce the number of questions and increase the explanation of the variable and response levels to ensure that respondents were interpreting variables as we had intended.

Some authors also commented on the difficulty of distilling a complex topic like land tenure into a three or four-level categorical variable. We felt it was important to match the response options provided to the authors as closely as possible to the ones used in our coding for the sake of minimizing our *interpretation* of author's responses and descriptions of the local conditions in the areas where they worked. If we had not, our interpretation of the author's responses would have been no different from our interpretation of the information presented in their publications. Nevertheless, it may be necessary to highlight this goal in future research so that authors understand our need for their informed distillation of the concept into a categorical response.

While our coding matched authors' responses to the questionnaire in some cases, it differed in others. There are numerous explanations for deviation between researchers and authors. The researchers had extensive discussions about the coding protocol, tested it on a sample projects, and revised it based on an inter-coder reliability test. The authors did not have the benefit of this process. As such, the authors and researchers could have had different interpretations and understandings of a question. In fact, in several cases the authors provided information in open-ended comments that made it clear that their selection from the categories provided did not match how the researchers would have coded that variable. Future efforts may benefit from greater care in crafting questions for author responses.

Similarly, the researchers became familiar with the different categories and arrived at a common classification scheme to differentiate between those categories (e.g. the difference between "moderate" and "high" levels of market access). Without such interactions, it is conceivable that the author's perceptions of the differences between categories differed from those of the researchers, despite efforts to clarify those differences within the questionnaire.

Finally, there is a discrepancy between the information available to the researchers and the information that can be drawn upon by the authors. While the researchers limited themselves to the information about a project presented in the articles, the authors were able to draw on their full experience with the project. In fact, many authors commented within the survey that there was more to the project than they were able to present in the associated article.

We do think that the questionnaire was adequate for the limited purpose of filling in missing values in the database. With modification, in the future this approach may be a more useful tool for more fully assessing the validity of coding projects derived from a variety of research methods and analytical techniques and written by authors from diverse disciplinary backgrounds.

3. Exploring interactions between key predictors to generate more detailed hypotheses

In this study, we lacked the data to explore the ways in which variables within the same domain and variables from different domains may interact to affect conservation outcomes. It is possible, and even likely, that some of the effects we observed are dependent upon, or mediated by the effects of other predictors. For instance, the effectiveness of local institutions (a part of the Engagement variable, which was significantly related to behavioral success) may be related to the size of the population or sociocultural or economic heterogeneity in the community [32]. One could also imagine that equitable distribution of benefits may be a function of the degree of participation in project design and implementation. We believe that a crucial next step is to explore these, and other, potential interactions among variables. Qualitative case study analyses can assist with this, but more and better data are required before such a step can be taken using a quantitative comparative approach.

General conclusion

This systematic review provided further evidence that CBC can be an effective conservation tool, particularly when studies are well designed and initiated in favorable local contexts. This study aimed to be the first systematic review to explore the effect of national-level variables on the outcomes of CBC projects. Using multivariate analysis we find little support for the prediction that the measured national-level variables would be associated with CBC project outcomes. There were, however, some significant effects in bivariate studies, and these associations warrant further research. In future studies, it will be important to use meaningful measures of a range of outcome-types. Our other results broadly confirm the patterns of association between predictors and project outcomes that have been reported by some previous systematic reviews, as well as more generally in the conservation literature. The multivariate results provide strong indication that well-design projects may be able to overcome challenges imposed by national contexts (e.g. low human development, or rampant corruption) as well as local contexts (e.g. ineffective institutions or unsupportive cultural traditions). This is surely heartening to conservation policy and practitioners. However, understanding more about how factors from multiple scales affect project success, and thus improving conservation practice, is dependent on evidence. As such, improving the quantity and quality conservation monitoring and reporting is still a priority.

Endnotes

[a]This disparity between the bivariate results and multivariate results can be explained by a statistical phenomenon known as Simpson's paradox [120].

[b]See Methods section, multivariate analysis for treatment of separation problems for Attitudinal and Ecological analyses.

Additional files

Additional file 1: Appendix A.

Additional file 2: Appendix B. Correlation matrix using Spearman's *r* for predictors used in the multivariate analysis for all five imputed datasets.

Additional file 3: Appendix A-D. Results of proportional odds logistic regression after stepwise, forward AIC model selection technique. The estimates and standard errors for each variable from each of the five imputed datasets are shown. Pooled estimates, standard errors and confidence intervals are also included.

Additional file 4: Appendix D. Projects and associated publications included in the study along with study quality assessment results and the type of monitoring and outcome reported for each project (quant= quantitative, qual=qualitative, auth = author's judgment).

Competing interests

The authors have no financial competing interests. Monique Borgerhoff Mulder is on the Editorial Board of Environmental Evidence and is on the Board of Directors of Savannas Forever, a Tanzania-based NGO that emphasizes conservation, communities, business and wildlife. No projects associated with Savannas Forever were included in the study.

Authors' contributions

JSB contributed to the conception of the study conducted the search, organized articles, helped develop the hypotheses, helped create and revise the coding protocol, helped write the draft protocol, coded all projects, helped develop the author questionnaire, contacted corresponding authors with a request to complete the questionnaire, entered all data, conducted the analyses, created all tables and figures, participated in writing the systematic review, and trained and oversaw research assistants who participated in the search, contacting authors and conducting the study quality assessment. KAW contributed to the conception of the study, helped create and revise the coding protocol, helped develop the hypotheses, helped write the draft protocol, coded about half of the projects, commented on author questionnaires and analyses, and helped write the systematic review. MBM contributed to the conception of the study, helped develop the hypotheses, helped write the draft protocol, commented on the analyses, and helped write the systematic review. All authors read and approved the final manuscript.

Acknowledgements

We would like to thank Pete Brosius and his lab at the Center for Integrative Conservation Research at the University of Georgia for hosting JSB and KAW, which greatly helped this project get up and running. We would like to thank research assistants Shandel Brown, Lauren Weisenfluh, and David Samuel Kim for their tremendous help with ACSC library search, delivery of the author questionnaire, coding for the quality assessment, and coder validity analysis. We would also like to thank all of the authors who were kind enough to respond to our emails and respond to the questionnaire. Finally, we would also like to acknowledge support for JSB from the Arnold and Mabel Beckman Foundation through a postdoctoral fellowship at the Beckman Institute, University of Illinois Urbana-Champaign.

Author details

[1]The Ohio State University School of Environment and Natural Resources, 210 Kottman Hall, 2021 Coffey Rd, Columbus, OH 43212, USA. [2]The James Hutton Institute, Social, Economic, and Geographic Sciences Group Cragiebuckler, Aberdeen, Scotland AB15 8QH, UK. [3]Anthropology

Department, University of California, Davis, 210 Young Hall, One Shields Ave., Davis, CA 95616, USA.

References

1. Robinson JG: **Ethical pluralism, pragmatism and sustainability in conservation practice.** *Biol Conserv* 2011, **144**(3):958–965.
2. Western D, Wright M (Eds): *Natural connections: Perspectives in community-based conservation.* Washington, DC: Island Press; 1994.
3. Horwich RH, Lyon J: **Community conservation: practitioners' answer to critics.** *Oryx* 2007, **41**(3):376–385.
4. Agrawal A, Gibson CC: **Enchantment and disenchantment: the role of community in natural resource conservation.** *World Develop* 1999, **27**(4):629–649.
5. Ribot J, Agrawal A, Larson AM: **Recentralizing while decentralizing: how national governments reappropriate forest resources.** *World Develop* 2006, **34**(11):1864–1886.
6. Nelson F (Ed): *Community Rights, Conservation, and Contested Land.* Washington, DC: Earthscan; 2010.
7. Hulme D, Murphree M: **Communities, wildlife, and the "new conservation" in Africa.** *J Int Dev* 1999, **11**(2):277–285.
8. Barrett CB, Arcese P: **Are integrated conservation-development projects (ICDPs) sustainable? On the conservation of large mammals in Sub-Saharan Africa.** *World Develop* 1995, **23**(7):1073–1084.
9. McShane T, Wells MP (Eds): *Getting Biodiversity Projects to Work: Towards More Effective Conservation and Developmeny.* New York: Columbia University Press; 2004.
10. Brockington D, Scholfield K: **Expenditure by conservation nongovernmental organizations in sub-Saharan Africa.** *Conserv Lett* 2010, **3**(2):106–113.
11. Dressler W, *et al*: **From hope to crisis and back again? A critical history of the global CBNRM narrative.** *Environ Conserv* 2010, **37**(1):5–15.
12. Chan KMA, *et al*: **When agendas collide: human welfare and biological conservation.** *Conserv Biol* 2007, **21**(1):59–68.
13. Brooks JS, Franzen MA, Borgerhoff Mulder M, Holmes CH, Grote M: *Development as a conservation tool: Evaluating ecoological, economic, behavioural, and attitudinal outcomes. in Systematic Review No. 20.* Birmingham, UK: Centre for Evidence-Based Conservation; 2006.
14. Waylen KA, Fischer A, McGowan PK, Thirgood SJ, Milner-Gulland EJ: *The effect of local cultural context on community-based conservation interventions: evaluating ecological, economic, attitudinal, and behavioural outcomes. in Systematic Review No. 80.* Birmingham, U.K: Collaboration for Environmental Evidence; 2010.
15. Garnett ST, Sayer J, du Toit JT: **Improving the effectiveness of interventions to balance conservation and development: a conceptual framework.** *Ecology and Society* 2007, **12**(1):2.
16. Tole L: **Reforms from the ground up: a review of community-based forest management in tropical developing countries.** *Environ Manage* 2010, **45**(6):1312–1331.
17. Borgerhoff Mulder M, Coppolillo P: *Conservation: Linking Ecology, Economics and Culture.* Princeton, NJ: Princeton University Press; 2005.
18. Brooks JS, Franzen MA, Holmes CM, Grote M, Borgerhoff Mulder M: **Testing hypotheses for the success of different conservation strategies.** *Conserv Biol* 2006, **20**(5):1528–1538.
19. Padgee A, Kim Y, Daugherty PJ: **What makes community forestry management successful: a meta-study from community forests throughout the world.** *Soc Natur Resour* 2006, **19**(1):33–52.
20. Waylen KA, Fischer A, McGowan PK, Thirgood SJ, Milner-Gulland EJ: **The effect of cultural context on the success of community-based conservation interventions.** *Conserv Biol* 2010, **24**(4):1119–1129.
21. Agrawal A, Chhatre A: **Explaining success on the commons: community forest governance in the Indian Himalaya.** *World Develop* 2006, **34**(1):149–166.
22. Tallis H, Kareiva P, Marvier M, Chang A: **An ecosystem services framework to suppport both practical conservation and economic development.** *Proc Nat Acad Sci USA* 2008, **105**(28):9457–9464.
23. Gutierrez NL, Hilborn R, Defeo O: **Leadership, social capital and incentives promote successful fisheries.** *Nature* 2011, **470**:386–390.
24. Struhsaker TT, Struhsaker PJ, Siex KS: **Conserving Africa's rain forests: problems in protected areas and possible solutions.** *Biol Conserv* 2005, **123**(1):45–54.
25. Ostrom E: **A diagnostic approach for going beyond panaceas.** *Proc Nat Acad Sci USA* 2007, **104**(39):15181–15187.
26. Nelson F, Agrawal A: **Patronage or participation? community-based natural resource management reform in Sub-Saharan Africa.** *Develop Change* 2008, **39**(4):557–585.
27. Smith RJ, Muir RDJ, Walpole MJ, Balmford A, Leader-Williams N: **Governance and the loss of biodiversity.** *Nature* 2003, **426**:67–70.
28. Dasgupta P: *Economics: A Very Short Introduction.* Oxford: Oxford University Press; 2007.
29. Sutherland WJ, Pullin AS, Dolman PM, Knight TM: **The need for evidence-based conservation.** *Trend Ecol Evolut* 2004, **19**(6):305–308.
30. Wollenberg E, Merino L, Agrawal A, Ostrom E: **Fourteen years of monitoring community-managed forests: learning from IFRIs experience.** *Int For Rev* 2007, **9**(2):670–683.
31. Agrawal A, Ostrom E: **Collective action, property rights, and decentralization in resource use in India and Nepal.** *Politics and Society* 2001, **29**(4):485–514.
32. Poteete AR, Ostrom E: **Heterogeneity, group size, and collective action: the role of institutions in forest management.** *Develop Change* 2004, **35**(3):435–461.
33. Ruttan L: **Sociocultural heterogeneity and the commons.** *Curr Anthropol* 2006, **47**(5):843–853.
34. Ostrom E, Schlager E: **The formation of property rights.** In *Rights to nature.* Edited by Hanna SS, Foke C, Maler KG. Washington DC: Island; 1996:127–156.
35. Gibson C, Williams JT, Ostrom E: **Local enforcement and better forests.** *World Develop* 2005, **33**(2):273–284.
36. Gibson C, Doods D, Turner P: **Explaining community-level forest outcomes: salience, scarcity and rules in Eastern Guatemala.** *Conserv Soc* 2007, **5**(3):361–381.
37. Schwartzman S, Nepstad D, Moreira A: **Arguing tropical forest conservation: people versus parks.** *Conserv Biol* 2000, **14**(5):1370–1374.
38. Redford KH, Padoch C: *Conservation of neotropical forests: working from traditional resource use.* New York: Columbia University Press; 1992:475. xiii
39. Naughton-Treves L, Buck M, Brandon K: **The role of protected areas in conserving biodiversity and sustaining local livelihoods.** *Annu Rev Environ Resour* 2005, **30**(1):219–252.
40. Dietz T, Ostrom E, Stern PC: **The struggle to govern the commons.** *Science* 2003, **302**(5652):1907–1912.
41. Ostrom E: **A behavioral approach to the rational choice theory of collective action.** *Am Pol Sci Rev* 1998, **92**(1):1–22.
42. Alvard M: **Indigenous hunting in the neotropics: conservation or optimal foraging.** In *Behavioral Ecology and Conservation Biology.* Edited by Caro T. New York: Oxford University Press; 1998.
43. Agrawal A: **Common property institutions and sustainable governance of resources.** *World Develop* 2001, **29**(10):1649–1672.
44. Cox M, Arnold G, Tomas SV: **A review of design principles for community-based natural resource management.** *Ecology and Society* 2010, **15**(4):38–57.
45. Ostrom E, Burger J, Field CB, Norgaard RB, Policansky D: **Revisiting the commons: Local lessons, global challenges.** *Science* 1999, **284**(5412):278–282.
46. Baland J, Abraham A: **Participatory development and the preferences of endogenous community imperfections.** *J Dev Stud* 2002, **39**(2):104–136.
47. Godoy R: *Indians, Markets, & Rainforests: Theories, Methods & Analysis.* New York: Columbia University Press; 2001.
48. Ostrom E: *Governing the Commons: The Evolution of Institutions for Collective Action.* Cambridge: Cambridge University Press; 1990.
49. Peterson R, Russell D, West PC, Brosius JP: **Seeing (and doing) conservation through cultural lenses.** *Environ Manage* 2010, **45**(1):5–18.
50. Bruner AG, Gullison RE, Rice RE, da Fonseca GAB: **Effectiveness of parks in protecting tropical biodiversity.** *Science* 2001, **291**(5501):125–128.
51. Berkes F: **Re-thinking community-based conservation.** *Conserv Biol* 2004, **18**(3):621–630.
52. Lindsay WK: **Integrating parks and pastoralists: Some lessons from Amboseli.** In *Conservation in Africa: People, policies and practice.* Edited by Anderson D, Grove R. Cambridge: Cambridge University Press; 1987:149–167.
53. Caro TM, *et al*: **Consequences of different forms of conservation for large mammals in Tanzania: preliminary analyses.** *Afr J Ecol* 1998, **36**(4):303–320.

54. Ferraro PJ: **The future of payments for environmental services.** *Conserv Biol* 2011, **25**(6):1134–1138.
55. Bardhan P: **Decentralization of governance and development.** *J Econ Perspect* 2002, **16**(4):185–205.
56. Dasgupta A, Beard VA: **Community driven development, collective action and elite capture in indonesia.** *Develop Change* 2007, **38**(2):229–249.
57. Saito-Jensen M, Nathan I, Treue T: **Beyond elite capture? community-based natural resource management and power in Mohammed Nagar village, Andhra Pradesh, India.** *Environ Conserv* 2010, **37**(3):327–335.
58. Pretty J: **Social capital and the collective management of resources.** *Science* 2003, **302**(5652):1912–1914.
59. Janssen MA, Ostrom E: **Adoption of a new regulation for the governance of common pool resources by a heterogeneous populations.** In *Inequality, Cooperation, and Environmental Sustainability.* Edited by Baland J, Bardhan P, Bowles S. Princeton, NJ: Princeton University Press; 2007:60–96.
60. Salafsky N, *et al*: **A systematic test of an enterprise strategy for community-based biodiversity conservation.** *Conserv Biol* 2001, **15**(6):1585–1595.
61. Baland JM, Bardhan P, Bowles S (Eds): *Inequality, cooperation, and environmental sustainabllity.* Princeton, NJ: Princeton University Press; 2007.
62. Abbot JIO, Thomas DHL, Gardner AA, Neba SE, Khen MW: **Understanding the links between conservation and development in the Bamenda highlands, Cameroon.** *World Develop* 2001, **29**(7):1115–1136.
63. Brashares JS, Golden CD, Weinbaum KZ, Barrett CB, Okello GV: **Economic and geographical drivers of wildlife consumption in rural Africa.** *Proc Nat Acad Sci USA* 2011, **108**(34):13931–13936.
64. Sponsel LE, Headland TE, Bailey RC (Eds): *Tropical Deforestation: The Human Dimensions.* New York: Columbia University Press; 1996.
65. Sayer J, Campbell B: *The Science of Sustainable Development.* Cambridge, UK: Cambridge University Press; 2004.
66. Paciotti B, Borgerhoff Mulder M: **Sungusungu: the role of preexisting and evolving social instiutions among Tanzanian vigilante organizations.** *Hum Organ* 2004, **63**(1):112–124.
67. Carpenter JP: **Punishing free-riders: how group size affects mutual monitoring and the provision of public good.** *Games Econ Behav* 2007, **60**(1):31–51.
68. Marwell G, Oliver PE: *The Critical Mass in Collective Action: A Micro-Social Theory.* New York: Cambridge University Press; 1993.
69. Olson M: *The Logic of Collective Action: Public Goods and the Theory of Groups.* Cambridge, Mass: Harvard University Press; 1965:176. x.
70. Agrawal A, Goyal S: **Group size and collective action, third party monitoring in common-pool resources.** *Comp Pol Stud* 2001, **34**(1):63–93.
71. Agrawal A, Yadama GN: **How do local institutions mediate market and population pressures on resources? Forest Panchayats in Kumaon, India.** *Dev Change* 1997, **28**(3):435–465.
72. Ruttan L: **Economic heterogeneity and the commons: effects on collective action and collective goods provisioning.** *World Develop* 2008, **36**(5):969–985.
73. Baland J, Platteau JP: **The ambiguous impact of inequality on local resource management.** *World Develop* 1999, **27**(5):773–788.
74. Adhikari BD, Lovett JC: **Institutions and collective action: does heterogeneity matter in community-based resource management.** *J Dev Stud* 2006, **42**(3):426–445.
75. Andersson K, Agrawal A: **Inequalities, institutions, and forest commons.** *Glob Environ Chang* 2011, **21**(3):866–875.
76. ACSC (2010) http://www.tradeoffs.org/app/Public/Catalog. (Advancing Conservation in a Social Context Reference Library).
77. Persha L, Agrawal A, Chhatre A: **Social and ecological synergy: local rulemaking, forest livelihoods, and biodiversity conservation.** *Science* 2011, **331**(6024):1606–1608.
78. Oldekop JA, Bebbington AJ, Brockington D, Prieziosi RF: **Understanding the lessons and limitations of conservation and development.** *Conserv Biol* 2010, **24**(2):461–469.
79. Higgins J & Green S eds (2008) *Cochrane Handbook for Systematic Reviews of Interventions* (The Cochrane Collaboration, http://www.cochrane-handbook.org).
80. Gamer M, Lemon J, Fellows I: *The irr package for R: various coefficients of interrater reliability and agreement.* Vienna: R foundation for Statistical Computing; 2008.
81. Team RDC: *R: A language and environment for statistical computing. Version 2.13.0.* Vienna: R Foundation for Statistical Computing; 2011.
82. Cohen J: **Weighted kappa: nominal scale agreement with provision for scaled disagreement or partial credit.** *Psychol Bull* 1968, **70**(4):213–220.
83. Siegel S, Castellan NJJ: *Nonparametric statistics for the life sciences.* London: McGraw-Hill; 1988.
84. Schwartzman S, Moreira A, Nepstad D: **Rethinking tropical forest conservation: perils in parks.** *Conserv Biol* 2000, **14**(5):1351–1357.
85. Redford KH: **The empty forest.** *Bioscience* 1992, **42**(6):412–422.
86. Storey JD: **A direct approach to false discovery rates.** *J R Stat Soc B-Stat Methodol* 2002, **64**(3):479–498.
87. Roback PJ, Askins RA: **Judicious use of multiple hypothesis tests.** *Conserv Biol* 2005, **19**(1):261–267.
88. Little RJA, Rubin DB: *Statistical Analysis with Missing Data.* 2nd edition. New York: John Wiley; 2002.
89. Chatfield C: **Model uncertainty, data mining and statistical inference.** *J R Stat Soc A* 1995, **158**(3):419–466.
90. Babyak MA: **What you see may not be what you get: a brief, nontechnical introduction to overfitting in regression-type models.** *Psychosom Med* 2004, **66**(3):411–421.
91. Venables WN, Ripley BD: *Modern Applied Statistics with S. Fourth Edition.* NY: Springer Science New York; 2002.
92. Burnham KP, Anderson DR: *Model Selection and Multi-Model Inference: a Practical Information-Theoretic Approach.* New York, NY: Springer; 2002.
93. Steyerberg EW, Marinus JC, Eijkemans J, Habbema DF: **Stepwise selection in small data sets: a simulation study of bias in logistic regression analysis.** *J Clin Epidemiol* 1999, **52**(10):935–942.
94. Reineking B, Schroder B: **Constrain to perform: regularization of habitat models.** *Ecol Model* 2006, **19**(3–4):675–690.
95. Tibshirani R: **Regression shrinkage and selection via the lasso.** *J R Stat Soc B* 1996, **58**(1):267–288.
96. Harrell FE: *Regression Modeling Strategies: With Applications to Linear Models, Logistic Regression, and Survival Analysis.* New York: Springer; 2001.
97. Gelman A, Hill J: *Data Analysis Using Regression and Multilevel/Hierarchical Models.* New York, NY: Cambridge Univeristy Press; 2007.
98. Rubin DB: *Multiple imputation for nonresponse in surveys.* New York: Wiley; 1987.
99. Albert A, Anderson JA: **On the existence of maximum likelihood estimates in regression models.** *Biometrika* 1984, **71**(1):1–10.
100. Sponsel LE, Bailey RC, Headland TN: **Anthropological perspectives on the causes, consequences, and solutions of deforestation.** In *Tropical deforestation: The human dimension.* Edited by Sponsel LE, Headland TN, Bailey RC. New York: Columbia University Press; 1996:3–52.
101. Andam KS, Ferraro PJ, Pfaff A, Sanchez-Azofiefa GA, Robalino JA: **Measuring the effectiveness of protected areas networks in reducing deforestation.** *Proc Nat Acad Sci USA* 2008, **105**(42):16089–16094.
102. Andam KS, Ferraro PJ, Sims KRE, Healy A, Holland MB: **Protected areas reduced poverty in Costa Rica and Thailand.** *Proc Nat Acad Sci USA* 2010, **107**(22):9996–10001.
103. Landis JR, Koch CG: **The measurement of observer agreement for categorical data.** *Biometrics* 1977, **33**(1):159–174.
104. Mansuri G & Rao V (2003) Evaluating community-based and community-driven development: a crtical review of the evidence. (World Bank), pp 1–54.
105. Klein J, Reau B, Kalland I, Edwards M: **Conservation, development and a heterogeneous community: The case of the Ambohitantely Special Reserve.** *Soc Natur Resour* 2007, **20**(5):451–467.
106. Stevens S: *Conservation through cultural survival.* Washington DC: Island Press; 1997.
107. Van Rijsoort J, Jinfeng Z: **Participatory resource monitoring as a means for promoting social change in Yunnan, China.** *Biodivers Conserv* 2005, **14**(11):2543–2573.
108. Topp-Jorgensen E, Poulsen MK, Lund JF, Massao JF: **Community-based monitoring of natural resource use and forest quality in montane forests and miombo woodlands of Tanzania.** *Biodivers Conserv* 2005, **14**(11):2653–2677.
109. Campbell LM, Hallboom BJ, Trow J: **Sustainability of community-based conservation: sea turtle egg harvesting in Ostinal (Costa Rica) ten years later.** *Environ Conserv* 2007, **34**(2):122–131.
110. Scanlon LJ, Kull CA: **Untangling the links between wildlife benefits and community-based conservation at Torra Conservancy, Namibia.** *Dev South Afr* 2009, **26**(1):75–93.
111. Bajracharya SB, Furley PA, Newton AC: **Effectiveness of community involvement in delivering conservation benefits to the Annapurna Conservation Area, Nepal.** *Environ Conserv* 2005, **32**(3):239–247.

112. Aswani S, Albert S, Sabetian A, Furusawa T: **Customary management as precautionary and adaptive principles for protecting coral reefs in Oceania.** *Coral Reefs* 2007, **26**(4):1009–1021.
113. Porter-Bolland L, *et al*: **Community managed forests and forest protected areas: an assessment of their conservation effects across the tropics.** *Forest Ecol Manage* 2011, **268**(15):6–17.
114. Cao S, Zhing B, Yue H, Zeng H, Zeng J: **Development and testing of a sustainable environmental restoration policy on eradicating the poverty trap in China's Changting County.** *Proc Nat Acad Sci USA* 2009, **106**(26):10712–10716.
115. Tallis H, Goldman R, Uhl M, Brosi B: **Integrating conservation and development in the field: implementing ecosystem service projects.** *Front Ecol Environ* 2009, **7**(1).
116. Lele S, Wilhusen P, Brockington D, Seidler R, Bawa K: **Beyond exclusion: alternative approaches to biodiversity conservation.** *Curr Opin Environ Sustain* 2010, **2**(1–2):94–100.
117. Stem CJ, Lassoie JP, Lee DR, Deshler DD, Schelhas JW: **Comunity participation in ecotourism benefits: the link to conservation practices and perspectives.** *Soc Natur Resour* 2003, **64**:479–498.
118. Conservation Measures Partnership (http://www.conservationmeasures.org/).
119. Cambridge Conservation Forum (http://www.cambridgeconservationforum.org.uk/projects/measures/).
120. Simspon EH: **The interpretation of interaction in contingency tables.** *J R Stat Soc B* 1951, **13**(2):238–241.
121. Dudley N (Ed): *Guidelines for Applying Protected Area Management Categories.* Gland, Switzerland: IUCN; 2008.

3

A systematic review of phenotypic responses to between-population outbreeding

Raj Whitlock[1*], Gavin B Stewart[2], Simon J Goodman[3], Stuart B Piertney[4], Roger K Butlin[5], Andrew S Pullin[6] and Terry Burke[5]

Abstract

Background: The translocation of plants or animals between populations has been used in conservation to reinforce populations of threatened species, and may be used in the future to buffer species' ranges from the anticipated effects of environmental change. This population admixture can result in outbreeding, and the resulting "hybrid" offspring can be either fitter (heterosis) or less fit (outbreeding depression) than their parents. Outbreeding depression has the potential to undermine conservation plans that mix populations of declining or threatened species.

Methods: We searched for literature documenting phenotypic responses to intraspecific outbreeding between natural populations of animal and plant species. Outbreeding responses were summarised as log-response ratios that compared hybrid with mid-parent phenotypes (528 effect sizes from 98 studies). These data included effect sizes from both fitness components (survival, viability and fecundity traits) and other traits (e.g. morphological, physiological, defence), and were pooled using Bayesian mixed-effects meta-analysis.

Results: There was no overall effect of outbreeding on hybrid phenotypes (overall pooled effect = +2.61% phenotypic change relative to parents, 95% credible interval (CI) −1.03–6.60%). However, fitness component traits responded significantly more negatively to outbreeding than traits less directly linked with fitness. Our model predicted a significant 6.9% F1 generation benefit to outcrossing through non-fitness traits (CI 2.7–11.2%), but no significant benefit to these traits in the F2 (3.5%; CI −4.3–12.2%). Fitness component traits were predicted to suffer a cost (−8.8%) relative to parents in the F2 (CI −14.1– − 2.5%), but not in the F1 (+1.3%; CI −2.1–5.4%). Between-study variation accounted for 39.5% of heterogeneity in outbreeding responses, leaving 27.1% of heterogeneity between effect sizes within studies and 33.4% attributable to measurement error within effect sizes.

Conclusions: Our study demonstrates consistent effects of trait type on responses to intraspecific outbreeding, and indicates the potential for outbreeding depression in the F2. However, our analyses also reveal significant heterogeneity in outbreeding responses within and among studies. Thus, outbreeding costs will not always occur. Conservation practitioners may be able to anticipate when such outbreeding depression should arise using an existing decision-making framework that takes into account the context of hybridising populations.

Keywords: Outbreeding Depression, Heterosis, Phenotypic Response, Conservation Genetics, Line-cross Analysis, Outbreeding, Outcrossing, Intraspecific Hybridisation

* Correspondence: r.whitlock@liverpool.ac.uk
[1]Institute of Integrative Biology, University of Liverpool, The Biosciences Building, Crown Street, Liverpool L69 7ZB, UK
Full list of author information is available at the end of the article

Background

The need to ensure natural populations have sufficient genetic variation for their long-term sustainability has long been recognized [1,2]. Genetic diversity provides the raw genetic material upon which natural selection can act allowing adaption to changes in the environment (including current anthropogenic changes such as eutrophication and climate change). By definition, genetic diversity includes allelic and genotypic variants conferring both detrimental and beneficial effects on the phenotype and fitness. When sexual reproduction occurs among related individuals, detrimental variation can be exposed as a decline in fitness of the resulting offspring, an effect known as inbreeding depression [3,4]. Inbreeding becomes more likely as populations become smaller in size, and where this leads to inbreeding depression it may contribute to the extinction of small populations [5,6]. However, the natural movement or human-aided translocation of individuals into an inbred population from a separate population can result in fitness recovery, or "genetic rescue" within the inbred population [7-13].

Despite the potential benefits of creating hybrid populations, conservation practitioners remain cautious about applying population translocations widely as a tool to enhance population sustainability [14]. This caution may be well founded, since it has been shown that in some circumstances, population admixture can lead to a reduction in fitness of hybrid individuals within the mixed population, called "outbreeding depression" [8,15,16]. The basis of such a fitness reduction is that the populations being mixed may have become adapted to the specific environments in which they exist (extrinsic outbreeding depression), or they may have diverged genetically such that genomic incompatibilities are exposed upon population mixing (intrinsic outbreeding depression, also known as the break up of co-adapted gene complexes [15,17,18]). Hybrid viability can also be undermined by chromosomal rearrangements that differentiate parent populations (including inversions, translocations and centric fusions [19-21]). These reductions in fitness can be observed readily at one extreme of the inbreeding-outbreeding continuum, in hybrid offspring derived from mating between individuals of different species [20,22,23].

Extrinsic outbreeding depression is driven by adaptation of parent populations to different environments [18]. Hybrids individuals are expected on average to possess a phenotype intermediate to that of the parents (considering only additive gene effects [24]) and so may be unfit in either or both parental environments. The potential for extrinsic outbreeding depression is thought to increase with environmental distance between parental populations, and with duration of isolation of the parent populations in these separate environments [21]. Intrinsic outbreeding depression arises as a further consequence of evolution in allopatry. Fitness often depends on the integrated functioning of alleles at different gene loci ("co-adapted gene complexes"). When parent populations have been isolated from each other for a sufficient number of generations, they can evolve to "find" different solutions to the problem of adaptation that depend upon different sets of alleles [25]. While these co-adapted gene sets may work well within each population, their functioning can break down if mixed in hybrid individuals with alleles with which they have never been tested. Intrinsic outbreeding depression occurs when hybrid fitness falls below that of the mid-parent value [15]. Theory predicts that the potential for intrinsic outbreeding depression should be associated with genetic distance between parent populations [17,26]. However, there is little consistent empirical evidence for this [15,16,27].

Risks of both intrinsic and extrinsic outbreeding depression are expected to be low where migration among parental populations is high, or where the populations are recently isolated. This is because the genetic composition of the parent populations will be more homogeneous, limiting the extent of adaptive differentiation and development of the divergent genetic architectures that lead to outbreeding depression [21,28].

A growing body of evidence exists on the phenotypic effects of intraspecific outbreeding, and there have been several attempts to synthesise this literature. Edmands reviewed the relative risks of inbreeding and outbreeding [15]. She found that outbreeding responses in the F2 may be comparable to the effects of close inbreeding, and that F2 phenotypic responses to outbreeding were significantly more negative than F1 responses. McClelland and Naish [16] carried out a meta-analysis that focussed on the phenotypic consequences of outbreeding within and between fish species. They found overall positive responses to outbreeding, but were not able to identify factors that consistently explained variation in outbreeding responses. Together, these two papers have improved our understanding of responses to outbreeding, but both suffer from problems. Edmands' analysis was limited to study-systems for which data were available on both responses to inbreeding and responses to outbreeding, and did not consider or control for within-study measurement error. McClelland and Naish's work was limited in taxonomic scope. In addition, these articles neither reviewed the literature systematically nor addressed the problem of the occurrence of multiple outbreeding effect-size estimates within each study. Thus, there is a need for an expanded synthesis of the evidence on outbreeding responses, in order to guide future conservation management that results in outbreeding (including translocations).

We carried out a systematic review to determine whether there is a net change in fitness or phenotype in hybrid offspring arising from mating between individuals from separate populations. We also assessed whether outbreeding responses varied through the generations following outbreeding, varied for different trait types, or with geographical, high-level phylogenetic, environmental or life history variables.

Methods

Question formulation

The core question and scope of this systematic review were refined at a review group meeting, held at Losehill Hall, Derbyshire 28–29 October 2008. The review group included representation from UK conservation agencies, conservation charities, and scientists specialising in the field of conservation genetics. When a draft systematic review protocol was developed, the review group were able to feed back on the plans for the review. This a-priori review protocol [29] was also peer-reviewed and published on-line. This process improved the review structure and scope markedly, and helped to ensure relevance to stakeholders who represented end-users of the review's results.

Literature searches

Scope of literature searches

On 5 October 2010 we searched computer databases and online repositories (Table 1), in order to identify articles, studies and datasets relevant to the review question.

Literature search terms

We used the search strings listed in Table 2 to retrieve articles from online databases and repositories. Where database search facilities allowed, we used search strings in Table 2 as follows. Compound search strings were generated by combining each of the search strings in groups (i) and (ii) with each of the strings in group (iii) for a total of 9 + 15 = 24 search strings (Table 2). Search terms in group (iv) are catch-all phrases that were used to carry out searches without combination into more complex strings (Table 2). Some databases provided only limited support for compound search queries, or did not support Boolean operators. In these cases we employed a simplified version of the search (Additional file 1: Table S1).

Study inclusion criteria

Relevant subjects and scope

Relevant subjects were defined as individuals of any natural population of animal or plant (or experimental individuals derived directly therefrom), at any location globally, and their progeny. We defined natural populations as naturally occurring, naturalised or (re-) introduced populations that occupy natural habitat, and that persist in the absence of human intervention. Studies that described populations with alternative phraseology such as "provenance", "land-race" or "cultivar" were retained as potentially relevant until it could be ascertained whether they fitted the criteria given in this section.

Table 1 Electronic databases searched to identify literature on intraspecific outbreeding

Database name	Subset of content searched	Source
†BIOSIS previews	All	http://apps.webofknowledge.com/
†Copac National, Academic, & Specialist Library Catalogue	Search limited to theses only	http://copac.ac.uk/
§†Countryside Council for Wales Library Catalogue	All	https://www-library.ccw.gov.uk/
†Dissertation Abstracts (Proquest dissertations and theses)	Search limited to dissertations and theses only	http://search.proquest.com/index
†Electronic theses online service (ETHOS)	All	http://ethos.bl.uk/
†ISI proceedings	All	http://apps.webofknowledge.com/
ISI Web of Science	All	http://apps.webofknowledge.com/
§†Joint Nature Conservation Committee publications catalogue	All	http://jncc.defra.gov.uk/page-2183
§†Natural England Publications and Products Catalogue	All	http://publications.naturalengland.org.uk/
SciVerse Science Direct	All	http://www.info.sciverse.com/sciencedirect
SciVerse Scopus	All	http://www.scopus.com/home.url
Scirus	Searches limited to first 50 results from journals-only searches	http://www.scirus.com/srsapp/advanced/
§†Scottish Natural Heritage	All	http://www.snh.gov.uk/publications-data-and-research/publications/search-the-catalogue/

§ Databases or libraries held by UK public bodies (non-departmental governmental organizations).
† Database includes grey literature.

Table 2 Search strings used to query online databases

Group	Search string
(i) Outbreeding related strings	Outbreeding depression OR out-breeding depression OR outcrossing depression OR out-crossing depression OR out-mating depression OR outmating depression
(i)	(Hybridisation OR hybridization) AND (population OR interpopulation OR inter-population) NOT (interspecific OR inter-specific)
(i)	Heterosis AND population
(ii) Search strings related to the movement of individuals for conservation purposes	Translocation AND conservation
(ii)	Reinforcement AND conservation
(ii)	Augmentation AND conservation
(ii)	Restoration AND conservation
(ii)	Genetic rescue AND conservation
(iii) Fitness related strings	Fitness
(iii)	Surviv* OR mortality OR longevity
(iii)	Fecundity OR reproduc*
(iv) Catch-all search strings	Distance-dependent fitness
(iv)	Distance-dependent crossing success
(iv)	Distance-dependent mating success

Boolean syntax follows the ISI Web of Knowledge template, and was adapted as necessary for use within databases using other syntax.

In this review we focussed exclusively on outbreeding effects on phenotype occurring within the post-zygotic portion of the life cycle. This focus excluded pre-zygotic mating incompatibilities from consideration, but included, for example, progeny traits such as germination success and egg hatching rate. We excluded studies documenting outbreeding effects resulting from crosses amongst agricultural cultivars or strains, or populations under captive management (e.g. zoo populations). These populations have been subjected to a different selective regime (e.g. directional selection for yield and or disease resistance, adaptation to captivity) than may prevail in natural populations. We did, however, include studies that observed the effects of hybridisation between natural populations and farmed individuals of the same species. These hybridisation events are relevant to the conservation of natural populations and have been known to occur in aquaculture and fisheries contexts, for example. We excluded studies documenting outbreeding effects in microorganisms. This was because (1) many groups of microorganisms are poorly characterised from a taxonomic point of view (2) the meaning of outbreeding differs for some of these microorganisms whose genomes may be non-recombining, or that may engage in horizontal gene transfer, and (3) bacteria are not themselves the target of translocation between sites or populations within the context of conservation plans. Studies focussing on crosses or hybridisation among different recognised species (but not sub-species) were excluded. This is because the costs and benefits of interspecific hybridisation are better known than those stemming from intraspecific hybridisation. In addition, the conservation community has a greater awareness of hybridisation involving multiple species because conservation actions are often arranged around particular known species of conservation concern. Studies focussing on crosses or hybridisation among taxa with uncertain taxonomy at the species level were excluded, to avoid inadvertent inclusion of studies involving interspecific hybridisation.

Types of intervention

We considered the "interventions" listed below:

- Outbreeding resulting from experimental translocation of individuals from a donor population to a geographically separated recipient population.
- Outbreeding resulting from observed natural migration between geographically separated populations.
- Outbreeding resulting from experimental crosses between individuals from geographically separated populations.
- Outbreeding resulting from experimental crosses between individuals separated by known distances within a single population.

Types of comparator

We defined the treatment group and corresponding reference (comparator) group on the basis of pedigree information presented within individual studies. The treatment group common to the interventions above is composed of hybrid offspring arising from outcrossing between geographically defined populations or across space within a population. The corresponding comparator or reference group(s) are composed of less outbred (but not deliberately inbred) individuals stemming from either within-population crosses or such crosses occurring over a smaller physical distance than those in the treatment group. The comparison made in this review is between the phenotype or fitness of hybrid individuals and that of their less outbred parent lineages. Thus, regardless of the intervention type, the phenotype or fitness of parent individuals or non-hybrid offspring from the "source populations" and that of their hybrid offspring were required for inclusion of any study in the review.

Types of outcome

Relevant outcomes were measures of the phenotype or traits of individuals within "reference" parental lineages or "treatment" lineages resulting from outbreeding between these former. Pre-zygotic traits were excluded. We made the assumption that phenotypes were a function of the genes of the individuals measured and their immediate measurement environment. Thus, we also assumed that phenotypes were not influenced by maternal or other persistent environmental effects.

Types of study

Relevant studies were defined as those with treatment and reference groups that could be identified using pedigree information, and that also recorded appropriate outcome measures. We did not include studies that used mean d-squared to infer the extent of outbreeding using molecular markers. This is because mean d-squared may not provide a reliable estimate of the position of any individual on the inbreeding–outbreeding continuum [30]. We chose to exclude these studies because (1) there is no identifiable reference class to make a comparison with and (2) because doubt exists as to whether the measure faithfully reflects differences in the quantity of inbreeding or outbreeding that occurred to create an individual's genome [30]. Articles that reviewed or meta-analysed the existing literature were excluded from the review, although any missing records from their bibliographies were added to the project database and assessed against the criteria defined above.

Potential sources of heterogeneity

We investigated the relationship between outbreeding effect sizes and several sources of heterogeneity (i.e. effect modifiers; hereafter "explanatory variables"), as described below (see Additional file 1: Text S2 for full details).

- ***Taxon category*** High-level taxonomic category for species; one of amphibian, bony fish, crustacean, gastropod, insect, mammal, bivalve, nematode, plant, reptile, tunicate
- ***Lifespan category*** Description of the longevity of an organism e.g. short-, ≤ 2 years; or long-lived, > 2 years
- ***Physical distance*** Euclidian distance separating outcrossed populations, km
- ***Hybrid generation*** Generation following outbreeding in which phenotype was observed; F1, F2, etc.
- ***Trait type*** Categorical descriptor of trait; one of defence, development, fecundity, fitness, growth-rate, physiology, size, survival, viability, or other
- ***Fitness class*** Categorical description of whether the phenotypic measure was a component of fitness (fecundity, survival, viability, integrated/ compound measures of fitness), or was more distantly related to fitness (all other trait types)
- ***Trait timing*** Categorical description of the timing of expression or measurement of a trait (early, mid or late), relative to the life-history of the organism in question
- ***Mating system*** Predominant sexual reproductive mode of the study organism; highly inbreeding, mixed mating, highly outbreeding
- ***Observation environment*** Categorical description of the environment in which the phenotype was expressed and observed; one of lab, common garden or natural population or habitat

Article screening strategy

We assessed articles for inclusion in the review based on an hierarchical assessment of relevance. This was done by assessing article titles, followed by reading the abstract of articles with relevant titles, followed by reading the full-text of articles with relevant abstracts (Additional file 1: Text S3). The objectivity of the article assessment procedure was determined using the Kappa coefficient of agreement applied to independent assessments of article subsets ([31]; Additional file 1: Text S3).

Data extraction

We attempted to extract data from all articles assessed as relevant at the full-text article assessment stage. We approached the authors of articles to request raw data or data summaries in cases where articles did not present the data needed for meta-analysis.

Data for traits measured on a continuous scale were extracted as mean values for each outcrossing class (outbred treatment individuals, parental comparator lineages). We also extracted standard error estimates for these means and recorded the number of families contributing to each cross class and, where the information was available, the number of individuals measured per family and/or cross class. Where separate data were available for both parent lineages, these were summarised as a mean (mid-parent) value for the comparator group. Data for separate traits were extracted separately. In some studies that used more than two parental comparator lineages (populations) trait summaries were not available for each separate parental lineage. In these studies the comparator data were taken as mean values (within traits) across all parent lineages. Equivalently, in some cases, trait data from treatment individuals (outbred offspring) were available only as mean values across crosses or population pairs.

Data for traits measured as counts (e.g. survival) were extracted as successes (survival) and failures (e.g. non-survival, death), for each of the outcrossing classes. In cases where proportion data were presented, these were used to estimate the original counts by multiplication with the sample size (number of individuals) followed by rounding to a whole number. Where proportions were given as a mean across families with associated standard error we treated these in the same way as for the continuous data, described above. Data extraction was not duplicated but consistency of the extracted data was checked and extraction of data from difficult articles was discussed.

Study quality assessment

We assessed the quality of the relevant full-text articles by determining, for each article, a score based on the presence or absence of attributes that indicated its internal or external validity. The attributes that were scored are summarised in Table 3. We assigned points for each of the desired attributes possessed by each study and measured any studies' overall quality as the sum of points scored across the component attributes. The maximum possible quality score was 7 points, one for each of the seven attributes recorded. Assessments of study quality are not often presented in traditional reviews and meta-analyses, but their use is considered best practice in systematic reviews within ecology and evolution [32].

Data synthesis

Effect size metric

We summarised outbreeding effect sizes for trait data measured on a continuous scale (e.g. growth rate, seed mass production, lifespan) using the log response ratio [33]:

$$ES = \log(\mu_H/\mu_P) \tag{1}$$

where μ_H and μ_P are trait mean values for the outbred (hybrid) treatment group and parental comparator groups respectively. We estimated the study measurement error variance as [33]:

$$mev = S^2\left(1/\left(n_H\ {\mu_H}^2\right) + 1/\left(n_P\ {\mu_P}^2\right)\right) \tag{2}$$

where n_H and n_P are respective sample sizes (numbers of families) measured for the hybrid treatment and parental comparator groups, and S is the pooled standard deviation for the hybrid treatment group and the parental comparator group. We used n_H and n_P (number of families) to convert standard errors reported within articles to standard deviations, separately for the treatment and comparator groups. These were then pooled to yield S. ([33]; p. 22).

Where trait data were binomially distributed we calculated effect sizes as the log risk ratio [33]:

$$ES = \log\ ((a/n_H)/(c/n_P)) \tag{3}$$

Table 3 Attributes used to assess study quality

	Design feature	Study attribute
Internal validity of study	Appropriate comparator lineages used	Were both parental lineages represented for each hybrid cross? 1 point for yes, 0 points otherwise
	Temporal controls	Were outbred and less outbred individuals studied at the same time? 1 point for yes, 0 points otherwise
		Were all parental lineages observed via controlled crosses? E.g. F1 parental control cross for comparison to F1 treatment cross, F2 for F2 etc.? 1 point for yes, 0 points otherwise
	Sampling strategy	Were multiple independent pairs or sets of populations/lineages studied? 1 point for yes, 0 points for only a single pair of lineages
External validity of study	Environment for trait observations	Within what environmental context were progeny traits observed? Natural population or native habitat (1 point); environmentally realistic common garden (0.5 points); lab environment (0 points; includes glasshouses, labs, terraria, aquaria)
	Observation window for inbred/outbred individuals	Were phenotypic effects tracked until at least the F2 generation of progeny? 1 point for yes, 0 points otherwise
	Status of populations/lineages studied	Did the study include natural, naturalized, or cultivated lineages? 1 point where all populations were natural populations, 0 points where naturalized or cultivated lineages were included

In this case we estimated the study measurement error variance as [33]:

$$mev = 1/a - 1/n_H + 1/c - 1/n_P \quad (4)$$

In these equations, a and c are the numbers (counts) of successes (e.g. survival or reproduction of tested individuals) in treatment and reference groups. n_H and n_P in this case refer respectively to the total number of individuals tested in hybrid treatment and parental comparator groups. Our effect sizes represent measures of intrinsic outbreeding responses, i.e. deviation of hybrid from mid-parent phenotypes, due predominantly to dominance and epistatic gene effects [15,17,18].

Meta-analysis

Effect size estimates were meta-analysed using the R package MCMCGLMM [34], which provides functions for fitting generalised linear mixed models in a Bayesian framework, via a Markov chain Monte Carlo sampler. Models fitted by MCMCGLMM extend to fixed- and random-effects meta-analyses that model and take into account the study measurement error variance ([35]; equivalent to weighting the analysis by the inverse of the study measurement error variance in a traditional meta-analysis). All models were run for a total of 6×10^5 iterations, with a burn in of 10^5 iterations and a subsequent thin interval of 50 iterations. This sampling schema leads to a total MCMC sample size of 1000 for each fitted model. We used multivariate normal priors with mean 0 and variance 10^{10} for the fixed effects. Priors for random effects were uniform improper distributions on the standard deviation of the random effects as recommended in [36]. Measurement error variance estimates (*mev*) were fitted as a set of random effects, and these we assumed to be known without error. Study identities (*Study*) were fitted as random effects in all models, in order to capture and model variation in outbreeding contexts. Explanatory variables of interest were fitted as fixed effects. Note that the distinction between "fixed" and random effects is arbitrary in a Bayesian modelling framework such as used here, where all effects are random. However, we retain separate notation for fixed and random effects specifications to facilitate conceptual distinction between explanatory variables ("fixed" effects of primary interest) and study identities (random effects whose variation we wish to account for and estimate). Additional file 1: Text S4 gives further details on model specification.

In cases where individual model parameters represented pooled effects of interest, we used posterior means and 95% credible intervals for the parameters to derive predictive intervals for the effect sizes. In other cases, we predicted pooled effect sizes by summing over the posterior distributions of their component parameters, and then summarising these as above. The meta-effect sizes (parameter estimates) were considered statistically "significant" when their 95% credible interval did not bracket 0. Results of the meta-analyses were presented graphically using forest plots for effect sizes and pooled effect sizes. Model goodness of fit was assessed via the Deviance Information Criterion DIC; [37]. DIC is subject to variation between separate runs of identically specified MCMCGLMM models, due to Monte Carlo error, even in a well-fitted model. Therefore, we carried out three replicate model runs for each model fitted to ensure that we compared the goodness of fit of competing models fairly and consistently. Models were checked to assess the degree of mixing and convergence by visual inspection of the sampled MCMC chains of component parameters, by evaluating chain autocorrelation, and by determining the effective size of MCMC chains. MCMCGLMM initiates a single MCMC chain during model fitting. Therefore we also assessed MCMC chain convergence using Gelman-Rubin diagnostics applied to pairs of replicate model runs initiated independently from over-dispersed starting values.

We took two approaches to fitting explanatory variables using MCMCGLMM. In the first we fitted models with only a single fixed effect (one model for each explanatory variable). The aim of fitting these models was to explore variation in outbreeding responses with predictors of these responses. We considered variables to be potentially relevant in explaining outbreeding responses when any of their component parameters differed significantly from zero. Trait type and fitness class (fitness component and other traits) contained redundant information; levels of trait type were nested within levels of fitness class. Therefore, we fitted the fitness component vs. other traits comparison in two ways. First as a post-hoc orthogonal contrast within the trait.type predictor, and second using the fitness class predictor. Variation in outbreeding responses among trait types within the fitness component trait class was investigated using further post-hoc contrasts.

In the second approach to modelling outbreeding responses we fitted multiple explanatory variables within a single maximal model, which had a fixed-effects specification as follows:

$$\begin{aligned}&\text{Generation} + \text{Trait.type} + \text{Trait.timing}\\&\text{Taxon.category} + \text{Lifespan.category} + \text{Mating.system} +\\&\text{Quality.score}\end{aligned} \quad (5)$$

We omitted fitness class from this model because this predictor and the trait type predictor contained redundant information. It was also necessary to omit physical distance from the model reduction analysis, since we had incomplete information on this predictor across the dataset. After fitting the maximal model, we then defined a minimal model by elimination of those fixed effects that did not contribute to improving the model fit (as judged by changes in the deviance information criterion, DIC). The maximal model included main effects only (no interactions among fixed effects). We carried out three replicate model runs for the maximal model and each nested model derived from this, to ensure that fixed effects were eliminated or retained based on consistent changes in DIC. Only predictors whose exclusion resulted in a consistent cost to model fit were retained in the model.

We were also interested specifically in whether our minimal model implied either net costs or benefits to outbreeding in different generations. Therefore we ran an additional model that included an interaction between the explanatory variable identified by our minimal model and hybrid generation. We give results for similar models combining generation with each of the other explanatory variables in Additional file 1: Figure S12.

We evaluated the proportion of the heterogeneity in outbreeding responses attributable to variation among studies, and the proportion of heterogeneity associated with the residual variance component using the approach of Sutton *et al.* [38]. However, we took the median of the *mev* as our estimate of the typical measurement error variance, instead of equation 9 in [39], which gave a poor estimate of central tendency of the *mev* for our data.

Publication bias

We used the R package METAFOR [40] to create enhanced funnel plots as a graphical check for the presence of funnel-plot asymmetry (indicating publication bias). We used study-mean effect sizes to create study-level funnel plots, since publication bias is likely to operate at the level of studies rather than individual effect sizes within studies (effect sizes within studies are likely to be correlated). An additional reason for doing this was that the number of effect sizes per study in our data was unbalanced, undermining any assessment of bias based on the full dataset. We used the median measurement error variance for the effect sizes within each study as a "typical" study-level measurement error variance. We also used the Egger regression to test for the presence of funnel-plot asymmetry [41], using study-level data, as above.

Sensitivity analyses

In order to understand whether outbreeding responses were sensitive to study quality we included our study quality variable in both a single-predictor meta-analysis and in the model reduction analysis, as described above. We also trialled inverse gamma and "parameter expanded" proper Cauchy priors for the standard deviation of the random effects, as alternatives to the improper flat priors that we used. Variance component estimates were found to be insensitive to the choice of prior. Finally we tested whether our model and its underlying assumptions was consistent with the observed data, using posterior predictive simulation [42]. Full details and results for the posterior predictive simulation are given in Additional file 1: Text S7 and Additional file 1: Figure S8.

Results

Review descriptive statistics

The primary literature search (conducted 5 October 2010) yielded 9631 papers (Table 4; Additional file 2). The minimum set of data necessary for meta-analysis was obtained for a total of 93 articles published between 1987 and 2010 (Table 5; Additional file 2, and Additional file 3). Over this period, articles on phenotypic responses to between-population outbreeding increased in publication frequency roughly linearly with time (Figure 1). Most (58.1%) of the 93 articles in the review used plants as study taxa (Figure 2). Fish were also well represented in the review dataset. The numbers of effect size estimates contributed by different taxonomic groups corresponded approximately with the numbers of articles focussing on these groups in the review (Figure 2). 61.2% of effect sizes were for traits that were components of fitness (fecundity, survival, viability and compound measures of fitness; Figure 3).

Quantitative synthesis

Study-mean effect sizes were evenly distributed in funnel plots (Additional file 1: Figures S5, S6), and showed no obvious signs of asymmetry indicative of publication bias. The Egger test for funnel plot asymmetry was non-significant (intercept = 0.41, $p = 0.166$).

Meta-analyses with a single explanatory variable

On average, hybrid offspring experienced neither a phenotypic benefit nor a cost to outcrossing (+2.6% phenotypic change relative to parents; 95% credible interval −1.0–6.4%; *pMCMC* (Bayesian *p*-value) = 0.156; Figure 4a).

The trait type, fitness class (whether a trait was a component of fitness or not), trait timing, generation, taxon category and observation environment predictors contained individual factor levels with which outbreeding responses varied significantly (Figure 4; Additional file 1: Table S9). Of these, only trait type, fitness class and trait timing improved overall model goodness-of-fit, as judged by DIC (Table 6). Physical distance separating parent populations improved model fit, according to DIC, but parameters for this explanatory variable did not differ significantly from 0. No other

Table 4 Summary of the article assessment procedure

Stage of article assessment	Total number assessed	Subset assessed in duplicate (% of total)	Number passing assessment	Agreement level, κ (interpretation according to [43])
Title assessment	9631	600 (6.2)	905	0.61 (Substantial)
Abstract assessment	905	200 (22.1)	285	0.49 (Moderate)
Fulltext assessment	285	40 (14.0)	93*	0.83 (Almost perfect)

Objectivity of article assessment was determined using the Kappa coefficient of agreement via independent assessments of a subset of articles [31].
*17 articles could not be assessed and were excluded at this stage due to being written in a language other than English, because of unanswered data requests, or because the original source material could not be located or no longer existed. Data were either unavailable or not extractable for a further 13 articles.

variables were significant predictors of outbreeding responses (Figure 4; Table 6; Additional file 1: Table S9).

Outbreeding responses varied significantly with trait type. Growth-rate and "other" trait types showed the greatest hybrid benefit following outcrossing (Figure 4d; Additional file 1: Table S9). Defence, survival and viability trait types showed the most negative responses to outbreeding. Orthogonal contrasts within the trait type predictor indicated that fitness component traits (survival, viability, fecundity traits and compound measures of fitness) responded more negatively to outbreeding than all other traits ($pMCMC = 0.024$; Figure 5). This difference appeared to be driven by survival and viability traits, which responded significantly more negatively to outbreeding than the remaining fitness component traits ($pMCMC = 0.004$; Figure 5). The outbreeding responses of viability traits did not differ significantly from those of the remaining later acting survival traits ($pMCMC =$ 0.76; Additional file 1: Table S12).

The more negative response of fitness components to outbreeding was borne out by our fitness class predictor, which grouped trait types depending on whether they were components of fitness or not. Outbreeding responses were consistently less positive for traits that were components of fitness relative to other traits ($pMCMC < 0.001$; Figure 4c, Additional file 1: Table S9). Fitness component traits showed outbreeding responses that were close to the mid-parent value (0.0% phenotypic change relative to the mid-parent), and not consistently different from zero (Figure 4c, Additional file 1: Table S9). The remaining non-fitness component traits conferred a consistent benefit to hybrids on outcrossing (by 6.6%).

Traits that acted during the middle or later stages of the lifecycle showed significantly more positive responses to outbreeding compared with early acting traits ($pMCMC =$ 0.006 and 0.004 respectively). The absolute outbreeding response in late-acting traits was also significantly greater than zero (the mid-parent phenotype; Figure 4e).

F1 hybrids experienced a (non-significant) benefit to outbreeding (3.7% benefit in phenotype relative to the mid-parent value; $pMCMC = 0.064$;). In the F2, hybrids experienced a significantly lower phenotypic response to outbreeding than F1 hybrids (Figure 4b; $pMCMC <$ 0.001; Additional file 1: Table S9). F2 and F3 hybrids experienced a net cost to outbreeding (−4.7% and −15.9%), but these responses were not consistently different from the mid-parent phenotype (Figure 4b). Results for the F3 generation were supported by only a very small number of articles and effect sizes.

There was little evidence that outbreeding responses were explained by high-level taxonomy. Only mammals showed outbreeding responses that differed significantly from parent phenotypes ($pMCMC = 0.01$; Figure 4f), and this taxonomic group was represented by data from only four articles.

The association of physical distance with outbreeding responses was very small, and not significantly different from zero (−0.5% phenotypic change for each log unit of distance; $pMCMC = 0.368$). Only ~80% of articles reported useable data on physical distance.

Observation environment was not a consistent predictor of the outbreeding response. However, we found that phenotypic responses to outbreeding observed in lab environments were of consistently lower magnitude than those observed in natural populations or habitats (Figure 4k; $pMCMC = 0.002$; Additional file 1: Table S9). Neither population status, nor study quality score were consistent predictors of outbreeding responses (Figure 4; Table 6).

Meta-analysis with multiple explanatory variables

We used a model reduction approach to determine the set of explanatory variables that best predicted the data. The best-fitting minimal model contained only the trait type predictor (Figure 4d; Additional file 1: Tables S10 & S11).

Figure 6 shows outbreeding responses for different fitness classes (representing trait type; fitness components or not) in different generations. We could not fit a model including an interaction between trait type and generation because some trait types contained data from only one generation. F1 fitness component traits showed little response to outbreeding (+1.3% relative to parent lineages; Figure 6). However, fitness component traits showed a significantly negative response to outbreeding in the F2 (−8.8%). The remaining non-fitness component traits showed a consistently positive response to outbreeding during the F1 (+6.9%), and also a positive response during the F2 (+3.5%; Figure 6).

Table 5 Summary of species and studies included in this review

Database record number	Species	Taxon category	Source	No. effect sizes
24	*Litoria peronii*	Amphibian	[44]	1
195	*Rana temporaria*	Amphibian	[45]	2
4211	*Boeckella dilatata*	Crustacean	[46]	1
214	*Tigriopus californicus*	Crustacean	[47]	4
258	*Tigriopus californicus*	Crustacean	[48]	12
458	*Tigriopus californicus*	Crustacean	[49]	66
31	*Tigriopus californicus*	Crustacean	[50]	6
10135	*Gambusia holbrooki*	Bony fish	[51]	5
10138	*Heterobranchus longifilis*	Bony fish	[52]	10
10130	*Labeo rohita*	Bony fish	[53]	2
189	*Micropterus salmoides*	Bony fish	[54]	8
10150	*Micropterus salmoides*	Bony fish	[55]	4
209	*Micropterus salmoides*	Bony fish	[56]	4
802	*Oncorhynchus gorbuscha*	Bony fish	[57]	2
3933	*Oncorhynchus gorbuscha*	Bony fish	[58]	2
263	*Oncorhynchus gorbuscha*	Bony fish	[59]	2
10133	*Oncorhynchus mykiss*	Bony fish	[60]	10
239	*Oncorhynchus mykiss*	Bony fish	[61]	14
9192	*Oncorhynchus mykiss*	Bony fish	[62]	8
2400	*Oncorhynchus tshawytscha*	Bony fish	[63]	8
10097	*Poeciliopsis occidentalis occidentalis*	Bony fish	[64]	7
10896	*Salmo salar*	Bony fish	[65]	2
11269	*Salmo salar*	Bony fish	[66]	14
10834	*Salmo salar*	Bony fish	[67]	4
3269	*Salmo salar*	Bony fish	[68]	5
10094	*Salmo trutta*	Bony fish	[69]	2
10154	*Salvelinus alpinus*	Bony fish	[70]	4
11273	*Physa acuta*	Gastropod	[71]	4
33	*Physa gyrina*	Gastropod	[72]	1
771	*Oedipoda germanica*	Insect	[73]	2
141	*Ovis canadensis*	Mammal	[74]	3
2488	*Panthera leo*	Mammal	[12]	1
3471	*Peromyscus leucopus*	Mammal	[75]	4
11140	*Puma concolor*	Mammal	[7]	4
2412	*Argopecten circularis*	Bivalve	[76]	2
2413	*Argopecten circularis*	Bivalve	[77]	1
3514	*Caenorhabditis briggsae*	Nematode	[78]	2
424	*Agrostemma githago*	Plant	[79]	2
10860	*Antirrhinum majus*	Plant	[80]	4
9631	*Arnica montana*	Plant	[81]	1
10533	*Aster amellus*	Plant	[82]	4
118	*Avena barbata*	Plant	[83]	5
354	*Brassica cretica*	Plant	[84]	1
154	*Calylophus serrulatus*	Plant	[85]	12
54	*Campanula americana*	Plant	[86]	6

Table 5 Summary of species and studies included in this review *(Continued)*

9624	*Campanula americana*	Plant	[87]	3
10095	*Castilleja levisecta*	Plant	[88]	2
151	*Chamaecrista fasciculata*	Plant	[89]	4
3764	*Clarkia tembloriensis*	Plant	[90]	8
339	*Cochlearia bavarica*	Plant	[91]	3
197	*Dalechampia scandens*	Plant	[92]	6
3323	*Delphinium nelsonii*	Plant	[93]	12
22	*Digitalis purpurea*	Plant	[94]	8
11205	*Echinacea angustifolia*	Plant	[95]	2
3373	*Espeletia schultzii*	Plant	[96]	2
505	*Eucalyptus globulus*	Plant	[97]	5
502	*Eupatorium perfoliatum*	Plant	[98]	5
502	*Eupatorium resinosum*	Plant	[98]	5
3307	*Gentiana pneumonanthe*	Plant	[99]	3
515	*Gentianella germanica*	Plant	[100]	1
10096	*Gentianella germanica*	Plant	[101]	6
11035	*Geum urbanum*	Plant	[102]	8
10531	*Helianthus verticillatus*	Plant	[103]	7
123	*Hypochoeris radicata*	Plant	[104]	10
3369	*Ipomopsis aggregata*	Plant	[105]	8
1267	*Lobelia cardinalis*	Plant	[106]	1
488	*Lobelia siphilitica*	Plant	[107]	3
404	*Lotus scoparius*	Plant	[108]	1
3325	*Lychnis flos-cuculi*	Plant	[109]	5
11152	*Lychnis flos-cuculi*	Plant	[110]	3
10662	*Macadamia tetraphylla*	Plant	[111]	2
10567	*Mercurialis perennis*	Plant	[112]	1
47	*Panax quinquefolius*	Plant	[113]	4
424	*Papaver rhoeas*	Plant	[79]	1
4224	*Phlox drummondii*	Plant	[114]	3
1852	*Pinus banksiana*	Plant	[115]	2
1706	*Pinus pinaster*	Plant	[116]	3
172	*Piriqueta caroliniana*	Plant	[117]	12
11042	*Plantago lanceolata*	Plant	[118]	6
64	*Ranunculus reptans*	Plant	[119]	18
170	*Ranunculus reptans*	Plant	[120]	3
3363	*Sabatia angularis*	Plant	[121]	9
411	*Sarracenia flava*	Plant	[122]	2
231	*Scorzonera humilis*	Plant	[123]	1
9737	*Senecio pterophorus*	Plant	[124]	3
392	*Shorea cordifolia*	Plant	[125]	6
424	*Silene alba*	Plant	[79]	2
10585	*Silene latifolia*	Plant	[126]	4
73	*Silene latifolia*	Plant	[127]	6
167	*Silene vulgaris*	Plant	[128]	6
332	*Silene vulgaris*	Plant	[129]	12

Table 5 Summary of species and studies included in this review *(Continued)*

392	*Syzygium rubicundum*	Plant	[125]	6
278	*Thymus vulgaris*	Plant	[130]	3
182	*Viola pumila*	Plant	[131]	2
182	*Viola stagnina*	Plant	[131]	2
65	*Zostera marina*	Plant	[132]	2
4539	*Zootoca (Lacerta) vivipara*	Reptile	[133]	3
4060	*Botryllus schlosseri*	Tunicate	[134]	18

The fifth column refers to the number of effect size estimates contributed by each study.

Heterogeneity in outbreeding responses

The study variance component (describing heterogeneity in outbreeding response among studies) was 0.0145 in the minimal model (Table 6), and accounted for 39.5% of total heterogeneity in outbreeding responses. The within-study (between effect size) variance accounted for 27.1% of heterogeneity. The remaining heterogeneity (33.4%) was attributable to measurement error variance (variation within effect sizes).

Discussion

In this review we have shown that hybrids arising from intraspecific outbreeding express phenotypes that do not differ from those expected given parent phenotypes (+2.6% phenotypic change; $pMCMC = 0.156$). However, we also identified predictors with which outbreeding responses were associated significantly, and heterogeneity in responses among studies.

Reasons for heterogeneity

There was significant heterogeneity in outbreeding responses among studies. This among-study variation in outbreeding responses could be due to species-specific or population-specific effects, or other study-specific contexts. In other words, part of this heterogeneity may be attributable to unmeasured, but biologically significant case-specific variables. These might include population demographic history and isolation, genetic architecture and differential adaptation between hybridising populations [17,21]. This study-level contextual variation must be taken into account when interpreting the relationships between explanatory variables and outbreeding responses described in this review.

We found that trait type explained variation in outbreeding responses, and was the single best predictor of these responses. Fitness components (especially survival and viability) had consistently lower outbreeding responses than other traits. In the only other meta-analysis on outbreeding responses [16], McClelland and Naish found that overall positive responses to outbreeding in fish species were driven largely by the responses of morphological traits. This observation is consistent with our result that non-fitness traits show a significantly more positive response to outbreeding than fitness component traits. Our result implies that the traits most important in regulating population size, growth rate and persistence may be those affected most negatively by outbreeding. In animals, fitness traits are also those influenced most strongly by inbreeding depression [135]. This effect is thought to be driven by differences in the extent of dominance variance between fitness and other traits, or in the average directionality of dominance interactions [135]. However, these arguments cannot be extended easily to explain our result. In fact, they lead to the expectation that fitness traits should respond more positively to outbreeding than other traits, at least where the source populations are inbred. We suggest that our result implies that fitness component traits are underpinned to a greater extent by epistatic interactions among loci than is the case for other traits. This suggestion is consistent with the observation that fitness is less heritable than other traits, and the idea that this lower heritability is associated with a greater non-additive component to the phenotype (reviewed in [136]). More broadly, this finding suggests that it may be unrealistic

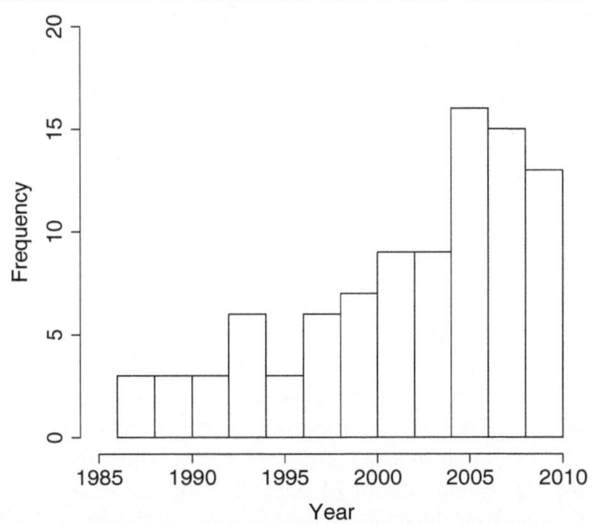

Figure 1 Publication frequency of articles presenting data on phenotypic responses to outbreeding over time. The articles shown are limited to those included in this systematic review.

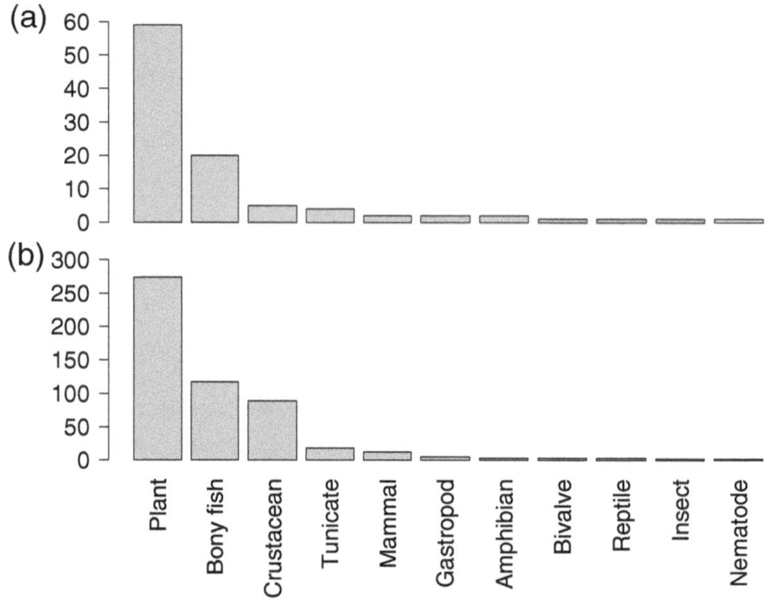

Figure 2 Representation of different high-level taxonomic categories within the systematic review. (**a**) Frequencies of effect sizes by taxonomic category. (**b**) Frequencies of studies by taxonomic category.

to understand fitness responses to outbreeding by considering only extrinsic outbreeding effects (i.e. taking into account only additive genetic variation).

We found that hybrid generation was also significantly associated with phenotypic responses to outbreeding. Specifically, there was a cost to outbreeding in the F2 and F3 relative to the F1. This shift in the hybrid outbreeding response over generations is expected from theory [17]. Between-population outbreeding creates between-population heterozygosity across the genome in the F1. This process can mask the expression of deleterious alleles that have drifted to high frequency in either of the parent populations, resulting in an improved hybrid phenotype (and potentially greater fitness, i.e. heterosis) relative to either parent population. In the F2 generation segregation reduces the between-source heterozygosity by half, while recombination breaks up favourable epistasis present in either source population. Thus we expect a decrease in the expected phenotype in F2 compared with F1 individuals. From a conservation perspective, the crucial question is whether the absolute fitness of F2 or later generation hybrids is lower or greater than the mid-parent value. If the hybrids are less fit than their parents (outbreeding depression), then there has been a true cost to hybridisation, and this may compromise conservation plans that result in population admixture and hybridisation. Our observation that F2 individuals have decreased phenotypic performance relative to F1 individuals is consistent with the synthesis presented by Edmands [15]. However, in their meta-analysis, McClelland and Naish found positive responses to outbreeding (relative to parental lineages) in both the F1 and F2 [16]. This result runs counter to our observation. We suggest that the difference arose because we were able to control for between-study variation in outbreeding responses, by fitting study-level random effects, allowing a more sensitive comparison of F1 and F2 responses. However it may be that McClelland and Naish's inclusion of outbreeding between species, inclusion of outbreeding between farmed lineages, or focus on fish could also explain the difference in outcomes.

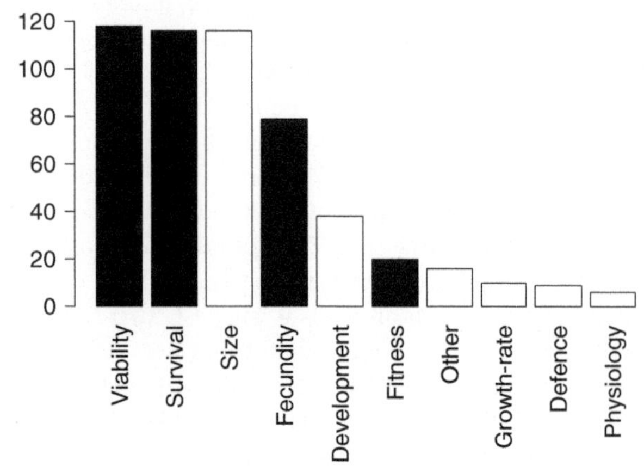

Figure 3 Frequency of trait types within our outbreeding response effect size data. Bars with black shading refer to components of fitness (viability, fecundity, survival, compound measures of fitness); open bars refer to other traits.

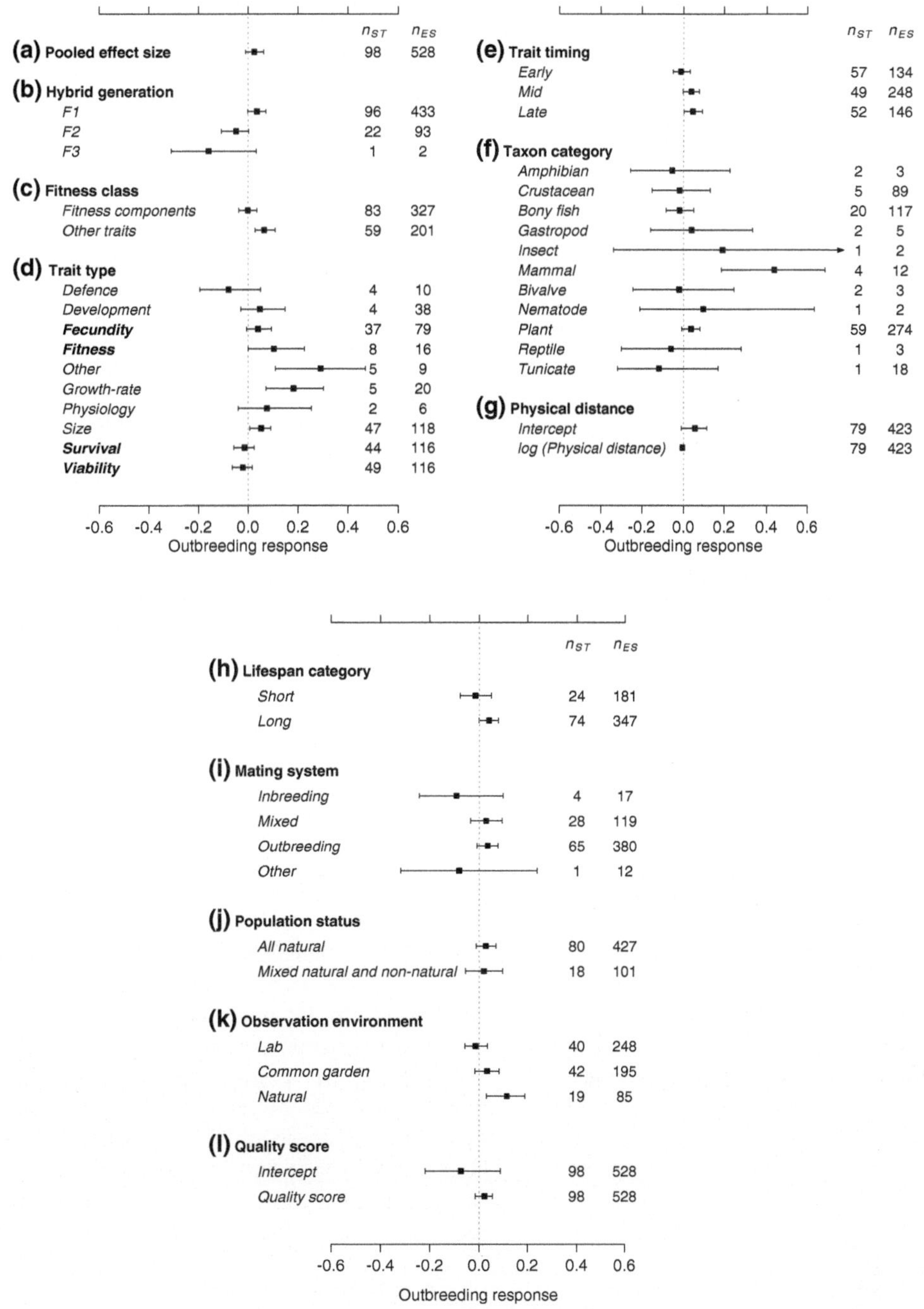

Figure 4 Variation of outbreeding responses with explanatory variables (effect modifiers). Outbreeding responses are given as proportion changes in phenotype relative to the mid-parent phenotype. An outbreeding response of zero means hybrid offspring have phenotype equivalent to the mid-parent value. Positive values represent phenotypic benefit relative to parent lineages. (**a**) Overall pooled outbreeding response (for a model fitting only an intercept). (**b**) – (**l**) Variation of outbreeding responses with individual explanatory variables. Points give posterior means for the pooled effect sizes (on the relative phenotypic scale). Error bars bracket the corresponding 95% credible interval. n_{ST} and n_{ES} give respectively the number of studies and the number of effect sizes underpinning each estimate. See Methods and Additional file 1: Text S2 for a description of the explanatory variables.

A breakdown of trait responses by generation indicated approximate parity of F1 fitness with the mid-parent phenotype (Figure 6). Other (non-fitness component) traits showed a consistent expected benefit following outbreeding during the F1. During the F2, fitness components suffered a significant net cost relative to parent fitness. These results suggest that there may be absolute fitness costs to outbreeding in the F2 (outbreeding depression). However, we must be cautious about this suggestion for two reasons. First, there were only 22 studies with

Table 6 Model fitting summaries for meta-analyses containing a single fixed-effects explanatory variable

Model	Fixed effects	DIC	Vs DIC	Δ DIC	Range in DIC	Study variance component	Residual variance component	Min. effective samples	Max. auto-correlation
1	~ Intercept	−576.1	-	-	1.7	0.0200	0.0110	889	0.058
2	~ Generation	−573.4	−576.1	2.7	1.7	0.0204	0.0110	1000	0.027
3	~ Fitness class	−610.9	−576.1	−34.8	1.8	0.0187	0.0102	1000	0.044
4	~ Trait type	−619.4	−576.1	−43.4	1.3	0.0145	0.0099	908	0.032
5	~ Trait timing	−586.6	−576.1	−10.5	0.4	0.0196	0.0107	1000	0.050
6	~ Taxon category	−571.0	−576.1	5.0	1.0	0.0187	0.0111	883	0.061
7	~ log(physical distance)	−467.8	−464.6	−3.2	0.6	0.0246	0.0107	856	0.077
8	~ Lifespan category	−575.6	−576.1	0.5	1.2	0.0194	0.0110	1000	0.028
9	~ Mating system	−574.5	−576.1	1.5	1.2	0.0205	0.0110	1000	0.044
10	~ Population status	−576.7	−576.1	−0.7	1.2	0.0201	0.0109	770	0.041
11	~ Observation environment	−573.6	−576.1	2.4	2.4	0.0178	0.0110	840	0.086
12	~ Quality score	−575.8	−576.1	0.3	3.4	0.0200	0.0109	1000	0.032

DIC (column two) gives the deviance information criterion. "Vs. DIC" gives the DIC for the corresponding model including only an intercept term. "Δ DIC" gives the difference in DIC between columns two and three. Range in DIC gives the range in DIC among three replicate model runs, as an indication of the extent of Monte Carlo error in DIC estimates. Study and residual variance components are variance estimates for study and residual random effects. "Min. effective samples" gives the minimum number of effective samples for the parameters in the model. "Max. auto-correlation" gives the maximum absolute sample autocorrelation at lag 1 for any parameter in the model.

observations on F2 hybrid performance. Second, we do not know whether these 22 studies made observations using species and population pairs that are representative of all possible intraspecific outbreeding contexts. The effects in Figure 6 strengthen our argument that epistasis has a relatively greater role in underpinning fitness component traits. We observed overall outbreeding depression (fitness costs) in the F2, but no such F2 costs in traits that were not fitness components. The former result is highly suggestive of a role for epistasis, whereas the generational shift in hybrid phenotype for non-fitness component traits may be consistent with the loss of between-source heterozygosity [17].

Outbreeding responses for early acting traits were significantly lower than those for mid- and late-acting traits. It is possible that this result was caused by maternal effects, such that F1 phenotypes are a function of maternal parent phenotypes, and F2 phenotypes are a function of maternal F1 phenotypes [24,137]. This response to the maternal environment during outbreeding can result in F2 phenotypes exceeding F1 phenotypes. Early acting traits showed an F2 decline in phenotype (relative to F1) closely similar to that of mid and late

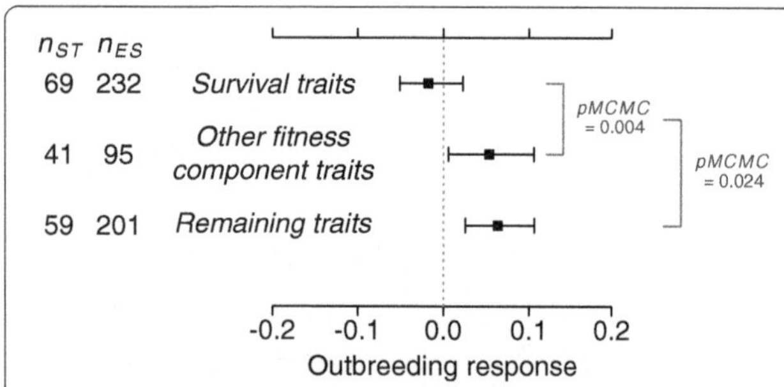

Figure 5 Outbreeding responses vary between fitness component and all other traits, and between survival traits (survival and viability) and other traits within the fitness component category (fecundity traits and integrated or multiplicative measures of fitness). Posterior means and credible intervals were estimated as contrasts within the trait type predictor. X-axis scale, point estimates, error bars and abbreviations as in Figure 4.

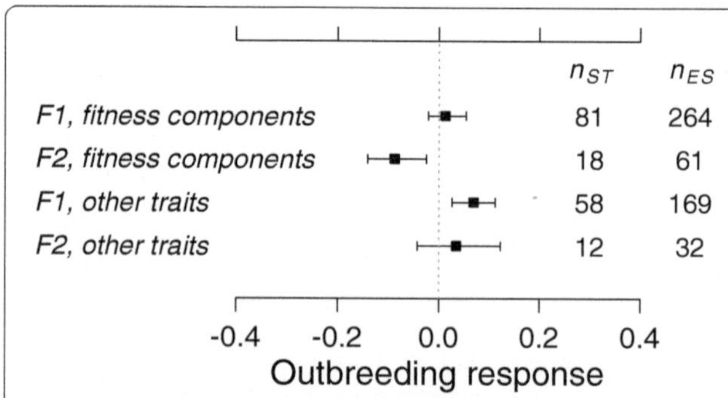

Figure 6 Fitness class (fitness component or remaining trait types) and generation (F1 or F2) predict variation in outbreeding responses. Fitness components were defined as viability, survival, fecundity traits or integrated/ multiplicative measures of fitness. Posterior means and credible intervals were estimated by fitting a fitness class × generation interaction. X-axis scale, point estimates, error bars and abbreviations as in Figure 4.

acting traits (Additional file 1: Figure S12). These results indicate that maternal effects may be weak relative to the effects of the decline of between-source heterozygosity in the F2, or the break-up of internal co-adaptation. Most (86.6%) of the effect sizes for early acting traits were viability traits (early survival). Outbreeding responses for these traits did not differ significantly from those of other (mid and late acting) survival traits. Thus, an alternative explanation for the more negative outbreeding responses observed in early acting traits may be that they were (almost all) survival traits.

The genetic divergence that underpins outbreeding responses depends in part on the extent of demographic connectivity between the populations by migration. We did not record migration rates per se, although we did record physical distance between crossed populations, as a proxy for this. We detected only very limited (and non-significant) variation in the outbreeding response with physical distance. This may be because physical distance scales differently with dispersal ability (and hence potential connectivity) for each species. Alternatively, there may be no relationship, or a non-linear relationship between outbreeding responses and physical separation of inter-mating populations.

Review limitations

Review scope

The scope of our systematic review was limited to intraspecific, post-zygotic outbreeding effects that involved at least one natural population, and excluded taxonomically complex outbreeding contexts. Thus, we cannot generalise our results to comment on taxonomically complex situations (hybrid swarms and speciation complexes), outbreeding between species, outbreeding exclusively between farmed populations, agricultural or horticultural cultivars or strains, or pre-zygotic outbreeding barriers.

Our effect size was a measure of intrinsic outbreeding responses [15,18], i.e. the deviation of hybrid phenotype from the mid-parent phenotype. Thus we cannot use our results to comment on the extent or magnitude of extrinsic responses to outbreeding (deviations in hybrid performance from either of the parents, due to additive gene effects).

Where populations mix under natural (uncontrolled) conditions, a broader range of crosses would be produced than the set of hybrid types that we considered. For example F1 hybrids could backcross with individuals that have a pure parental ancestry. Progeny of backcrosses between F1 and parental lineages are expected to possess between-source heterozygosity equivalent to F2 and later generation hybrids but with a reduced epistatic cost [17]. Thus, backcrossing may allow beneficial alleles to escape deleterious hybrid genetic backgrounds, and introgress into populations that have received managed immigration. However, where internal co-adaptation is between nuclear and cytoplasmic genomes, backcrossing may restore epistatic fitness loss only in one backcross direction [50].

Sources of bias

Visual inspection of funnel plots and a test of funnel plot asymmetry indicated little evidence for publication bias (either towards phenotypic outbreeding costs, or towards outbreeding benefits). Identification of publication bias from funnel plots may be prone to error in smaller meta-analyses (e.g. containing ~ 10 studies; [138]), but our data were based on 98 studies, and should be less sensitive to this issue of sample size. Bias would have most effect on our results and conclusions if any studies not included in our review (e.g. unpublished work, sources of grey literature that we did not consider) reported outbreeding responses that differed systematically from those in the studies that we did include. However, outbreeding responses may be biased in either of two directions: towards fitness costs (outbreeding depression), or towards fitness gains (heterosis), depending on assumptions made by the researcher. Thus, it is not obvious that any bias in outbreeding responses should influence our meta-analyses systematically.

The nested structure of our meta-analytic model precluded an assessment of the sensitivity of the results to the "file drawer problem" by traditional routes, e.g. by quantifying the number of missing non-significant studies required to make the observed pooled effect sizes non-significant ("fail-safe n"; [139]). The value of calculating a fail-safe n for the significance of fixed effects within our models could, in any case, be called into question, given the study-level heterogeneity in outbreeding responses.

Limitations of the primary research literature

We found relatively few studies that followed later generation responses to outbreeding (F3 and later hybrid generations). Thus the available evidence provides a poor basis for understanding the longer-term consequences of intraspecific outbreeding (i.e. the outbreeding responses of greatest potential interest to conservation practitioners). Only 20.4% of articles included in our review observed outbreeding responses within natural populations. This represents a shortcoming of the literature, given that the effects of outbreeding may differ between natural and lab environments (Figure 4k). Wherever possible, investigators should seek to conduct studies on phenotypic outbreeding effects either in natural populations or habitat, or under conditions that approximate as far as possible those within natural populations. Many articles also lacked clarity with regard to their crossing designs, their level of replication (attempted and realised), and to what hierarchical level in the sampling design measurement error referred (e.g. individual level, family level, treatment level).

Review conclusions

Implications for managing between-population outbreeding

The available evidence has implications for several sorts of conservation measures that result in between-population outbreeding. These measures include "range restoration", where individuals are translocated (re-introduced) within the existing native range of a species [140,141], "reinforcement" where individuals are translocated to a particular population that is in decline, in order to bolster that population demographically [141,142], and assisted migration e.g. [143], in cases where the translocated individuals come from multiple source populations. The conservation practitioner is interested in whether between-population outbreeding influences the outcome of these conservation measures, increasing or decreasing population growth or viability. The results of this review do not relate directly to the effect of outbreeding on demographic sustainability. However, our review contributes to this understanding by synthesising the effects of outbreeding on phenotypes of hybrid offspring relative to parent lineages. These phenotypes, including fitness components, are expected to contribute to the growth and viability of populations.

The available evidence indicates that between-population outbreeding may lead to phenotypic benefits in the F1, but could also lead to absolute costs to fitness in the F2 and later generations (the latter result is based on 18 studies; Figure 6). However, not all outbreeding will lead to outbreeding depression in the F2 generation. This is evident if we consider the study-level heterogeneity in outbreeding responses, which imply scatter in outbreeding responses around the typical F2 fitness response. In addition, there are good reasons for supposing that outbreeding depression may be more likely for some population pairs than for others. Previous studies have suggested that outbreeding depression is likely when outbreeding populations have been isolated for many generations (especially if they occupy divergent environments), are genetically divergent or are differentiated in chromosome structure or number [21,26]. These factors influence (or are measures of) the underlying basis of outbreeding depression, i.e. adaptive differentiation, and internal (epistatic) co-adaptation within the genome. Therefore, the available evidence suggests that conservation practitioners should avoid mixing population pairs that meet these conditions, since there is an enhanced risk that their admixture would lead to outbreeding depression [21]. Frankham *et al.* [21] give specific guidelines and a decision making tool to assist in identifying population pairs whose mixture may risk outbreeding depression.

Implications for research

The heterogeneity in outbreeding responses we observed among studies suggests that future research should strive to document as carefully and as fully as possible the factors that may lead to this variation. These could include the level of demographic isolation (present and historical) that a population has been exposed to, generation time, effective population size, the extent of cytogenetic differences between crossing populations, and the degree of environmental differentiation between hybridising populations (relative to other populations in a species' range). This would allow future reviews to model this heterogeneity and to begin to understand the causes of between-species and between-population differences in outbreeding responses. In addition, researchers should aim to observe hybrid offspring and parental lineages in environments that are as close to natural as possible. More generally, other aspects of study design that improve the internal and external validity of line-cross analyses and investigations into phenotypic outbreeding responses (Table 3) should be incorporated into future studies. For example, we suggest that prospective authors of outbreeding studies should aim to follow the fitness effects of between-population outbreeding into at least the F2 generation, and ideally to incorporate non-inbred backcrosses into their designs as well. Where resources allow, researchers should carry out control crosses within parental populations (e.g. F1, F2 etc. within population controls) as the gold standard for comparison with hybrid offspring. Ideally, researchers would also follow non-outbred crosses within both parent populations as comparators for hybrid performance. Wherever possible, investigators should measure total fitness rather than individual components of fitness (i.e. the contribution to future generations resulting from the aggregate effects of survival and reproduction). These improvements to study design should yield a clearer view of the conditions under which either heterosis, or outbreeding depression may occur following between-population outcrossing.

Additional files

Additional file 1: "Whitlock_outbreeding_review_supplementary_materials". This file contains supplementary materials (text, figures and tables).

Additional file 2: "Whitlock_outbreeding_review_audit_file.xlsx". This file contains details of all articles recovered through the literature searches, and documents the point at which articles were excluded (title assessment, abstract assessment, fulltext assessment), or whether they were, alternatively, included in the review.

Additional file 3: "Whitlock_outbreeding_review_dataset.xlsx". This file contains the dataset of effect sizes and other explanatory variables, along with article identifiers to link to Additional file 2.

Competing interests
The authors declared that they have no competing interests.

Author's contributions

RW lead the literature survey and assessment, carried out the data extraction, designed and carried out the statistical analysis and drafted the review report. All authors contributed to the design and focus of the study. ASP and GBS implemented the systematic review methodology as specifically employed in this review. GBS participated in literature searches, and contributed to the design of the statistical analysis. TB conceived of the study. All authors read and approved the final manuscript.

Acknowledgements

This work was supported by the UK Population Biology Network, through funding from the Natural Environment Research Council and Natural England. We thank Jack Brodie, Helen Hipperson, Marie Chadburn and Sophie Allen for assistance with literature searching, article assessment and data extraction. We also thank our review group for constructive criticism on the scope, development and structure of this review, and two peer reviewers for useful feedback on the review protocol. Finally we thank three peer reviewers who each provided constructive comments on this systematic review report.

Author details

[1]Institute of Integrative Biology, University of Liverpool, The Biosciences Building, Crown Street, Liverpool L69 7ZB, UK. [2]Centre for Reviews and Dissemination, University of York, York YO10 5DD, UK. [3]Faculty of Biological Sciences, University of Leeds, Leeds LS2 9JT, UK. [4]Zoology Building, School of Biological Sciences, University of Aberdeen, Aberdeen AB24 2TZ, UK. [5]UK Population Biology Network, Department of Animal and Plant Sciences, University of Sheffield, Sheffield S10 2TN, UK. [6]Centre for Evidence-Based Conservation, School of Environment and Natural Resources and Geography, Bangor University, Bangor LL57 2UW, UK.

References

1. Frankel OH: **Genetic conservation: our evolutionary responsibility.** *Genetics* 1974, **78:**53–65.
2. Frankel OH, Soulé ME: *Conservation and evolution.* Cambridge, UK: Cambridge University Press; 1981.
3. Crnokrak P, Roff DA: **Inbreeding depression in the wild.** *Heredity* 1999, **83:**260–270.
4. Keller LF, Waller DM: **Inbreeding effects in wild populations.** *Trends Ecol Evol* 2002, **17:**230–241.
5. O'Grady JJ, Brook BW, Reed DH, Ballou JD, Tonkyn DW, Frankham R: **Realistic levels of inbreeding depression strongly affect extinction risk in wild populations.** *Biol Conserv* 2006, **133:**42–51.
6. Reed DH, Frankham R: **Correlation between fitness and genetic diversity.** *Conserv Biol* 2003, **17:**230–237.
7. Hedrick PW, Fredrickson R: **Genetic rescue guidelines with examples from Mexican wolves and Florida panthers.** *Conserv Genet* 2010, **11:**615–626.
8. Tallmon DA, Luikart G, Waples RS: **The alluring simplicity and complex reality of genetic rescue.** *Trends Ecol Evol* 2004, **19:**489–496.
9. Westemeier RL, Brawn JD, Simpson SA, Esker TL, Jansen RW, Walk JW, Kershner EL, Bouzat JL, Paige KN: **Tracking the long-term decline and recovery of an isolated population.** *Science* 1998, **282:**1695–1698.
10. Hogg JT, Forbes SH, Steele BM, Luikart G: **Genetic rescue of an insular population of large mammals.** *Proc R Soc Lond, Ser B: Biol Sci* 2006, **273:**1491–1499.
11. Madsen T, Shine R, Olsson M, Wittzell H: **Restoration of an inbred adder population.** *Nature* 1999, **402:**34–35.
12. Trinkel M, Ferguson N, Reid A, Reid C, Somers M, Turelli L, Graf J, Szykman M, Cooper D, Haverman P, *et al*: **Translocating lions into an inbred lion population in the Hluhluwe-iMfolozi Park, South Africa.** *Anim Conserv* 2008, **11:**138–143.
13. Vilà C, Sundqvist A, Flagstad Ø, Seddon J, Björnerfeldt S, Kojola I, Casulli A, Sand H, Wabakken P, Ellegren H: **Rescue of a severely bottlenecked wolf (*Canis lupus*) population by a single immigrant.** *Proc Roy Soc Lond B Biol Sci* 2003, **270:**91–97.
14. Gregory A, Burke T, Ferris R, Robson J, Smithers R, Whitlock R: *The conservation of genetic diversity: science and policy needs in a changing world.* Peterborough, UK; 2006.
15. Edmands S: **Between a rock and a hard place: evaluating the relative risks of inbreeding and outbreeding for conservation and management.** *Mol Ecol* 2007, **16:**463–475.
16. McClelland EK, Naish KA: **What is the fitness outcome of crossing unrelated fish populations? A meta-analysis and an evaluation of future research directions.** *Conserv Genet* 2007, **8:**397–416.
17. Lynch M: **The genetic interpretation of inbreeding depression and outbreeding depression.** *Evolution* 1991, **45:**622–629.
18. Allendorf FW, Leary RF, Spruell P, Wenburg JK: **The problems with hybrids: setting conservation guidelines.** *Trends Ecol Evol* 2001, **16:**613–622.
19. Rieseberg LH: **Chromosomal rearrangements and speciation.** *Trends Ecol Evol* 2001, **16:**351–358.
20. White MJD: *Modes of speciation.* San Francisco, California: W. H. Freeman; 1978.
21. Frankham R, Ballou JD, Eldridge MDB, Lacy RC, Ralls K, Dudash MR, Fenster CB: **Predicting the probability of outbreeding depression.** *Conserv Biol* 2011, **25:**465–475.
22. Arnold ML, Hodges SA: **Are natural hybrids fit or unfit relative to their parents?** *Trends Ecol Evol* 1995, **10:**67–71.
23. Coyne JA, Orr HA: *Speciation.* Sunderland, MA, USA: Sinauer Associates; 2004.
24. Falconer DS, Mackay TFC: *Quantitative Genetics.* Harlow, England: Prentice Hall; 1996.
25. Orr HA: **The population genetics of speciation: the evolution of hybrid incompatibilities.** *Genetics* 1995, **139:**1805–1813.
26. Edmands S, Timmerman CC: **Modeling factors affecting the severity of outbreeding depression.** *Conserv Biol* 2003, **17:**883–892.
27. Edmands S: **Does parental divergence predict reproductive compatibility.** *Trends Ecol Evol* 2002, **17:**520–527.
28. Storfer A: **Gene flow and endangered species translocations: a topic revisited.** *Biol Conserv* 1999, **87:**173–180.
29. Whitlock R, Stewart GB, Pullin AS, Goodman SJ, Piertney SB, Butlin RK, Burke T, Brodie J, Hipperson H: *Does translocation of individuals between populations result in net cost to fitness in offspring of mixed populations?.* Bangor, UK: Collaboration for Environmental Evidence; 2009.
30. Hedrick PW: **Evaluation of D2, a microsatellite measure of inbreeding and outbreeding, in wolves with a known pedigree.** *Evolution* 2001, **55:**1256–1260.
31. Cohen J: **A coefficient of agreement for nominal scales.** *Educ Psychol Meas* 1960, **20:**37–46.
32. Pullin AS, Stewart GB: **Guidelines for systematic review in conservation and environmental management.** *Conserv Biol* 2006, **20:**1647–1656.
33. Borenstein M, Hedges LV, Higgins JPT, Rothstein HR: *Introduction to meta-analysis.* Chichester, UK: John Wiley & Sons; 2009.
34. Hadfield JD: **MCMC methods for multi-response generalized linear mixed models: the MCMCglmm R package.** *J Stat Softw* 2010, **33:**1–22.
35. Hadfield JD, Nakagawa S: **General quantitative genetic methods for comparative biology: phylogenies, taxonomies and multi-trait models for continuous and categorical characters.** *J Evol Biol* 2010, **23:**494–508.
36. Gelman A: **Prior distributions for variance parameters in hierarchical models.** *Bayesian Analysis* 2006, **1:**515–533.
37. Spiegelhalter DJ, Best NG, Carlin BP, van der Linde A: **Bayesian measures of model complexity and fit.** *J R Stat Soc Lond B* 2002, **64:**583–639.
38. Sutton JT, Nakagawa S, Robertson BC, Jamieson IG: **Disentangling the roles of natural selection and genetic drift in shaping variation at MHC immunity genes.** *Mol Ecol* 2011, **20:**4408–4420.
39. Higgins JPT, Thompson SG: **Quantifying heterogeneity in a meta-analysis.** *Stat Med* 2002, **21:**1539–1558.
40. Viechtbauer W: **Conducting meta-analyses in R with the metafor package.** *J Stat Softw* 2010, **36:**1–48.
41. Egger M, Smith G, Schneider M, Minder C: **Bias in metaanalysis detected by a simple, graphical test.** *Br Med J* 1997, **315:**629–634.
42. Gelman A, Carlin JB, Stern HS, Rubin DB: *Bayesian Data Analysis.* London, New York, Washington, D. C.: Chapman & Hall/CRC; 2004.
43. Landis JR, Koch GG: **The measurement of observer agreement for categorical data.** *Biometrics* 1977, **33:**159–174.
44. Sherman CDH, Wapstra E, Uller T, Olsson M: **Male and female effects on fertilization success and offspring viability in the Peron's tree frog, *Litoria peronii*.** *Austral Ecol* 2008, **33:**348–352.
45. Sagvik J, Uller T, Olsson M: **Outbreeding depression in the common frog, *Rana temporaria*.** *Conserv Genet* 2005, **6:**205–211.

46. Byrom AE, Burns CW, Wallis GP: **Experimental hybridization of alpine and lowland forms of *Boeckella dilata*, a calanoid copepod.** *Heredity* 1993, **71**:508–515.
47. Edmands S, Feaman HV, Harrison JS, Timmerman CC: **Genetic consequences of many generations of hybridization between divergent copepod populations.** *J Hered* 2005, **96**:114–123.
48. Edmands S, Deimler JK: **Local adaptation, intrinsic coadaptation and the effects of environmental stress on interpopulation hybrids in the copepod *Tigriopus californicus*.** *J Exp Mar Biol Ecol* 2004, **303**:183–196.
49. Edmands S: **Heterosis and outbreeding depression in interpopulation crosses spanning a wide range of divergence.** *Evolution* 1999, **53**:1757–1768.
50. Ellison CK, Burton RS: **Interpopulation hybrid breakdown maps to the mitochondrial genome.** *Evolution* 2008, **62**:631–638.
51. Leberg PL: **Strategies for population reintroduction - effects of genetic-variability on population growth and size.** *Conserv Biol* 1993, **7**:194–199.
52. Nguenga D, Teugels GG, Ollevier F: **Fertilization, hatching, survival and growth rates in reciprocal crosses of two strains of an African catfish *Heterobranchus longifilis* Valenciennes 1840 under controlled hatchery conditions.** *Aquacult Res* 2000, **31**:565–573.
53. Gjerde B, Reddy P, Mahapatra KD, Saha JN, Jana RK, Meher PK, Sahoo M, Lenka S, Govindassamy P, Rye M: **Growth and survival in two complete diallele crosses with five stocks of Rohu carp (*Labeo rohita*).** *Aquaculture* 2002, **209**:103–115.
54. Cooke SJ, Philipp DP: **Influence of local adaptation and interstock hybridization on the cardiovascular performance of largemouth bass *Micropterus salmoides*.** *J Exp Biol* 2005, **208**:2055–2062.
55. Cooke SJ, Kassler TW, Phillipp DP: **Physiological performance of largemouth bass related to local adaptation and interstock hybridization: implications for conservation and management.** *J Fish Biol* 2001, **59**:248–268.
56. Goldberg TL, Grant EC, Inendino KR, Kassler TW, Claussen JE, Philipp DP: **Increased infectious disease susceptibility resulting from outbreeding depression.** *Conserv Biol* 2005, **19**:455–462.
57. Gharrett AJ, Smoker WW, Reisenbichler RR, Taylor SG: **Outbreeding depression in hybrids between odd- and even-broodyear pink salmon.** *Aquaculture* 1999, **173**:117–129.
58. Gharrett AJ, Smoker WW: **Two generations of hybrids between even-year and odd-year pink salmon *Oncorhynchus gorbuscha* a test for outbreeding depression.** *Can J Fish Aquat Sci* 1991, **48**:1744–1749.
59. Gilk SE, Wang IA, Hoover CL, Smoker WW, Taylor SG, Gray AK, Gharrett AJ: **Outbreeding depression in hybrids between spatially separated pink salmon, *Oncorhynchus gorbuscha*, populations: marine survival, homing ability, and variability in family size.** *Environ Biol Fishes* 2004, **69**:287–297.
60. Johnsson JI, Clarke WC, Withler RE: **Hybridization with domesticated rainbow trout (*Oncorhynchus mykiss*) reduces seasonal variation in growth of steelhead trout (*O. mykiss*).** *Can J Fish Aquat Sci* 1993, **50**:480–487.
61. Miller LM, Close T, Kapuscinski AR: **Lower fitness of hatchery and hybrid rainbow trout compared to naturalized populations in Lake Superior tributaries.** *Mol Ecol* 2004, **13**:3379–3388.
62. Tymchuk WE, Sundström LF, Devlin RH: **Growth and survival trade-offs and outbreeding depression in rainbow trout (*Oncorhynchus mykiss*).** *Evolution* 2007, **61**:1225–1237.
63. Bryden CA, Heath JW, Heath DD: **Performance and heterosis in farmed and wild Chinook salmon (*Oncorhynchus tshawyacha*) hybrid and purebred crosses.** *Aquaculture* 2004, **235**:249–261.
64. Sheffer RJ, Hedrick PW, Velasco AL: **Testing for inbreeding and outbreeding depression in the endangered Gila topminnow.** *Anim Conserv* 1999, **2**:121–129.
65. Fraser DJ, Houde ALS, Debes PV, O'Reilly P, Eddington JD, Hutchings JA: **Consequences of farmed-wild hybridization across divergent wild populations and multiple traits in salmon.** *Ecol Appl* 2010, **20**:935–953.
66. Fraser DJ, Cook AM, Eddington JD, Bentzen P, Hutchings JA: **Mixed evidence for reduced local adaptation in wild salmon resulting from interbreeding with escaped farmed salmon: complexities in hybrid fitness.** *Evolutionary Applications* 2008, **1**:501–512.
67. Houde ALS, Fraser DJ, Hutchings JA: **Reduced anti-predator responses in multi-generational hybrids of farmed and wild Atlantic salmon (*Salmo salar* L.).** *Conserv Genet* 2010, **11**:785–794.
68. McGinnity P, Prodohl P, Ferguson A, Hynes R, Maoileidigh NO, Baker N, Cotter D, O'Hea B, Cooke D, Rogan G, *et al*: **Fitness reduction and potential extinction of wild populations of Atlantic salmon, *Salmo salar*, as a result of interactions with escaped farm salmon.** *Proc R Soc B-Biol Sci* 2003, **270**:2443–2450.
69. Dahl J, Pettersson E, Dannewitz J, Järvi T, Löf AC: **No difference in survival, growth and morphology between offspring of wild-born, hatchery and hybrid brown trout (*Salmo trutta*).** *Ecol Freshwat Fish* 2006, **15**:388–397.
70. Nilsson J: **Arctic charr strain crosses - effects on growth and sexual maturity.** *J Fish Biol* 1993, **43**:163–171.
71. Escobar JS, Nicot A, David P: **The different sources of variation in inbreeding depression, heterosis and outbreeding depression in a metapopulation of *Physa acuta*.** *Genetics* 2008, **180**:1593–1608.
72. McCarthy TM, Sih A: **Relatedness of mates influences mating behaviour and reproductive success of the hermaphroditic freshwater snail *Physa gyrina*.** *Evol Ecol Res* 2008, **10**:77–94.
73. Reinhardt K, Kohler G: **Conservation of the red-winged grasshopper, *Oedipoda germanica* (Latr.): the influence of reproductive behaviour.** *Biol Conserv* 2002, **107**:221–228.
74. Hogg JT, Forbes SH, Steele BM, Luikart G: **Genetic rescue of an insular population of large mammals.** *Proc R Soc B-Biol Sci* 2006, **273**:1491–1499.
75. Keane B: **Dispersal and inbreeding avoidance in the white-footed mouse *Peromyscus leucopus*.** *Anim Behav* 1990, **40**:143–152.
76. Cruz P, Ramirez JL, Garcia GA, Ibarra AM: **Genetic differences between two populations of catarina scallop (*Argopecten ventricosus*) for adaptations for growth and survival in a stressful environment.** *Aquaculture* 1998, **166**:321–335.
77. Cruz P, Ibarra AM: **Larval growth and survival of two catarina scallop (*Argopecten circularis*, Sowerby, 1835) populations and their reciprocal crosses.** *J Exp Mar Biol Ecol* 1997, **212**:95–110.
78. Dolgin ES, Felix MA, Cutter AD: **Hakuna Nematoda: genetic and phenotypic diversity in African isolates of *Caenorhabditis elegans* and *C. briggsae*.** *Heredity* 2008, **100**:304–315.
79. Keller M, Kollmann J, Edwards PJ: **Genetic introgression from distant provenances reduces fitness in local weed populations.** *J Appl Ecol* 2000, **37**:647–659.
80. Andalo C, Cruzan MB, Cazettes C, Pujol B, Burrus M, Thébaud C: **Post-pollination barriers do not explain the persistence of two distinct *Antirrhinum* subspecies with parapatric distribution.** *Plant Syst Evol* 2010, **286**:223–234.
81. Luijten SH, Kery M, Oostermeijer JGB, Den Nijs HJCM: **Demographic consequences of inbreeding and outbreeding in *Arnica montana*: a field experiment.** *J Ecol* 2002, **90**:593–603.
82. Raabová J, Münzbergová Z, Fischer M: **Consequences of near and far between-population crosses for offspring fitness in a rare herb.** *Plant Biol* 2009, **11**:829–836.
83. Johansen-Morris AD, Latta RG: **Fitness consequences of hybridization between ecotypes of *Avena barbata*: hybrid breakdown, hybrid vigor, and transgressive segregation.** *Evolution* 2006, **60**:1585–1595.
84. Rao GY, Andersson S, Widen B: **Developmental stability in *Brassica cretica*: the effect of crossing distance on fluctuating asymmetry in cotyledon morphology.** *Heredity* 2002, **88**:197–202.
85. Heiser DA, Shaw RG: **The fitness effects of outcrossing in *Calylophus serrulatus*, a permanent translocation heterozygote.** *Evolution* 2006, **60**:64–76.
86. Etterson JR, Keller SR, Galloway LF: **Epistatic and cytonuclear interactions govern outbreeding depression in the autotetraploid *Campanulastrum americanum*.** *Evolution* 2007, **61**:2671–2683.
87. Galloway LF, Etterson JR: **Population differentiation and hybrid success in *Campanula americana*: geography and genome size.** *J Evol Biol* 2005, **18**:81–89.
88. Kaye TN, Lawrence B: *Fitness effects of inbreeding and outbreeding on golden paintbrush (Castilleja levisecta): implications for recovery and reintroduction.* Corvallis, Oregon: Institute for Applied Ecology; 2003.
89. Erickson DL, Fenster CB: **Intraspecific hybridization and the recovery of fitness in the native legume *Chamaecrista fasciculata*.** *Evolution* 2006, **60**:225–233.
90. Holtsford TP: **Variation in inbreeding depression among families and populations of *Clarkia tembloriensis* (Onagraceae).** *Heredity* 1996, **76**:83–91.

91. Paschke M, Abs C, Schmid B: **Effects of population size and pollen diversity on reproductive success and offspring size in the narrow endemic *Cochlearia bavarica* (Brassicaceae).** *Am J Bot* 2002, **89:**1250–1259.
92. Pelabon C, Carlson ML, Hansen TF, Armbruster WS: **Effects of crossing distance on offspring fitness and developmental stability in *Dalechampia scandens* (Euphorbiaceae).** *Am J Bot* 2005, **92:**842–851.
93. Waser NM, Price MV: **Crossing-distance effects in *Delphinium nelsonii*: outbreeding and inbreeding depression in progeny fitness.** *Evolution* 1994, **48:**842–852.
94. Grindeland JM: **Inbreeding depression and outbreeding depression in *Digitalis purpurea*: optimal outcrossing distance in a tetraploid.** *J Evol Biol* 2008, **21:**716–726.
95. Wagenius S, Hangelbroek HH, Ridley CE, Shaw RG: **Biparental inbreeding and interremnant mating in a perennial prairie plant: fitness consequences for progeny in their first eight years.** *Evolution* 2010, **64:**761–771.
96. Sobrevila C: **Effects of distance between pollen donor and pollen recipient on fitness components in *Espeletia schultzii*.** *Am J Bot* 1988, **75:**701–724.
97. Hardner CM, Potts BM, Gore PL: **The relationship between cross success and spatial proximity of *Eucalyptus globulus* ssp. *globulus* parents.** *Evolution* 1998, **52:**614–618.
98. Byers DL: **Effect of cross proximity on progeny fitness in a rare and a common species of *Eupatorium* (Asteraceae).** *Am J Bot* 1998, **85:**644–653.
99. Oostermeijer JGB, Altenburg RGM, Den Nijs HCM: **Effects of outcrossing distance and selfing on fitness components in the rare *Gentiana pneumonanthe* (Gentianaceae).** *Acta Bot Neerl* 1995, **44:**257–268.
100. Fischer M, Matthies D: **Mating structure and inbreeding and outbreeding depression in the rare plant *Gentianella germanica* (Gentianaceae).** *Am J Bot* 1997, **84:**1685–1692.
101. Paland S, Schmid B: **Population size and the nature of genetic load in *Gentianella germanica*.** *Evolution* 2003, **57:**2242–2251.
102. Vandepitte K, Honnay O, Jacquemyn H, Roldán-Ruiz I: **Effects of outcrossing in fragmented populations of the primarily selfing forest herb *Geum urbanum*.** *Evol Ecol* 2010, **24:**1353–1364.
103. Ellis JR, McCauley DE: **Phenotypic differentiation in fitness related traits between populations of an extremely rare sunflower: conservation management of isolated populations.** *Biol Conserv* 2009, **142:**1836–1843.
104. Becker U, Reinhold T, Matthies D: **Effects of pollination distance on reproduction and offspring performance in *Hypochoeris radicata*: experiments with plants from three European regions.** *Biol Conserv* 2006, **132:**109–118.
105. Waser NM, Price MV: **Optimal outcrossing in *Ipomopsis aggregata* seed set and offspring fitness.** *Evolution* 1989, **43:**1097–1109.
106. Schlichting CD, Devlin B: **Pollen and ovule sources affect seed production of *Lobelia cardinalis* (Lobeliaceae).** *Am J Bot* 1992, **79:**891–898.
107. Mutikainen P, Delph LF: **Inbreeding depression in gynodioecious *Lobelia siphilitica*: among-family differences override between-morph differences.** *Evolution* 1998, **52:**1572–1582.
108. Montalvo AM, Ellstrand NC: **Nonlocal transplantation and outbreeding depression in the subshrub *Lotus scoparius* (Fabaceae).** *Am J Bot* 2001, **88:**258–269.
109. Hauser TP, Loeschcke V: **Inbreeding depression and mating-distance dependent offspring fitness in large and small populations of *Lychnis flos-cuculi* (Caryophyllaceae).** *J Evol Biol* 1994, **7:**609–622.
110. Leimu R, Fischer M: **Between-population outbreeding affects plant defence.** *PLoS One* 2010, **5.**
111. Pisanu PC, Gross CL, Flood L: **Reproduction in wild populations of the threatened tree *Macadamia tetraphylla*: interpopulation pollen enriches fecundity in a declining species.** *Biotropica* 2009, **41:**391–398.
112. Vandepitte K, Roldán-Ruiz I, Honnay O: **Reproductive consequences of mate quantity versus mate diversity in a wind-pollinated plant.** *Acta Oecol* 2009, **35:**548–553.
113. Mooney EH, McGraw JB: **Effects of self-pollination and outcrossing with cultivated plants in small natural populations of American ginseng, *Panax quinquefolius* (Araliaceae).** *Am J Bot* 2007, **94:**1677–1687.
114. Levin DA, Bulinska-Radomska Z: **Effects of hybridization and inbreeding on fitness in *Phlox*.** *Am J Bot* 1988, **75:**1632–1639.
115. Bolstad SB, Kang H, Guries RP, Marty TL: **Performance of interprovenance and intraprovenance crosses of jack pine in central Wisconsin.** *Silvae Genet* 1991, **40:**124–130.
116. Harfouche A, Kremer A: **Provenance hybridization in a diallel mating scheme of maritime pine (*Pinus pinaster*). I. Means and variance components.** *Can J For Res-Rev Can Rech For* 2000, **30:**1–9.
117. Rhode JM, Cruzan MB: **Contributions of heterosis and epistasis to hybrid fitness.** *Am Nat* 2005, **166:**E124–E139.
118. Crémieux L, Bischoff A, Müller-Schärer H, Steinger T: **Gene flow from foreign provenances into local plant populations: fitness consequences and implications for biodiversity restoration.** *Am J Bot* 2010, **97:**94–100.
119. Willi Y, Van Kleunen M, Dietrich S, Fischer M: **Genetic rescue persists beyond first-generation outbreeding in small populations of a rare plant.** *Proc R Soc B-Biol Sci* 2007, **274:**2357–2364.
120. Willi Y, Fischer M: **Genetic rescue in interconnected populations of small and large size of the self-incompatible *Ranunculus reptans*.** *Heredity* 2005, **95:**437–443.
121. Dudash MR: **Relative fitness of selfed and outcrossed progeny in a self-compatible protandrous species *Sabatia-angularis* L. Gentianaceae a comparison in three environments.** *Evolution* 1990, **44:**1129–1139.
122. Sheridan PM, Karowe DN: **Inbreeding, outbreeding, and heterosis in the yellow pitcher plant, *Sarracenia flava* (Sarraceniaceae), in Virginia.** *Am J Bot* 2000, **87:**1628–1633.
123. Colling G, Reckinger C, Matthies D: **Effects of pollen quantity and quality on reproduction and offspring vigor in the rare plant *Scorzonera humilis* (Asteraceae).** *Am J Bot* 2004, **91:**1774–1782.
124. Caño L, Escarré J, Blanco-Moreno JM, Sans FX: **Assessing the effect of inbreeding and long-distance gene flow on the invasive potential of *Senecio pterophorus* (Asteraceae).** *Aust J Bot* 2008, **56:**539–549.
125. Stacy EA: **Cross-fertility in two tropical tree species: evidence of inbreeding depression within populations and genetic divergence among populations.** *Am J Bot* 2001, **88:**1041–1051.
126. Hathaway L, Andersson S, Prentice HC: **Experimental crosses within European *Silene latifolia* (Caryophyllaceae): intraspecific differentiation, distance effects, and sex ratio.** *Botany* 2009, **87:**231–240.
127. Wolfe LM, Blair AC, Penna BM: **Does intraspecific hybridization contribute to the evolution of invasiveness?: an experimental test.** *Biol Invasions* 2007, **9:**515–521.
128. Bailey MF, McCauley DE: **The effects of inbreeding, outbreeding and long-distance gene flow on survivorship in North American populations of *Silene vulgaris*.** *J Ecol* 2006, **94:**98–109.
129. Emery SN, McCauley DE: **Consequences of inbreeding for offspring fitness and gender in *Silene vulgaris*, a gynodioecious plant.** *J Evol Biol* 2002, **15:**1057–1066.
130. Thompson JD, Tarayre M, Gauthier P, Litrico I, Linhart YB: **Multiple genetic contributions to plant performance in *Thymus vulgaris*.** *J Ecol* 2004, **92:**45–56.
131. Eckstein RL, Otte A: **Effects of cleistogamy and pollen source on seed production and offspring performance in three endangered violets.** *Basic Appl Ecol* 2005, **6:**339–350.
132. Billingham MR, Simoes T, Reusch TBH, Serrao EA: **Genetic sub-structure and intermediate optimal outcrossing distance in the marine angiosperm *Zostera marina*.** *Mar Biol* 2007, **152:**793–801.
133. Arrayago M, Bea A, Heulin B: **Hybridization experiment between oviparous and viviparous strains of *Lacerta vivipara*: A new insight into the evolution of viviparity in reptiles.** *Herpetologica* 1996, **52:**333–342.
134. Grosberg RK: **Limited dispersal and proximity-dependent mating success in the colonial Ascidian *Botryllus-schlosseri*.** *Evolution* 1987, **41:**372–384.
135. DeRose MA, Roff DA: **A comparison of inbreeding depression in life-history and morphological traits in animals.** *Evolution* 1999, **53:**1288–1292.
136. Merilä J, Sheldon BC: **Genetic architecture of fitness and non-fitness traits: empirical patterns and development of ideas.** *Heredity* 1999, **83:**103–109.
137. Roberts RC: **The effects on litter size of crossing lines of mice inbred without selection.** *Genet Res* 1960, **1:**239–252.
138. Terrin N, Schmid CH, Lau J: **In an empirical evaluation of the funnel plot, researchers could not visually identify publication bias.** *J Clin Epidemiol* 2005, **58:**894–901.
139. Orwin RG: **A fail-safe n for effect size in meta-analysis.** *J Educ Stat* 1983, **8:**157–159.
140. Falk DA, Millar CI, Olwell M: *Restoring diversity: strategies for reintroduction of endangered plants.* Washington, USA: Island Press; 1996.

141. Seddon PJ: **From reintroduction to assisted colonization: moving along the conservation translocation spectrum.** *Restor Ecol* 2010, **18:**796–802.
142. Allen WH: **Reintroduction of endangered plants.** *Bioscience* 1994, **44:**65–68.
143. McLachlan JS, Hellmann JJ, Schwartz MW: **A framework for debate of assisted migration in an era of climate change.** *Conserv Biol* 2007, **21:**297–302.

Have wet meadow restoration projects in the Southwestern U.S. been effective in restoring geomorphology, hydrology, soils, and plant species composition?

Karissa M Ramstead[1], James A Allen[1*] and Abraham E Springer[2]

Abstract

Background: Wet meadows occur in numerous locations throughout the American Southwest, but in many cases have become heavily degraded. Among other things they have frequently been overgrazed and have had roads built through them, which have affected the hydrology of these wetland ecosystems.

Because of the important hydrologic and ecological functions they are believed to perform, there is currently significant interest in wet meadow restoration. Several restoration projects have been completed recently or are underway in the region, sometimes at considerable expense and with minimal monitoring. The objective of this review was to evaluate the effects of wet meadow restoration projects in the southwestern United States on geomorphology, hydrology, soils and plant species composition. A secondary objective was to determine the effects of wet meadow restoration projects on wildlife.

Methods: Electronic databases, internet search engines, websites and personal contacts were used to find articles of relevance to this review. Articles were filtered by title, abstract and full text. Summary information for each of the articles remaining after the filtering process was compiled and used to assess the quality of the evidence presented using two different approaches.

Results: Our searches yielded 48 articles, of which 25 were published in peer-reviewed journals, 14 were monitoring or project reports, and 9 were published in conference proceedings or are unpublished theses or manuscripts.

A total of 26 operational-scale restoration projects were identified. A wide range of restoration techniques were employed, ranging from small-scale manipulations of stream channels (e.g., riffle structures) to large scale pond-and-plug projects. Other common restoration techniques included fencing to exclude livestock (and sometimes also native ungulates), other forms of grazing management, seeding, and transplanting seedlings.

Most of the articles reported that restoration was fully or partially effective, at least in the short-term. However, the relative lack of high quality quantitative data, and especially data extending more than two years after project implementation, greatly limits our ability to determine how effective restoration has truly been in practice.

(Continued on next page)

* Correspondence: James.Allen@nau.edu

[1]School of Forestry, Northern Arizona University, P.O. Box 15018, Flagstaff, AZ 86011, USA

Full list of author information is available at the end of the article

(Continued from previous page)

Conclusions: While caution is warranted due to data quality limitations, progress has been made over the past 20 years in wet meadow restoration. In particular, important contributions have been made in restoring the highly degraded wet meadow systems that are characterized by deep, wide and relatively straight gullies. There is evidence, for example, that the pond-and-plug approach is an effective technique for restoring many aspects of these systems, albeit at the cost of creating new features (ponds) that are not necessarily natural features of wet meadows.

There is a need to allocate additional effort to project documentation, including better-designed and longer-lasting monitoring programs. One approach that might help is for practitioners to work with scientists from government agencies, local universities and colleges, and other organizations. When this type of collaboration has happened in the past it appears to have been effective. Many important lessons could have been learned, and mistakes avoided, if more effort had been put into documenting both successes and failures of past projects.

Keywords: Systematic review, Wetland, Riparian, Biodiversity

Background

Introduction

High elevation streamside and groundwater-fed meadows occur in numerous locations throughout the American Southwest. They are referred to by a variety of names, such as riparian meadows, montane meadows, subalpine meadows, sedge meadows, or wet meadows. For the remainder of this review the term wet meadow will be used.

Where wet meadows have not been excessively altered, sedges (*Carex* spp.), rushes (*Juncus* spp.), spikerushes (*Eleocharis* spp.) and obligate or facultative wetland grasses (e.g., *Deschampsia* spp. and *Calamagrostis* spp.) are often the dominant species [1-4]. Willow (*Salix* spp.) and alder (*Alnus* spp.) species may also occur in or adjacent to these meadows [4-7].

Wet meadows in the American Southwest frequently occur along elevation gradients that include a stream or other water body at the lower end. The upland side may contain a vegetation community gradient including some of these types: mesic meadows, dry meadows, sagebrush (*Artemesia* spp.), pine (*Pinus* spp.) dominated forest, or mixed conifer forest. These elevation and vegetation gradients are closely associated with differences in flooding, depth to water table/soil moisture content, and other soil characteristics. Wet meadows typically occur on sites with water table levels near or slightly above the ground surface throughout the spring and early summer [8-11].

Distribution of wet meadows in the American Southwest

The extent of wet meadows in the American Southwest - which we define as the states of Arizona, California, Colorado, Nevada, New Mexico and Utah - is not well documented. When considered within the context of the entire southwestern landscape, we know that wet meadows are relatively rare. Less than 1% of the landscape in the region is characterized as wetlands [12], and wet meadows are just one of several wetland types that occur. Within certain subregions, however, wet meadows can be locally common. Approximately 133,000 ha of meadows are reported to occur in the Sierra Nevada Mountains of California [13]. That acreage is distributed among more than 10,000 individual meadows [13]; however, it is not clear what proportion of those meadows would be classified as wet meadows. Patton and Judd [1] reported that approximately 17,700 ha of wet and transitional (moist) meadows occur on national forests in Arizona and New Mexico, together accounting for 0.2% of the total area.

Ecological roles of wet meadows

Wet meadows are key habitats for many species. Elk (*Cervus elaphus*) and other large ungulates, for example, have been shown to make extensive use of wet meadows as foraging sites [1,14]. Small mammals that use wet meadows include meadow voles (*Microtus* spp.), pocket gophers (*Thomomys* spp.), field mice (*Peromyscus* spp.), shrews (*Sorex* spp.), and American minks (*Mustela vison*). In the Sierra Nevada Mountains, 82 terrestrial vertebrate species are dependent on riparian and meadow habitats [13].

In addition to the common wildlife species that are dependent on wet meadows, several threatened, endangered and sensitive species also occur. For example, threatened, endangered and sensitive species in the Sierra Nevada that use wet meadows for at least a portion of their life cycle requirements include the Yosemite toad (*Bufo canoru*), the mountain yellow–legged frog (*Rana muscosa*), the willow flycatcher (*Empidonax traillii*), and the great gray owl (*Strix nebulosa*;13). In east central Arizona populations of Apache trout (*Onocorhynchus gilae* var. *apache*) endemic to high elevation streams have dwindled due to habitat degradation, resulting in its listing as an endangered species [15].

Although generally less well-documented than the roles described above, wet meadows also perform other important ecosystem functions that are commonly attributed to wetlands, such as water-quality improvement, attenuation of peak flows, water storage, and carbon sequestration [13,16,17].

Alterations of wet meadow ecosystems

Wet meadows are subject to a variety of natural disturbances and are also one of the ecosystem types in the American Southwest that has been most altered by humans. Among other things, they have been used extensively for livestock grazing, are the sites of many small dams and stock tanks, have had roads built through or adjacent to them, and have been impacted by wild ungulate grazing, invasive species and wildfires in the surrounding uplands. These types of alterations have in turn had important impacts on the hydrology of wet meadows, most notably by lowering of their water tables due to stream channel incision, surface water diversions, or groundwater withdrawal [16,18-20]. In the Sierra Nevada Mountains, an estimated 52,630 to 80,970 ha of wet meadows have been impacted by these types of disturbances over the past 150 years [13]. Impacts include severe erosion, soil desiccation, shrub and tree encroachment, and changes in plant and animal species composition and diversity [13,21,22].

Objectives

The primary objective of this review is to address the question "Have wet meadow restoration projects in the southwestern United States been effective in restoring geomorphology, hydrology, soils, and plant species composition?" A secondary objective is to address the question "Have wildlife species increased in diversity and/or abundance following wet meadow restoration?"

Methods

Question formulation

The subject of wet meadow restoration was proposed by the reviewers primarily because of our observation that an increasing number of wet meadow restoration projects have been implemented in the region in recent years. Based on our discussions with resource managers involved in wet meadow restoration projects, it was apparent to us that the outcomes of these projects have been mixed. Furthermore, relatively little information on the successes and failures of wet meadow restoration projects seemed to be published or otherwise shared among managers across the region. We conducted a survey of managers [23] that helped refine our research questions and also provided evidence that many more wet meadow restoration projects are likely to be implemented over the next decade.

Search strategy

Electronic databases available through Northern Arizona University's Cline Library were the primary source. The following databases were used:

- Academic Search Premier
- BioOne
- CSA
- Forest Science Database (Ovid)
- JSTOR
- Northern Arizona University 360
- ProQuest
- Wilson OmniFile

Additional sources of information included:

- Google Scholar (a search engine)
- Government and university websites and libraries (e.g., USDA Forest Service's TreeSearch)
- Unpublished reports (e.g., project monitoring reports) were sought directly from individuals and organizations responsible for restoration projects.

Searches included the following English language search terms:

(*denotes = wet meadow, riparian meadow, sedge meadow, <u>or</u> montane meadow)

1. Wet Meadow* AND Restoration
2. Wet Meadow* AND Hydrology
3. Wet Meadow* AND Grazing
4. Wet Meadow* AND Erosion
5. Wet Meadow* AND Sedimentation
6. Wet Meadow* AND Channel
7. Wet Meadow* AND Biodiversity
8. Wet Meadow* AND *Carex*
9. Wet Meadow* AND *Juncus*

The majority of the database searches were conducted between July and October 2009, although additional searches were made as late as October 2011. The latter searches relied primarily on other sources than databases, including Google Scholar and organizational websites. They were also aided by personal contacts with restoration practitioners that had developed over the course of the review. Google Scholar was used to search for new articles published after the date of the database searches. Several searches were made using the custom date range feature; only the first ten pages of search results were examined.

Study inclusion criteria

Inclusion criteria were broken down into the four categories described below.

Relevant subject(s) and geographic area

- Herbaceous and mixed herbaceous/scrub-shrub dominated riparian ecosystems that are clearly wetlands (e.g., dominated by obligate or facultative wetland species in genera such as *Carex* spp., *Juncus* spp., *Salix* spp., and *Alnus* spp.); meadows adjacent to these wetlands that are more typically characterized as mesic and that often have a diverse flora characterized by facultative or facultative wetland species, including a greater number of grass and forb species than typically found in wet meadows. Only studies conducted in Arizona, California, Colorado, Nevada, New Mexico or Utah were used for data extraction, although other articles were retained through the full text filtering stage for possible use in the assessment of study limitations and future research needs.

Types of intervention

- Geomorphological and/or soil restoration techniques (e.g., pond-and-plug, channel relocation, site re-contouring, topsoil placement or removal).
- Hydrologic restoration techniques (e.g., check dams, artificial riffle formations). However, articles that focused entirely on stream or fisheries habitat restoration, even if the stream flowed through a wet meadow, were not included.
- Vegetation restoration techniques (e.g., seeding, planting, herbivore exclusion).
- Modifications of adjacent areas (e.g., thinning of adjacent upland forests).

Types of study

- All primary experimental and observational studies.

Types of outcome

- Geomorphological and soil outcomes such as channel stability, presence/movement of nick points, and development of redoximorphic soil properties.
- Hydrologic outcomes such as changes in water table levels and flooding depth, duration and/or timing.
- Vegetation outcomes such as species composition, percent cover and biomass, survival of planted material, and presence or absence of invasive species.
- Biodiversity outcomes such as change in species richness, evenness, or abundance due to restoration treatments or grazing management changes.

Using a progressive filtering approach, Karissa Ramstead first eliminated all articles that clearly did not meet our inclusion criteria based solely on their titles. She next read the abstracts (when available) of the remaining articles and eliminated those that upon closer inspection also did not meet the inclusion criteria. Finally, the articles remaining after the abstract filter were read in full to determine which were suitable for data extraction. All articles identified by Ramstead for potential inclusion were then reviewed by a second reviewer (James Allen).

Study quality assessment

After the articles were selected for review, Ramstead rated them using Pullin and Knight's Hierarchy of Quality of Evidence [24]. A Kappa test was then performed using 15 randomly selected articles to assess the two primary reviewers' (Allen and Ramstead) level of agreement about their quality. In addition to this rating approach, which is widely used in systematic reviews on conservation biology-related topics, we also developed the more specific assessment approach that is described in the Data synthesis and presentation section.

Data extraction

All articles that remained after full text filtering were summarized in a master spreadsheet (Additional file 1: Appendix A). Information extracted included study location, outcome (in terms of how it related to our primary and secondary questions), restoration intervention(s), research/monitoring methods used to assess the project, and results.

Data synthesis and presentation

Information on study location and on intervention(s) employed was used to examine overall patterns across the region in terms of where restoration has been done to date and what practices have been most commonly used. A two-part analysis was then done to assess the evidence of restoration effectiveness.

For the first part of the analysis of restoration effectiveness, we compiled a list of six characteristics of functional wet meadow ecosystems, which we also considered to be useful criteria for assessing the success of wet meadow restoration projects (Table 1). Our list was modified from a previous one developed for southwestern riparian ecosystems by Medina [16] and includes at least one characteristic that is directly related to each of the five outcomes addressed by our primary or secondary questions. The descriptions of each characteristic were modified from the previous list to be more specific to wet meadows and the new list was expanded to include the types of evidence that we considered to be useful for the assessment of each characteristic. Using the descriptions and types of evidence as guidelines, we then determined if each particular characteristic was addressed and, if so, whether there was sufficient

Table 1 Characteristics of functional wet meadow ecosystems (modified from [16])

Characteristic	Description	Assessment
Suitable Stream Morphology	A low gradient stream with a low width:depth ratio (<12) and medium to high sinuosity (>1.2), except in meadows believed to have no natural channel (e.g. [25])	Best evidence: Survey data on pre- and post-project stream channel morphology
		Other evidence: Photo points, high resolution aerial photography, qualitative descriptions
Stable Streambanks	Streambanks with the capacity to withstand repeated high water/flooding events without significant loss of bank material (except in meadows believed to have no natural channel (e.g. [25])	Best evidence: Survey data on post-project stream channel morphology following repeated hydrologic events
		Other evidence: Photo points taken after several seasons, qualitative descriptions of the effects of hydrologic events
High Water Table	High water tables are present within the meadow for long enough periods to result in reduced soil conditions near the soil surface (i.e. water table <30cm from soil surface for a period of at least 14 consecutive days during the growing season) and that allows for the support of native wetland plant species	Best evidence: Data from water table wells and stream hydrographs; redox potential measurements taken within the top 30 cm of the soil column.
		Other evidence: Visual observations of water levels, evidence of flooding, and soil redoximorphic characteristics; evidence of a transition to hydrophytic vegetation
Organic Matter Assimilation	Characterized by a high percentage of organic matter in diagnostic soil horizons and/or a high rate of organic matter accumulation (the latter may be especially important in newer restoration sites)	Best evidence: Measurements of soil organic matter content, with comparison to suitable reference sites
		Other evidence: Descriptions of soil horizons, soil color assessments using Munsell soil color charts
Perennial Native Vegetation	Perennial native vegetation covers ≥75% of the streambanks and ≥ 50% of the floodplain and is comprised predominantly of native, perennial species, especially wetland graminoid species with extensive and highly fibrous root systems (e.g., *Carex* spp.)	Best evidence: Quantitative data on species composition, cover and biomass using accepted vegetation measurement techniques and appropriate sample sizes
		Other evidence: Quantitative data from small samples; photo points; qualitative descriptions
Presence of Native Fauna	The presence and relative abundance of both aquatic and terrestrial organisms is similar to comparable reference sites	Best evidence: Quantitative data on target species using accepted techniques and sampling intensities
		Other evidence: Visual evidence of species presence and abundance

evidence presented to determine if the project was successful in achieving the desired characteristic. For this analysis, we used the five-point rating system shown in Table 2 and assessed only those articles that presented data for the operational-scale restoration projects, as opposed to articles that presented data collected primarily for research purposes. While we believe that most wet meadow restoration practitioners would agree that these six characteristics are important, they represent our view of what constitutes successful restoration and weren't always listed as explicit goals in the articles we assessed.

The second part of our analysis of restoration effectiveness consists of a narrative that synthesizes the results of all the articles in relation to our primary and secondary questions, with separate subsections of the results being devoted to the effects of restoration on (1) geomorphology, (2) hydrology, (3) soils, (4) vegetation and (5) wildlife/biodiversity.

Results

Review statistics

More than 27,500 publications were identified through our initial searches (Additional file 2: Appendix B), the vast majority of which were eliminated during the title filtering stage. After full text review, we retained 40 articles (Table 3). In Additional file 3, we provide a list of 170 additional articles that we retained through the full

Table 2 Rating system used for assessing evidence for restoration effectiveness

Assessment of evidence presented	Rating
Characteristic not mentioned	N/A
Characteristic mentioned; evidence presented to demonstrate failure	0
Characteristic mentioned; no evidence presented to demonstrate success	1
Characteristic mentioned; insufficient evidence presented to demonstrate success	2
Characteristic mentioned; sufficient evidence presented to demonstrate partial success	3
Characteristic mentioned; sufficient evidence presented to demonstrate full success	4

"Characteristic" refers to those listed in Table 1; "mentioned" refers to whether or not the characteristic was listed as specific goal of the restoration project.

text review, but that we excluded from the data extraction phase for the reasons provided. We believe these articles may be a useful resource for those with an interest in wet meadow restoration outside of the American Southwest. As we continued to work on the review, we located an additional 8 articles that we included, bringing the total to 48. All of these new articles were "grey literature" that we became aware of through Google Scholar searches or personal contacts.

Study description

The articles selected for inclusion in this review are diverse. Twenty-five of the included articles reported on research that was directly or indirectly related to wet meadow restoration and was published in peer-reviewed journals. Another 14 articles are better described as monitoring or project reports. The remaining 9 articles are mostly conference proceedings and unpublished theses or manuscripts.

Many articles only addressed one or two of our four main outcomes (i.e. effects on geomorphology, hydrology, soils, and vegetation). Geomorphology was addressed in 20 of the articles, hydrology in 18, soils in 8, and vegetation in 38. We found 13 articles that addressed our secondary question regarding the effects of restoration on wildlife/biodiversity.

Of the 48 articles selected for inclusion in this review, a total of 26 operational-scale restoration projects (as opposed to projects done strictly for research purposes and generally implemented at smaller scales) were identified. Of the 26 projects, 10 were in Arizona, 3 were in New Mexico, 12 were in California, and 1 was in Nevada (Figure 1).

At least 32 different types of restoration or research interventions were identified, with most being reported in only 1 or 2 articles (Figure 2, note some types of interventions are lumped into categories such as Other and "Greenhouse" Studies). The most common interventions were transplanting of vegetation, various forms of grazing management (including exclosures), and use of the pond-and-plug method for restoration of geomorphology and hydrology.

Because of the wide variation in study design, study quality, restoration/research interventions and the way that intervention effects were reported (which in many cases was primarily qualitative), we determined that the use of formal statistical meta-analysis techniques would not be practical. Two additional reasons why we opted not to conduct a meta-analysis include: (1) the relatively short-term nature of the studies, such that the true effects of restoration treatments had not yet been fully tested in many cases and (2) wide year-to-year variation in environmental conditions, especially related to precipitation and its effect on flooding and water table levels, that confounded treatment effects in otherwise well-designed studies e.g., [26,27].

Study quality assessment

The variation in types of study referred to in the previous section was reflected in their quality ratings (Figure 3). The Kappa analysis conducted by Ramstead and Allen showed 100% agreement between the two reviewers based on the ratings initially assigned by Ramstead to 15 randomly selected articles (Cohen's Kappa statistic = 1.0). However, we did experience difficulties deciding how to classify some articles, especially ones that were primarily about modelling studies and to a lesser degree some of the monitoring reports, that didn't seem to fit cleanly into the Pullin and Knight hierarchy.

Evidence of effectiveness

A summary of the evidence of restoration effectiveness is presented in Table 4 for the operational-scale restoration projects. Out of 162 individual ratings (27 studies x 6 characteristics), 56 (34.6%) were rated "N/A." Evidence was presented more frequently for the effects of restoration on perennial vegetation, stream morphology, and streambank stability (i.e. for vegetation and geomorphology) than for the other three characteristics. The characteristic examined least frequently was organic matter assimilation, which was mentioned in just two of the articles, only one of which presented quantitative data.

None of the characteristics had an average rating above 3 – the amount of evidence needed to demonstrate partial success. Of greater importance than the average ratings, however, are the individual ratings of either 3 or 4. These ratings indicate where restoration

Table 3 Articles included during filtering stages

Review stage	Number of articles
Publications captured from electronic databases (excluding duplicates)	~27,500
Publications captured by other sources	50
Publications remaining after title filter	~550
Publications remaining after abstract filter	210
Publications remaining after full text filter that were retained for data extraction	40
Publications added after initial searches in 2009	8

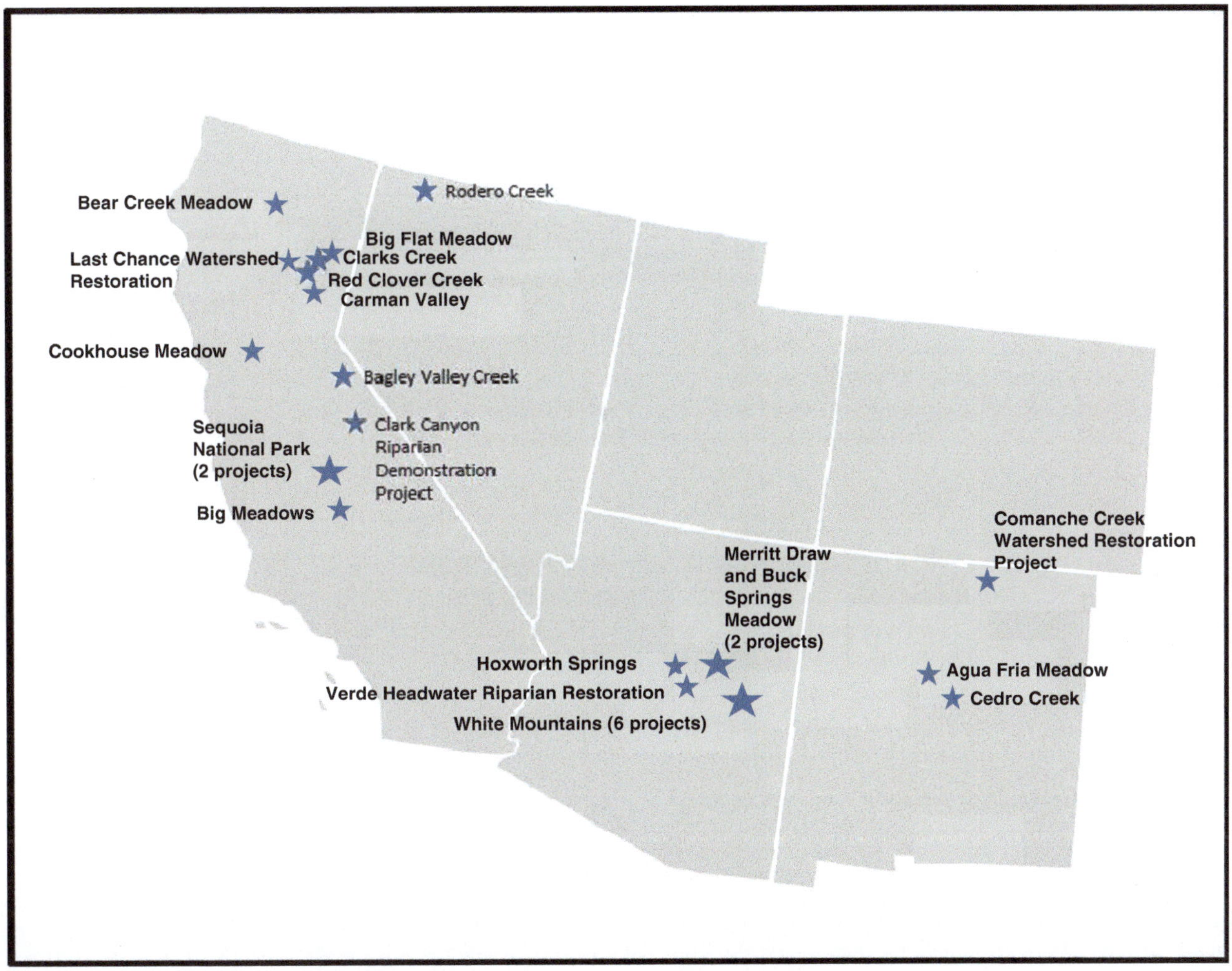

Figure 1 Approximate locations of the 26 operational-scale restoration projects included in this review. Because of their close proximity, some stars indicate more than one project.

techniques have been effective in improving specific wet meadow characteristics. Out of the total of 162 individual ratings in Table 4, there were 12 ratings of 4 (7.4% of the total) and 51 ratings of 3 (31.5% of the total). In the following subsections, we rely heavily on the evidence with ratings of 3 or 4, along with supporting evidence from the smaller, research-scale studies not included in Table 4, to provide a synthesis of the evidence of restoration effectiveness.

Geomorphology

Restoration of project site geomorphology was of central importance in the majority of restoration projects we reviewed. In most cases, this involved measures to restore or recreate a stream channel with a gradient, sinuosity, width-to-depth and entrenchment ratios, and other characteristics that were similar to what existed in the meadow prior to stream incision or other alteration. Only in one case [25] did we find that the goal was to eliminate a channel from a meadow and to establish a sheet flow hydrologic regime.

The restoration practices employed varied depending on the pre-project condition of the stream channel. In meadows where the stream was not deeply incised, relatively simple and inexpensive practices such as using locally available rocks and gravel to construct riffle structures e.g., [38] were typically employed. Where streams were more deeply incised, approaches such as "re-channelization" (i.e., reconfiguring the existing channel or constructing a new one) have been used [28,29,32].

In some of the most highly altered meadows, where the original shallow and meandering stream had been replaced by a deep, wide, and relatively straight gully, the pond-and-plug technique has been employed. This technique involves creating a series of earthen plugs in the gully by using fill material from within and adjacent to the gully (thereby creating alternating ponds and plugs). Streamflow is then diverted into either a remnant

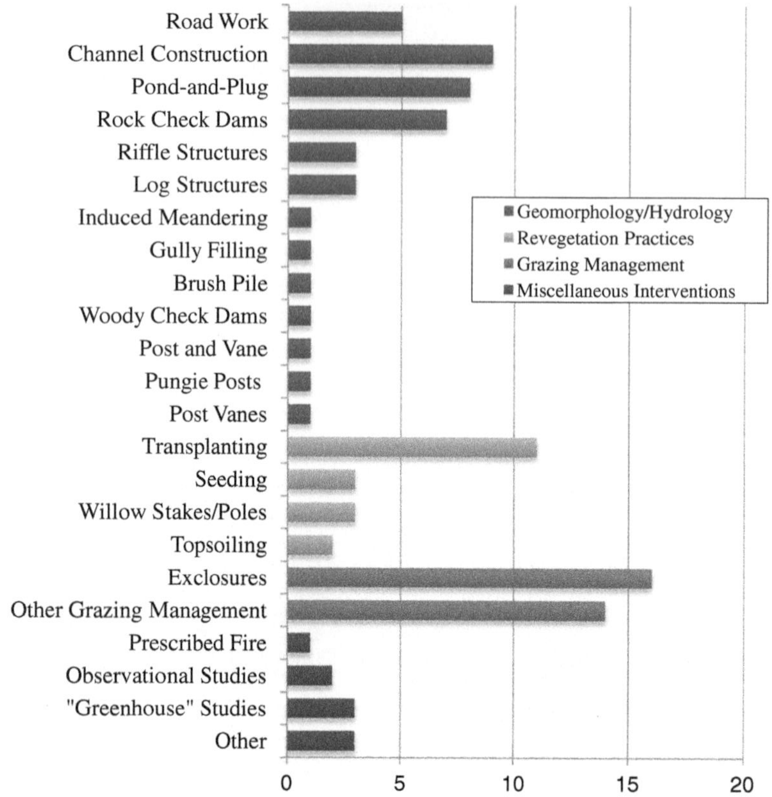

Figure 2 Types of restoration interventions (treatments) reported. The category "Greenhouse" Studies includes research projects that involved the manipulation of growing conditions such as light, temperature, or water table level in artificial environments.

stream channel or a newly created one on the original floodplain surface. This approach recreates part of the geomorphology of the meadow, but also results in a number of ponds that were not a feature of the original landscape. The pond-and-plug approach was first employed approximately 15 years ago and has been used with increasing frequency, particularly in the Sierra Nevada region of California [51].

The articles that we reviewed present a considerable amount of evidence that geomorphological restoration techniques can be effective. Pre- and post-project longitudinal profiles, stream and/or floodplain cross-sections, characterization of streambed particle size, and qualitative evidence (e.g., photo points) were the most frequently used types of evidence. These types of evidence, especially the stream profiles and cross-sections, can

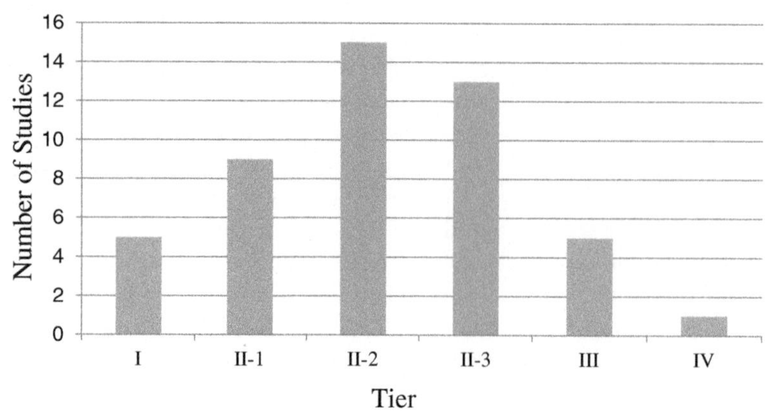

Figure 3 The number of articles by Pullin and Knight's [24] hierarchy of evidence category.

Table 4 Evidence ratings of the success of wet meadow restoration projects

	Characteristics of a functional wet meadow ecosystem						
Article	**Suitable stream morphology**	**Stable streambanks**	**High water table**	**Organic matter assimilation**	**Perennial vegetation**	**Presence of native fauna**	**Total evidence rating**
Anderson et al. 2003 [28]	4	3	2	N/A	3	N/A	12
Borgmann et al.[29]	1	1	3	N/A	3	3	11
Chambers and Lesh[29]	N/A	N/A	N/A	N/A	3	3	7
Godwin 2004 [30]	N/A	N/A	N/A	N/A	3	N/A	3
Hammersmark et al. 2009 [31]	3*	N/A	3*	N/A	4	N/A	10
Herbst and Kane 2009 [32]	3	3	N/A	N/A	3	3	12
Holmquist et al. 2010 [33]	N/A	N/A	N/A	N/A	3	3	6
Key and Gish 1989 [34]	3	2	2	N/A	3	2	12
Long and Endfield 2000 [35]	2	2	2	N/A	2	1	9
Long et al. 2004a [36]	1	1	1	N/A	1	1	5
Long et al. 2004b [37]	4	3	2	N/A	3	3	15
Medina and Long 2004 [38]	3	3	2	N/A	3	2	13
Medina and Steed 2002 [39]	3	3	2	N/A	3	3	14
Norman and Immeker 2009 [40]	4	4	4	N/A	3	N/A	15
Plumas Corporation 2004 [41]	3	2	2	N/A	2	N/A	9
Plumas Corporation 2011 [42]	3	2	N/A	N/A	2	3	10
Ramstead 2011 [23]	1	3	N/A	4	4	N/A	12
Rosen et al. 1999 [43]	3	3	3	N/A	1	N/A	10
Sierra Valley Resource Conservation District 2004 [44]	2	2	2	N/A	1	3	10
State of California 2005 [45]	1	1	1	N/A	2	4	9
Swanson et al. 1988 [46]	3	3	1	N/A	2	2	11
Szewczak 2004 [47]	N/A	N/A	N/A	N/A	N/A	3	3
Vrooman 2004 [48]	3	3	N/A	N/A	3	N/A	9
Vrooman 2005 [49]	3	2	1	N/A	3	N/A	9
Wilcox 2010 [50]	4	4	N/A	N/A	N/A	N/A	8
Wildlife Fish Habitat Initiative 2008 [29]	2	3	2	N/A	3	3	13
Wolf and Cooper 2011 [25]	3**	3**	4	1	3	N/A	14
Average Rating	2.7	2.5	2.2	2.5	2.6	2.6	10.0

*These are addressed more completely in a related study.
**The desired floodplain morphology in this case was a stable floodplain surface *without* a defined stream channel.
The six characteristics assessed are described in Table 1 and the rating system is described in Table 2.

effectively demonstrate that the post-project stream or floodplain configuration was similar to specified construction targets or to reference sites in relatively undisturbed meadows e.g., [25].

Probably the most important question regarding the effectiveness of geomorphological restoration techniques is how well they work in the long-term. Because most of the articles and project reports we reviewed covered no more than five years of post-project results, considerable caution is warranted. The wet meadows of the region contain many examples of failed efforts to restore stream channels [52, personal observations], most of which are believed to have failed during extreme flooding events. During the relatively short post-project periods covered by most of the articles and project reports included in this review, there were several examples of flood-related damage [25,35,53,54]. We have personally observed post-project damage at some sites that has not yet been reported in print, such as headcuts that have developed at the Hoxworth Springs restoration site [30] and failed log structures and gabions at Buck Springs Meadow and Merritt Springs Meadow [23] and elsewhere on the Coconino National Forest in northern Arizona.

While even seemingly well-designed projects have been damaged or failed completely, there is also encouraging evidence that others have been able to withstand truly extreme events. The best example of this may be the large flooding event withstood by a pond-and-plug project in 2009. The flood, which was determined to be in the 500-year return interval category, impacted the

Big Meadows project [50]. Virtually no changes in channel morphology or to the plugs were observed despite the flashy nature of this flood and the fact that it occurred at a time of year (October) when the area was dry and the plugs and channel may have been most vulnerable to damage.

Hydrology

The efforts to restore geomorphic processes described in the previous section are primarily designed to return streamflow, overbank flooding, ground water levels and other aspects of a site's hydrology back to close approximations of their former patterns. The best evidence of success in achieving hydrologic restoration comes from direct measurements of these variables pre- and post-project, through the use of stream gauges, groundwater monitoring wells and similar techniques. Relatively few of the articles we reviewed used one or more of these techniques or monitored for sufficient periods of time to demonstrate substantial, long-term improvements in hydrology. Another challenge we faced was that, in some cases where it appears that adequate hydrological data were collected, the documents we reviewed did not present the data in detail.

Most of the highest quality data we could find on hydrology came from pond-and-plug projects in California, as well as the channel elimination project. One example is the first-ever pond-and-plug project in California, at Big Flat Meadow [29]. Monitoring of this project included stream gauges above and below the project site and a network of ground water monitoring wells along transects perpendicular to the stream. Interpretation of the data is somewhat confounded by the wide year-to-year variation in precipitation, but the data appear adequate to demonstrate that flow duration increased, peak flows below the project site decreased, groundwater levels increased, and stream temperature decreased as a result of the project, all of which are positive results.

A clearer and more dramatic effect on groundwater levels was reported for the Upper Halstead Meadow project. This project sought to eliminate (fill in completely) a gully and to restore a nearly level topography and a sheet flow hydrologic regime to the meadow [25]. Monitoring wells in this meadow showed wide variations in groundwater levels prior to restoration, with levels commonly dipping below 1.5 m during the growing season. Groundwater levels post-restoration fluctuated very little and remained within 20 cm of the surface for most of the three growing seasons that were monitored. Post-project groundwater levels were very similar to those recorded at five reference sites.

While other articles or project reports presented evidence of improvements in hydrology, most tended to do so with less data and sometimes relied on a combination of quantitative and qualitative evidence. An example is Medina and Long [38], whose primary treatment was the installation of riffle structures within a small creek channel. They reported that the riffles reduced the average distance from the streambed to the bank from 63 cm to 45 cm. This in turn reduced the channel capacity and promoted overbank flooding, while also creating a streambank environment more conducive to the establishment of sedges. Accompanying pre- and post-project photos, showing the development of lush vegetation, provided supporting evidence of the project's success in restoring both geomorphology and hydrology.

Several efforts have been made to examine the effects of restoration projects on hydrology using numerical models [11,31,55-57]. These models have been developed and/or tested using data from project sites, and are promising tools for improving both our understanding of the effects of restoration and our ability to design better projects in the future. In one such modeling study, Hammersmark et al. [56] documented three types of hydrological responses to a pond-and-plug project, including (1) increased groundwater levels and subsurface storage, (2) increased frequency and duration of floodplain inundation, and (3) decreased annual runoff (by 1-2%) and duration of base flow (by 2 weeks). The decreased runoff and duration of base flow was driven by the raised channel bed and higher evapotranspiration rates. In the simulations, additional water exited the site as groundwater flow as opposed to stream flow.

Soils

Other than the earthmoving associated with geomorphic restoration, we found little evidence of any direct manipulation of soils as part of restoration projects. Only in the case of small-scale research projects e.g., [26,27] did we encounter the use of soil treatments such as fertilization or aeration. Overall, little work appears to have been done on the effects of restoration on soil; most of what has been done focuses primarily on the effects of eliminating or reducing grazing.

Grazing clearly has caused considerable damage to wet meadows in the region and elsewhere in the western U.S. [58]. Therefore studies aimed at both quantifying the impacts of grazing and understanding how soils recover once it is reduced or eliminated can be very useful. Examples of studies that have quantified impacts of grazing are those of Flenniken et al. [59], Wheeler [60], and Blank et al. [61]. Flenniken et al. [59] demonstrated that grazing, even the very short-term grazing they examined, can significantly decrease microchannel sinuosity and drainage density, thereby increasing flow depths, total runoff and the potential for erosion. Wheeler [60] showed that short-term but moderately heavy grazing caused significant impacts to soil physical properties, but recovery was rapid once grazing was eliminated. Blank et al. [61] demonstrated grazing effects on soil nutrients, many of which increased in response to grazing. While potentially important, we did not attempt to include all the articles of this type on grazing effects, since they are only indirectly related to the central questions addressed in this review.

Studies such as [23,33] were designed to address more directly the effects of reducing or eliminating grazing on wet meadow soils, since protection from overgrazing is one of the most common restoration measures taken in the region. Holmquist et al. [33] found no differences in soil compaction between sites subject to light grazing by pack animals and ungrazed sites. Ramstead [23] examined two areas that had been subject to heavier grazing and found that approximately 20 years of protection from grazing resulted in generally less compacted soils and greater organic matter content.

A final type of study of relevance to soils and wet meadow restoration is more purely ecological in nature. Studies of this type are designed to help determine the relationship between site characteristics, including soils, and the occurrence of plant species and/or community types. Examples of such studies in the American Southwest include those by Benedict, Castelli et al. and Long et al. [2,9,62]. While these studies do not address the effects of restoration projects directly, they have the potential to be very useful for the design of future projects.

Vegetation

Of the 38 articles or reports that addressed vegetation in some way, 25 focused directly on operational-scale projects. Similar to what we have reported above, the type and quality of the data collected, as well as the length of post-project monitoring data collection, is highly variable. The evidence suggests, however, that generally good revegetation success can be obtained if the hydrology has been restored, the site is protected sufficiently from grazing and, where applicable, proper revegetation practices were used.

Wolf and Cooper [25] presented some of the most detailed information we found for an operational-scale project. Their results indicated good short-term survival (~97%) of the 34,250 seedlings that were initially planted at the site. Multivariate analyses showed a clear trend in the overall plant community towards dominance by wetter-site species and towards a greater resemblance to reference sites. Their results or observations also demonstrated some of the problems that can be encountered, however, including a substantial loss of planted seedlings following a large flooding/erosional event and slow growth of seedlings planted in the filled gully, where the fill had been excessively compacted. Both of these issues have reportedly been addressed successfully by refilling the damaged areas with a lighter mix that consisted of 70% soil and 30% sawdust.

When combined with data collected in the field, the modeling studies described above also provide good quality evidence of changes in vegetation caused by restoration projects. Hammersmark et al. [31], for example, used data from 170 vegetation plots established post-project, combined with their hydrologic model, to demonstrate that the habitat suitability for wetland plant species increased throughout a large portion of the project site.

Some of the evidence from restoration projects is qualitative but nonetheless compelling, including observations by experienced personnel and photo points that depict the same sites pre- and post-project. Good examples of photo point evidence can be seen for the Red Clover/McReynolds Creek [42]; Photos 2 and 3] and Pacheta Creek projects [38]; Figures 1 and 5), both of which show impressive development of wet meadow vegetation in the post-project photos.

In cases where the physical site conditions are suitable and grazing has been eliminated or reduced, active revegetation practices (e.g., transplanting of plugs or seeding) may not be required. Ammon and Stacey [6], for example, found that long-term (~30 year) protection of a riparian meadow from grazing resulted in a substantial recovery of willows (*Salix* spp.) and greater vertical vegetational diversity as opposed to a portion of the same meadow that is still subject to grazing. No other restoration practices other than exclusion of livestock grazing were reported for this meadow. Similar results were reported by Shulz and Leininger [5] who noted that, after 29 years of rest from grazing, the riparian vegetation was so dense that it concealed the stream from view. In their review papers, both and Platts and Rinne [63] and Elmore and Kaufmann [64] stated that protection of riparian sites from grazing has repeatedly been shown to result in the recovery of vegetation in western U.S. Wet meadows appear to be no exception.

Where they have been used, the results obtained using active revegetation practices have been mixed. The successes and problems encountered with planted seedlings in one project have already been mentioned above [25]. In a project that involved both substantial earth-moving and active revegetation [28] all project goals were reportedly met except for those associated with revegetation. While the approximately 1/3 of the project area located furthest downstream was beginning to show successful Nebraska sedge (*Carex nebrascensis*) and Baltic rush (*Juncus balticus*) establishment, there was still an overall average of 64% bare ground 3 years post- restoration.

Restorationists could benefit from a number of studies that have been conducted in the region that focus either on plant species propagation or species-site relationships, several of which we have included in this review. Steed et al. [65] and Steed and DeWald [66], for example, provide practical information on three common *Carex* species based on their research on the species' responses to differing water table levels and transplanting methods. Articles such as those already mentioned in the soils section provide information on species-site relationships that can be used to guide species selection or to establish targets for restoring hydrology.

Wildlife/biodiversity

The response of wildlife and biodiversity to wet meadow restoration efforts has been investigated both directly, through changes in species presence and/or abundance, and indirectly, through changes in habitat characteristics. While generally positive effects have been noted, in some cases the data are confounded in ways that limit the ability to draw clear conclusions and in other situations the effects may be at least partially detrimental.

Studies that reported clear increases in animal species presence and/or abundance following restoration have been published for benthic macroinvertebrates [64], avifauna [42] and bats [47].

Changes in habitat quality have been reported for taxa as diverse as birds, small mammals, and fish. Ammon and Stacey [6] investigated the effects of protection of a wet meadow site from grazing on nest predation. The recovery of willows provided better cover, resulting in substantial declines in nest predation for both ground and above-ground nesting birds. The biggest impact was on ground nests, which had an 83% success rate on the restored site and only 36% in the portion of the same meadow that was still subject to grazing. Chambers and Lesh [unpublished manuscript[29]] found that Mogollon voles (*Microtus mogollonensis*) were captured more frequently and runway densities were higher inside exclosures located within several wet meadows in northern Arizona. Other articles have reported positive effects of restoration on habitat quality for trout, such as decreased water temperature [13,42] and improved channel bed characteristics [36].

An example of a study with confounding effects is the Clarks Creek Project [45]. Increases in small mammals and avifauna were noted at the site after project implementation, but similar increases were noted at a nearby control site. Also, a decrease in the number of fish in the restored channel was observed post-project, but the decrease appears to have been caused by beaver dams that reduced access to the channel, rather than a decline in habitat quality of the restored reach. Borgmann et al. [67] reported confounding effects on butterflies, which increased in richness and abundance at the project site the year after implementation but also increased at nearby control sites. Confounding effects were also reported for another project in the Sierra Nevada region, in this case due to the difficulty of separating changes in bird captures at the restoration site from similar changes noted at a much larger network of capture sites in the area [44].

Potential negative impacts were noted for at least one pond-and-plug project, where two non-native species had become established in some of the ponds [44]. The two species (green sunfish; *Lepomis cyanellus* and bullfrog: *Rana catesbeiana*) are both known to prey on native fish and amphibians. Although no specific detrimental impacts were reported, this may be one aspect of the pond-and-plug approach that would benefit from additional research. Permanent ponds are not a natural feature of most wet meadows and their presence may alter animal species communities in unintended ways.

Reasons for variation in effectiveness

We cannot pinpoint any instances where there was variation in the effectiveness of projects that were truly comparable in terms of ecological setting, disturbance history, and treatments applied. In reality each project was unique. The fact that each meadow and stream complex is unique - and therefore requires site-specific treatments - was mentioned by several of the restoration practitioners and researchers that we contacted during the course of this review.

In a more general sense, however, there are many factors that can be identified that cause variation in project effectiveness. In the Upper Halstead Meadow Project, for example, the initial effectiveness of the project was reduced by the choice of fill material and by compacting the material excessively. This resulted in both increased susceptibility to erosion and to reduced growth of seedlings planted in the fill material [25]. It is clear from the literature on restoration of other types of ecosystems that variations in project design, quality of construction, choice of planting material, level of training and supervision of the personnel implementing the project, and many other variables associated with project implementation can have a large effect on the results e.g., [68].

In addition to factors related to project design and implementation, year-to-year variation in environmental conditions such as precipitation, temperature, timing of snowmelt, and flooding can also have a large impact on projects. The two main effects of this type of variation are damage to projects caused by major events such as large storms/floods e.g., [25,50,54,69] and difficulties in interpreting research or monitoring results caused by more typical variation among years in water table levels or other environmental conditions [26,27].

Review limitations

Key limitations of this review include (1) the scarcity of studies that were designed to address our specific questions in a rigorous, quantitative fashion, (2) the inaccessibility of data (versus general descriptions of the results) for most of the projects that included monitoring components, (3) wide variation in methods of data collection, analysis and presentation, which effectively ruled out a meta-analysis, (4) year-to-year environmental variation that confounded the results in some cases, and (5) the relatively short duration of post-project data collection for most of the studies we found.

Another critical concern is that many restoration projects appear to have been implemented over the years, going back at least to the 1930's [50], yet little or no information is available on many (and probably a majority) of them. Our review would have been stronger if the effectiveness of more wet meadow projects had been investigated over the years or, in cases where they were investigated, if the results had been made more widely available. Although speculative, we suspect that the results of projects that failed are less likely to have been published, and yet much can be learned from failed projects. Also, if most failed projects were not reported in accessible formats, then our results may be biased by a tendency to report more frequently on successful projects. None of the articles we found reported a total failure to achieve the stated objectives, although one reported on damage that may have resulted in failure if it hadn't been effectively repaired [25].

A broader search strategy may have yielded additional articles that would have been useful to us. In particular, a search strategy that encompassed more completely the literature on grazing management in riparian areas e.g., [64] and the restoration of riparian habitat for fisheries e.g., [63] may have been valuable in the context of this review. We also recognize that important work has been done on wet meadow restoration in other regions, much of which could be relevant to restoration efforts in the American Southwest.

Reviewers' conclusions

While much of the evidence is not of the highest quality and therefore considerable caution is warranted, it is nevertheless apparent that progress has been made over the past 20 years in wet meadow restoration. In particular, especially important contributions have been made in restoring the highly degraded wet meadow systems that are characterized by deep, wide and relatively straight gullies. There is substantial evidence that the pond-and-plug approach, which was first implemented approximately 15 years ago, is an effective technique for restoring most aspects of these systems, albeit at the cost of creating new features (ponds) that are not necessarily natural features of wet meadows.

Important lessons continue to be learned about wet meadow restoration practices, including those aimed at establishing suitable floodplain and stream channel morphology, a more natural hydrologic regime, and more effective revegetation. The damage that continues to plague many projects, such as erosion and structure damage caused by floods, demonstrates that we still have much to learn about project design and implementation. It also suggests we have to recognize that post-project vigilance and a commitment to ongoing maintenance will be needed to ensure long-term success.

Implications for management

One of the most important implications for managers is that not enough information on the projects carried out to date has been documented and shared. During the course of this review, we encountered several projects in the field or on websites for which no substantive documentation of any kind could be located; we expect that this is common throughout the region. Many opportunities to learn both from successes and failures have undoubtedly been lost over the years due to this failure to document projects and to make the information widely available. Making more project information available through websites can be a powerful tool. The Feather River Coordinated Resource Management group's website (http://www.feather-river-crm.org/) is perhaps the best example in the region of this approach to sharing information on wet meadow restoration.

We understand that there are serious constraints on managers' time and resources, but allocating additional effort to project documentation, including more formal and longer-lasting monitoring programs, is an important need that should be addressed. One approach that might help in this regard is for practitioners to work with scientists from government agencies, local universities and colleges, and other organizations. When this type of collaboration has happened in the past it appears to have been effective e.g., [47].

Implications for research

More research on wet meadow ecosystem processes, the ecology of individual plant and animal species, and specific restoration and management techniques clearly would be beneficial. We have developed a list of research questions that would be worthwhile to address. While many of these have been investigated to some degree already, there is much more to learn. Although they are listed within each section in order of priority from our perspective, we realize that research priorities may vary depending on the predominant types of disturbance being addressed within a given area, as well as the management priorities of government agencies and other landowners.

Geomorphology and Hydrology

- How do the pond-and-plug technique, re-channelization, and smaller-scale channel restoration

techniques perform in the long-term, especially in terms of stream channel/floodplain stability?
- How can we improve upon current practices to make them more resistant or resilient to disturbances such as flooding?
- It would be valuable to investigate the impact and development of created ponds (when using the pond-and-plug technique) more than has been done to date – what impact do they have on meadow hydrology and do they show signs of filling in over time?
- How do individual restoration projects affect water storage, streamflow and water supplies downstream? What is the cumulative effect of a large number of projects on the water budget in a watershed?
- How might climate change impact hydrology and the likelihood of extreme events? Should we be planning for such changes in the design of wet meadow restoration projects?

Soils

- How do soils at restoration sites of different ages compare to relatively undisturbed meadows?
- When are soil treatments needed during project implementation to increase survival and growth of vegetation? What treatments are most effective?

Vegetation

- When can we rely on natural revegetation and when do we need to use techniques such as planting or seeding?
- What improvements do we need to make to wet meadow plant species propagation and establishment techniques?

Wildlife/Biodiversity

- What is the impact of the pond-and-plug approach (and the ponds in particular) on wildlife? Do the ponds harbor species that otherwise would not occur in wet meadows? If so, what are their impacts?
- What is the impact of restoration practices on individual species that have endangered, threatened, or sensitive status (e.g., Apache trout, mountain yellow–legged frog?

Other

- Would improved wet meadow classification systems help guide restoration project design?
- Are there viable alternatives to fencing that require less maintenance?
- What types of livestock grazing regimes are compatible with wet meadow restoration?
- Many wet meadows are on tribal lands. What is the cultural importance of wet meadows to various tribes? What can we learn from the way they managed wet meadows in the past?
- What is the true scope of the need for wet meadow restoration? How many degraded or destroyed meadows are there, where are they, and how many are realistic candidates for restoration?

Additional files

Additional file 1: Appendix A. Summary of included articles [70-75].

Additional file 2: Appendix B. Summary of database search results.

Additional file 3: Provided with this submission: Articles excluded at full text assessment.

Competing interests
The Ecological Restoration Institute (ERI) of Northern Arizona University provided financial and editorial support for this project. The reviewers are not aware of any conflicts of interest that might result from this support.

Authors' contributions
KR conducted all of the literature searches, except those done after 2009 using Google Scholar. She also conducted the title, abstract, and full text filtering (the latter in cooperation with JA) and wrote the first draft of the manuscript. JA initiated this review, designed the protocol, conducted the Google Scholar searches, and provided extensive assistance with the analysis and writing. AS provided input into the protocol and reviewed earlier drafts of the manuscript. All authors read and approved the final manuscript.

Acknowledgements
We would first like to thank the Ecological Restoration Institute of Northern Arizona University for funding this project. Dick Fleishman of the United States Forest Service, Coconino National Forest helped keep us inspired by taking us on several field trips of local wet meadow restoration projects. The feedback from Jonathan Long of the Pacific Southwest Research Station has been invaluable. Jonathon also helped us get in contact with several other people involved in wet meadow restoration. We would like to thank all the practitioners and researchers who participated in a survey that was helpful in looking at the different benefits and disadvantages of individual restoration techniques. Some of them also provided us with literature for this review. Finally, we would like to thank Liz Kalies, of the Ecological Restoration Institute, as well as two anonymous reviewers and the *Environmental Evidence* editorial staff, for their guidance and comments on earlier drafts of this review.

Author details
[1]School of Forestry, Northern Arizona University, P.O. Box 15018, Flagstaff, AZ 86011, USA. [2]School of Earth Sciences and Environmental Sustainability, Northern Arizona University, P.O. Box 4099, Flagstaff, AZ86011, USA.

References
1. Patton DR, Judd BI: **The role of wet meadows as wildlife habitat in the Southwest.** *J Range Manage* 1970, **23:**272–275.
2. Benedict NB: **Plant associations of subalpine meadows, Sequoia National Park, California.** *Arctic Alpine Res* 1983, **15:**383–396.
3. Hendrickson DA, Minckley WL: **Ciénegas – vanishing climax communities of the American Southwest.** *Desert Plants* 1984, **6:**131–175.
4. Muldavin E, Durkin P, Bradley M, Stuever M, Mehlhop M: *Handbook of wetland vegetation communities of New Mexico, Volume I: Classification and community descriptions.* Albuquerque, NM, USA: New Mexico Natural Heritage Program, Biology Department, University of New Mexico; 2000.

5. Schulz TA, Leininger WC: **Differences in riparian vegetation structure between grazed areas and exclosures.** *J Range Manage* 1990, **43:**295–299.
6. Ammon EM, Stacey PB: **Avian nest success in relation to past grazing regimes in a montane riparian system.** *Condor* 1997, **99:**7–13.
7. Long JW: **Evaluating recovery of riparian wetlands on the White Mountain Apache Reservation.** In *Ph.D. dissertation.* Northern Arizona: Northern Arizona University, School of Forestry; 2002.
8. Judd BI: **Vegetation zones around a small pond in the White Mountains of Arizona.** *Great Basin Nat* 1972, **32:**91–96.
9. Castelli RM, Chambers JC, Tausch RJ: **Soil-plant relations along a soil-water gradient in Great Basin riparian meadows.** *Wetlands* 2000, **20:**51–266.
10. Dwire KA, Kauffman JB, Baham JE: **Plant species distribution in relation to water-table depth and soil redox potential in montane riparian meadows.** *Wetlands* 2006, **26:**31–146.
11. Loheide SP, Deitchman RS, Cooper DJ, Wolf EC, Hammersmark CT, Lundquist JD: **A framework for understanding the hydroecology of impacted wet meadows in the Sierra Nevada and Cascade Ranges, California, USA.** *Hydrogeol J* 2009, **17:**229–246.
12. Dahl TE: *Wetlands losses in the United States, 1780's to 1980's.* Washington, D. C.: U.S. Department of the Interior, Fish and Wildlife Service; 1990.
13. National Fish and Wildlife Foundation: *Business Plan: Sierra Nevada Meadow Restoration.* Washington, D.C.: National Fish and Wildlife Foundation; 2010. http://www.nfwf.org/Content/ContentFolders/NationalFishandWildlifeFoundation/GrantPrograms/Keystones/WildlifeandHabitat/Sierra_Meadow_Restoration_business_plan.pdf.
14. Dodd NL, Gagnon JW, Boe S, Schweinsburg RE: **Assessment of elk highway permeability by using global positioning system telemetry.** *J Wildlife Manage* 2007, **71:**1107–1117.
15. Arizona Game and Fish Department: *Apache Trout abstract compiled and edited by the Heritage Data Management System.* Phoenix, Arizona: Arizona Game and Fish Department; 2001. http://www.azgfd.gov/w_c/edits/documents/Oncoapac.fo_000.pdf.
16. Medina AL: *Native aquatic plants and ecological condition of southwestern wetlands and riparian areas.* Albuquerque, New Mexico: USDA Forest Service, Rocky Mountain Forest and Range Experiment Station, General Technical Report RM-GTR-272; 1996:329–335.
17. Richardson DM, Holmes PM, Eisler KJ, Galatowitsch SM, Stromberg JC, Kirkman SP, Pysek P, Hobbs RJ: **Riparian vegetation: Degradation, alien plant invasions, and restoration prospects.** *Divers Distrib* 2007, **13:**126–139.
18. Chambers JC, Miller JR (Eds): *Geomorphology, hydrology, and ecology of Great Basin meadow complexes – implications for management and restoration. USDA Forest Service General Technical Report RMRS-GTR-258.* Fort Collins, Colorado: Rocky Mountain Research Station; 2011.
19. Neary DG, Medina AL: **Geomorphic response of a montane riparian habitat to interaction of ungulates, vegetation, and hydrology.** In *Desired Future Conditions for Southwestern Riparian Ecosystems: Bringing Interests and Concerns Together,* USDA Forest Service General Technical Report RM-GTR-272. Edited by Shaw DW, Finch DM. Fort Collins, Colorado: Rocky Mountain Forest and Range Experiment Station; 1996:143–147.
20. Gage E, Cooper DJ: *Historic range of variation assessment for wetland and riparian ecosystems, US Forest Service Region 2.* Golden, Colorado: USDA Forest Service, Region 2; 2008.
21. Allen-Diaz BH: **Water table and plant species relationships in Sierra Nevada meadows.** *Am Midl Nat* 1991, **126:**30–43.
22. Brown DE: *Biotic communities of the Southwestern United States and Northwestern Mexico.* Salt Lake City: University of Utah Press; 1994.
23. Ramstead KM: **Wet meadow restoration in the Southwestern United States.** In *MSc thesis.* Northern Arizona: Northern Arizona University, School of Forestry; 2011.
24. Pullin AS, Knight TM: **Support for decision making in conservation practice: An evidence-based approach.** *J Nat Conserv* 2003, **11:**83–90.
25. Wolf E, Cooper D: *Restoration of Geomorphic Structure, Hydrologic Regime, and Vegetation in Upper Halstead Meadow.:* Colorado State University, Fort Collins, CO; 2011. http://www.cfc.umt.edu/CESU/NEWCESU/Assets/Individual%20Project%20Reports/NPS%20Projects/CSU/2005/05_06_08Cooper_SEKI_Halstead%20Meadow_frpt.pdf.
26. Martin DW, Chambers JC: **Effects of water table, clipping, and species interactions on *Carex nebrascensis* and *Poa pratensis* in riparian meadows.** *Wetlands* 2001, **21:**422–430.
27. Martin DW, Chambers JC: **Restoring degraded riparian meadows: biomass and species responses.** *J Range Mgmt* 2001, **54:**284–291.
28. Anderson D, Springer AE, Kennedy J, Odem W, DeWald L, Fleishman D: *Verde River Headwaters Restoration Demonstration Project, Final Report.* Arizona: Water Protection Fund, Grant No. 98–059; 2003.
29. Wild Fish Habitat Initiative: *Big Flat Meadow Re-Watering Project.* Bozeman, MT: Montana Water Center, University of Montana-Bozeman; 2007. http://wildfish.montana.edu/Cases/browse_details.asp?ProjectID=43.
30. Godwin TN: **Evaluation of streambed restoration and occurrence of Hoxworth Springs, Coconino County, Arizona.** In *MSc thesis.* Northern Arizona: Northern Arizona University, Geology Department; 2004.
31. Hammersmark CT, Rains MC, Wickland AC, Mount JF: **Vegetation and water-table relationships in a hydrologically restored riparian meadow.** *Wetlands* 2009, **29:**785–797.
32. Herbst DB, Kane JM: **Responses of aquatic macroinvertebrates to stream channel reconstruction in a degraded rangeland creek in the Sierra Nevada.** *Ecol Restor* 2009, **27:**76–88.
33. Holmquist JC, Schmidt-Gengenbach J, Haultain SA: **Does long-term grazing by pack stock in subalpine wet meadows result in lasting effects on arthropod assemblages?** *Wetlands* 2010, **30:**252–262.
34. Key JW, Gish MA: **Clark Canyon (Mono County) riparian demonstration area.** In *In Proceedings the California Riparian Systems Conference, September 22–24, 1998, Davis, California; USDA Forest Service General Technical Report PSW-110.* Albany, CA: USDA Forest Service, Pacific Southwest Research Station; 1988:127–134.
35. Long JW, Endfield D: *Restoration of White Springs, Proceedings of Land Stewardship in the 21st Century: The contributions of watershed management.* U.S.D.A. Forest Service Proceedings RMRS-P-13. Fort Collins, Colorado, USA: Rocky Mountain Research Station; 2000:358.
36. Long JW, Burnette BM, Medina AL, Parker JL: *Restoration of Soldier Spring: an isolated habitat for native Apache trout,* Proceedings of the 16th International Conference, Society for Ecological Restoration, August 24–26, 2004. Victoria, Canada: 2004.
37. Long JW, Burnette BM, Medina AL: *Restoring wetlands after the Rodeo-Chediski wildfire,* Proceedings of the 16th International Conference, Society for Ecological Restoration, August 24–26, 2004. Victoria, Canada: 2004.
38. Medina AL, Long JW: **Placing formations to restore stream functions in a wet meadow.** *Ecol Restor* 2004, **22:**120–125.
39. Medina AL, Steed JE: *West Fork Allotment riparian monitoring study 1993–1999.* Fort Collins, CO: USDA Forest Service, Rocky Mountain Research StationFinal Project Report Volume I; 2002.
40. Norman S, Immeker D: *Interim Monitoring Report for the Big Meadow Creek in Cookhouse Meadow Restoration Project.* South Lake Tahoe, CA: USDA Forest Service, Lake Tahoe Basin Management Unit; 2009. http://www.fs.usda.gov/Internet/FSE_DOCUMENTS/stelprdb5158607.pdf.
41. Plumas River Corporation: *Last Chance Watershed Restoration Project CalFed Agreement #2000-E01 Final Report.* 2004. http://www.feather-river-crm.org/images/pdfs/calfed_final.pdf.
42. Plumas Corporation: *Red Clover/McReynolds Creek Restoration Project Monitoring Report.* 2010. http://feather-river-crm.org/pdf/RedCloverMonitoringRpt2010.pdf.
43. Rosen J, Jemison R, Pawelek D, Neary D: **Using GIS technology to analyze and understand wet meadow ecosystems.** In *USDA Forest Service Proceedings RMRS-P-7.* Fort Collins, CO: Rocky Mountain Research Station; 1999:175–199.
44. Sierra Valley Resource Conservation District: *Carman Valley Watershed Restoration Project, Final Report.* 2004. http://www.feather-river-crm.org/pdf/CarmanFinalReport%2011_2004.pdf.
45. State of California: *Clarks Creek Stream/Meadow Restoration Project Fish and Wildlife Monitoring Report: District Report.* 2004. http://www.feather-river-crm.org/images/pdfs/clarks_report.pdf.
46. Swanson S, Franzen D, Manning M: **Rodero Creek: Rising water on the high desert.** *J Soil Water Conserv* 1987, **42:**405–407.
47. Szewczak JM: *Bats of Carman Valley, Report 1997-2004 Survey and Monitoring.* Quincy, CA: Feather River Coordinated Resource Management, Plumas Corporation; 2004. http://www.feather-river-crm.org/pdf/CarmanBatReport2004.pdf.
48. Vrooman S: *Comanche Creek Morphology and Riparian Vegetation Monitoring Report*; 2005. http://comanchecreek.org/images/links/30-Morphology_and_Riparian_Monitoring_Report_2005.pdf.
49. Vrooman S: *Cedro Creek Monitoring Report Final Report, 2005–2007*; 2008. http://quiviracoalition.org/images/global/52-Pedro_Creek_2008_Monitoring_Report.pdf.

50. Wilcox J: *Big Meadows Restoration Project, October 14, 2009 Flood Event. Technical Report #1, Plumas Corporation*; 2010. http://www.feather-river-rm.org/pdf/BigMeadowsTechReport.pdf.
51. Lindquist DS, Wilcox J: *New concepts for wet meadow restoration in the Northern Sierra Nevada*, Proceedings of the International Erosion Control Association, Conference 31. California: Palm Springs; 2000:145–152.
52. Wilcox J, Benoit T, Mink L: *Evaluation of geomorphic restoration techniques applied to fluvial systems*. Quincy, CA: Feather River Coordinated Resource Management Group; 2001. http://www.feather-river-crm.org/project-files/georest/cover.html.
53. Jemison R: *Forest road reengineering to restore riparian meadow conditions in the Zuni Mountains of New Mexico*, Proceedings of the 2003 International Conference on Ecology and Transportation. Raleigh, NC, USA: Center for Transportation and the Environment, North Carolina State University; 2003:99–106.
54. Quivira Coalition: *Comanche Creek Watershed Restoration Project-Restoring Habitat for the Rio Grande Cutthroat Trout, Part 2~ EPA 319(h) Final Report*. Santa FE, NM: Quivira Coalition; 2008. http://comanchecreek.org/images/links/206-Comanche_II_Final_Report_9-19-08.pdf.
55. Loheide SP, Gorelick SM: **Riparian hydroecology: a coupled model of the observed interactions between groundwater flow and meadow vegetation patterning.** *Water Resour Res* 2007, **43:**WO7414. 16 pp.
56. Hammersmark CT, Rains MC, Mount JF: **Quantifying the hydrological effects of stream restoration in a montane meadow, northern California, USA.** *River Res Appl* 2008, **24:**735–753.
57. Hammersmark CT, Dobrowski SZ, Rains MC, Mount JC: **Simulated effects of stream restoration on the distribution of wet-meadow vegetation.** *Restor Ecol* 2010, **18:**882–893.
58. Kaufmann JB, Krueger WC: **Livestock impacts on riparian ecosystems and streamside management implications: a review.** *J Range Mgmt* 1984, **37:**430–438.
59. Flenniken M, McEldowney RR, Leininger WC, Frasier GW, Trlica MJ: **Hydrologic responses of a montane riparian ecosystem following cattle use.** *J Range Mgmt* 2001, **54:**567–574.
60. Wheeler MA, Trlica MJ, Frasier GW, Reeder JD: **Seasonal grazing affects soil physical properties of a montane riparian community.** *J Range Manage* 2002, **55:**49–56.
61. Blank RR, Svejcar T, Riegel G: **Soil attributes in a Sierra Nevada riparian meadow as influenced by grazing.** *Rangeland Ecol Manage* 2006, **59:**321–329.
62. Long JW, Tecle A, Burnette BM: **Geologic influences on recovery of riparian wetlands on the White Mountain Apache Reservation.** *J Arizona-Nevada Acad Sci* 2003, **36:**46–60.
63. Platts WS, Rinne JN: **Riparian and stream enhancement management and research in the Rocky Mountains.** *N Am J Fish Manage* 1985, **5:**115–125.
64. Elmore W, Kaufmann B: **Riparian and watershed systems: degradation and restoration.** In *Ecological Implications of Herbivory in the West*. Edited by Vavra M, Laycock WA, Pieper RD.; 2004:212–231.
65. Steed JE, DeWald LE, Kolb TE: **Physiological and growth responses of riparian sedge transplants to groundwater depth.** *Int J Plant Sci* 2002, **163:**925–936.
66. Steed JE, DeWald LE: **Transplanting sedges (*Carex spp.)* in Southwestern riparian meadows.** *Restor Ecol* 2003, **11:**247–256.
67. Borgmann KL, Groce J, Morrison ML: *Chapter II: Cookhouse Meadow Restoration Project*. South Lake Tahoe, CA: USDA Forest Service, Lake Tahoe Basin Management Unit; 2007. http://www.fs.usda.gov/Internet/FSE_DOCUMENTS/stelprdb5150406.pdf.
68. Kusler JA, Kentula ME: *Wetland Creation and Restoration: Status of the Science*. Washington, D.C: Island Press; 1990.
69. Jemison R: **Forest road reengineering to restore riparian meadow conditions in the Zuni Mountains of New Mexico**. In *Proceedings of the 2003 International Conference on Ecology and Transportation*. Edited by Irwin Cl, Garrett P, McDermott KP. Raleigh, North Carolina: Center for Transportation and the Environments, North Carolina State University; 2003:99–106.
70. Beck JL, Peek JM: **Herbage productivity and ungulate use of northeastern Nevada mountain meadows.** *J Range Manage* 2004, **57:**376–383.
71. Mount JF, Hammersmark CT: **Ecohydrologic effects of stream restoration.** In *UC Water Resources Center Technical Completion Report Project No. WR-995*. Berkeley, CA: University of California; 2007:46.
72. Jones KL, Roundy BA, Shaw NL, Taylor JR: **Environmental effects on germination of *Carex utriculata* and *Carex nebrascensis* relative to riparian restoration.** *Wetlands* 2004, **24:**467–479.
73. Mullen RM, Springer AE, Kolb TE: **Complex effects of prescribed fire on restoring the soil water content in a high-elevation riparian meadow, Arizona.** *Restor Ecol* 2006, **14:**242–250.
74. Norton JB, Bowannie F Jr, Peynetsa P, Quandelacy W, Siebert SF: **Native American methods for conservation and restoration of semiarid ephemeral streams.** *J Soil Water Conserv* 2002, **57:**250–258.
75. Sarr DA, Dudley TL: **Survival and restoration potential of beaked sedge (*Carex utriculata*) in grazed riparian meadows of the southern Sierra Nevada (California).** *Ecol Restor* 2008, **26:**186–188.

5

Are alternative livelihood projects effective at reducing local threats to specified elements of biodiversity and/or improving or maintaining the conservation status of those elements?: a systematic review protocol

Dilys Roe[1*], Mike Day[1,2], Francesca Booker[1,2], Wen Zhou[2], Sophie Allebone-Webb[3], Noëlle Kümpel[3], Nicholas A O Hill[3,7], Juliet Wright[3,4], Niki Rust[1,8], Terry CH Sunderland[2], Kent Redford[5] and Gillian Petrokofsky[2,6]

Abstract

Background: Alternative livelihood projects are used by a variety of organisations as a tool for achieving conservation results. Yet these interventions, including their objectives, vary a great deal, and there is no single accepted definition of what constitutes an alternative livelihood project. In addition, very little is known about what impacts, if any, alternative livelihoods projects have had on biodiversity conservation, as well as what determines the success or failure of these interventions. Reflecting this concern, a resolution was passed at the IUCN World Conservation Congress in 2012 calling for a critical review of the benefits to biodiversity of alternative livelihood projects. This systematic review is intended to contribute to this resolution.

Methods/Design: This protocol details the methodology for exploring the research question: *Are alternative livelihood projects effective at reducing local threats to specified elements of biodiversity and/or improving or maintaining the conservation status of those elements?* The aim of this proposed systematic review is to provide an overview for researchers, policy makers and practitioners of the current state of the evidence base. To undertake this, the systematic review will explore peer-reviewed research from the bibliographic databases SCOPUS, Web of Science, CAB Abstracts, AGRIS and AGRICOLA using search terms and Boolean search operators. The systematic review will also explore grey literature sources by conducting searches on Google, subject specific websites and institutional websites. Additionally, the review will use calls for papers and snowballing techniques to further identify literature. In order to identify relevant evidence, inclusion criteria will be used to screen the titles and abstracts of the captured literature. Data will be extracted from the final list of relevant documents by using a questionnaire established through literature review and a stakeholder workshop. The key output of this review will be a report on the state of the evidence, and, if the data permits, an estimate of the effectiveness of alternative livelihood interventions.

Keywords: Alternative livelihoods, Biodiversity, Conservation, Community attitudes and conservation threats

* Correspondence: dilys.roe@iied.org
[1]International Institute for Environment and Development (IIED), 80-86 Grays Inn Road, London WC1X 8NH, UK
Full list of author information is available at the end of the article

Background

Strategies designed to alleviate threats to biodiversity have been applied at a variety of scales from local to global. Recent attention has focused on global and regional scales due to concerns about the impacts of climate change and large-scale habitat loss. However, there has also been considerable investment in strategies designed for implementation at the local level. Many of these strategies have targeted the actions of people living in and around protected areas or other areas of biodiversity importance, and have consisted of both negative and positive incentives to change people's attitudes and behaviour. These incentives are intended to reduce or cease people's exploitation of components of biodiversity that are of conservation concern. Commonly-used disincentives include resource access restrictions, with penalties imposed if restrictions are not adhered to; increased law enforcement; and land/resource use zoning. Frequently-used positive incentives include payments for ecosystem services and other compensatory mechanisms, and so-called "alternative livelihoods projects", which are the focus of this review.

Alternative livelihood projects are poorly defined in the literature – indeed there is no common definition. In general, however, they can be understood to be an approach to achieving biodiversity conservation by substituting a livelihood strategy that is causing harm to a biodiversity target (for example, through unsustainable use), for one that has a more positive impact. In other words, they seek to reduce locally-driven threats to biodiversity. In some cases this might mean providing an *alternative resource* to the one that is being exploited. For example, encouraging local people to farm cane rats as a source of protein, rather than hunting bushmeat (e.g. Development d'Alternatives au Bracconage en Afrique Centrale project [1]).

In other cases, the focus of the project may be on providing an *alternative occupation or source of income* so as to reduce the need to exploit the biodiversity target. Commonly used alternative occupations include craft making [2], bee-keeping [3] or butterfly farming [4] as a substitute for expanding subsistence agriculture around protected areas, or seaweed farming as an alternative to artisanal fishing [5,6]. Finally, a third approach involves encouraging an *alternative method of exploiting a resource* that has a lower impact than the original method. Examples of such interventions include the promotion of fuel-efficient stoves to reduce the demand for firewood [7]; or the introduction of a mariculture programme to ameliorate pressure on a certain fish species [8].

Alternative livelihoods projects are sometimes standalone initiatives and at other times part of a broader integrated conservation and development (ICD) programme. In all cases, the alternative livelihoods projects or components share a common objective: *"to provide an alternative means of making a living that reduces pressure on exploited resources"* [3]. The use of alternative livelihood projects is not limited to the conservation sector. In Afghanistan, the promotion of alternative livelihoods is a central pillar of the government's counter narcotics strategy [9] while in Ghana, alternative livelihoods projects such as cassava farming, poultry farming, and grass-cutter or snail rearing have been promoted in order to combat illegal artisanal mining [10].

The on-going commitment to alternative livelihood approaches within the conservation sector is demonstrated by the United States Agency for International Development (USAID) - funded Central African Regional Program for the Environment (CARPE). This initiative includes an objective to *"expand alternative livelihood opportunities for rural communities"* as a goal in its latest (2012–2020) phase [11]. Yet despite such investment, very little is known about what has worked, what has not worked and why. Furthermore, both conservation and development practitioners have expressed concern that the alternative livelihoods approach may be flawed. In particular, with the emergence of the sustainable livelihoods framework in the late 1990s [12], it is better recognised that poor households are likely to engage in a combination of diverse livelihood activities [13,14]. Commentators have suggested, therefore, that because of this multiplicity of activities the introduction of an alternative livelihood activity may simply be incorporated into the overall mix of activities rather than acting as a genuine replacement [15].

Reflecting this concern, a motion was passed at the IUCN World Conservation Congress in 2012 calling for a critical review of biodiversity benefits of alternative livelihood projects and the development of *"best practice guidelines for alternative livelihood projects to ensure sustainable benefits to species, ecosystems and people"* [16]. This systematic review is a response to that call.

Objective of the review

The objective of this systematic review is to critically examine the evidence on the effectiveness of alternative livelihoods projects in order to inform the future design of such interventions. The goal is to conduct an overview of the state of the evidence base and illuminate the strengths, weaknesses and prevailing knowledge gaps. This review arose from a meeting convened by the Zoological Society of London (ZSL) in May 2012, which challenged a small group of participants to identify case studies of alternative livelihoods interventions that have experienced success, specifically in achieving positive conservation outcomes. This in turn stimulated passing of the aforementioned resolution to IUCN calling for a critical review in order to help guide future donor investments.

The timing of the IUCN resolution coincided with a move by the Center for International Forestry Research (CIFOR) to promote evidence-based approaches to science-policy dialogue. As part of this, CIFOR established an *"Evidence-Based Forestry Initiative"* to act as a platform for promoting systematic reviews of key policy questions, including this review [15]. CIFOR, the Institute for International Environment and Development (IIED) and ZSL are collaborating on this review and form the core review team.

The primary research question for the review (summarised in Table 1) is: *"Are alternative livelihood projects effective at reducing local threats to specified elements of biodiversity and/or improving or maintaining the conservation status of those elements?"* Within this, *'alternative livelihood projects'* are defined as interventions that *"seek to alleviate a human threat to biodiversity through providing, or encouraging the use of an alternative resource; an alternative occupation; or an alternative method (lower impact) of exploitation"*; and "effectiveness is defined as: 1) threat alleviation through changes in human attitude, behaviour and/or wellbeing; and ultimately 2) improvements in the conservation status of the specified element of biodiversity under threat. The secondary question is: What is the state of the evidence base for the effectiveness of alternative livelihood projects for biodiversity conservation?

Our definition of alternative livelihoods projects distinguishes the review's scope from having a broader focus on *'livelihood projects',* which can include any conservation intervention that addresses the economic needs of local people. This includes, for example, community-based conservation, payments for ecosystem service (PES) schemes and integrated conservation and development programmes (although we would include such approaches if they include an alternative livelihoods intervention as a component of a wider scheme). Specifically, our definition implies that there is some kind of substitution of an unsustainable activity for one that is considered as sustainable – or even has a positive effect -on the conservation status of a particular biodiversity target (for example species, group of species, habitat).

Methods/Design

Literature searching

Developing and testing a search strategy

Initial literature searching was tested in Elsevier's Scopus [17] and Thomson Reuter's Web of Knowledge (Science) [18]. A list of search terms was compiled for the population and intervention elements that construct the research question. Search terms included synonyms for alternative livelihoods, examples of specific types of alternative livelihoods interventions (such as beekeeping or cane rat farming) and terms related to biodiversity or conservation. This list was developed iteratively as more terms were identified from the literature. Final search terms were selected based on the number of search results they generated and the presence of publications from a test library of 15 relevant studies. This is an evolving test library with further relevant papers added as they are identified to pass our primary inclusion criteria (see 'Additional file 1: A list of Search Terms used and the Test Library').

A scoping search was also conducted in CABI's database CAB Abstracts [19] using the OvidSP platform [20]. This search used the CAB Abstracts Thesaurus to identify search terms and generated 12,389 search results (see 'Additional file 2: The Search Strategy used for Bibliographic Databases'). The search was intended to be broad to allow for an initial review of the literature, paying attention to the terminology used to refer to alternative livelihoods in the titles and abstracts. It is expected that this process will help to further refine the search strategy generated from testing Scopus and Web of Knowledge.

In addition to the bibliographic databases, search terms were tested in the search engine platform Google Scholar [21]. The terms that produced relevant hits were combined to produce six search strings of alternative livelihood synonyms and eight search strings for specific types of alternative livelihoods interventions (see 'Additional file 3: The Search Strategy used for Internet Search Engines').

Search terms and languages

Search terms are detailed in 'Additional file 1: A list of Search Terms used and the Test Library'. Searches will be carried out using the listed English terms in the first instance, and replicated in French and Spanish. Although, this will not be necessary for the bibliographic searches as titles and abstracts for all foreign language papers are included in English in the selected databases. The team's linguistic abilities and time limitations preclude any further language coverage. Terms listed in three relevant multilingual subject-specialist thesauri will also be incorporated: AGROVOC [22], CAB Thesaurus (English and Spanish terms) [23] and NAL Thesaurus [24].

Table 1 Elements of the systematic review question

Population	Intervention	Comparators	Outcomes
Biodiversity target	Alternative livelihoods project	Prior to alternative livelihoods project intervention/ without alternative livelihood intervention	Reduced threat and/or improved conservation status of specified biodiversity target

The generic search strategy is shown in more detail across Additional files 1, 2, 3 and 4. The strategy will vary slightly between each bibliographic database according to their particular field structures use of Boolean operators, wildcards, proximity searches and the exclusion of irrelevant subject areas; full descriptions of the final search strategies will be documented.

Sources of publications

(a) Bibliographic databases

The following bibliographic databases will be searched for publications:

- Scopus [17]
- Web of Science [18]
- CAB Abstracts [19]
- AGRIS [25]
- AGRICOLA [26]

(b) Individual journals

A number of individual journals will be targeted. The Directory of Open Access Journals (DOAJ) [27] will be used to identify freely accessible publications in as well as INASP's (International Network for the Availability of Scientific Publications) [28] regional journal platforms such as African Journals Online (AJOL) [29], Asia Journals Online [30] and the Poverty and Conservation Learning Group (PCLG) journal digest [31]. For searches in all these platforms, simple search terms will be used as follows: conservation AND development, alternative livelihoods, or just livelihood where characters are limited. The equivalent French and Spanish terms will also be used where appropriate.

(c) Internet searching

Google Scholar [21] will be searched using a total of 14 search strings. The first 150 search results, when organised by relevance, will be captured for the 6 alternative livelihood synonym strings, while the first 100 results will be captured for those 8 strings using the specific alternative livelihood intervention terms. This is because the latter 8 strings give a narrower range of results when compared with the 6 strings that use broader alternative livelihood synonyms. These Google Scholar strings will also be translated into French and Spanish languages and the results captured. Searches for grey literature will also be conducted using simple search terms in the search engine Google [32]. The first 100 results, when ordered by relevance, will be reviewed (see 'Additional file 3: The Search Strategy used for Internet Search Engines').

(d) Specialist searching

Grey literature will be searched through the websites of key donor, implementer and research organisations. 'Additional file 4: The Search Strategy to be used for Specialist Websites' details the list of organisations that will be targeted. These organisations have been selected according to their prominence in conservation and development, as well as the ability to search their websites. A number of subject specific websites with potential relevant publications will also be searched. Searches on these websites will use the simple search terms that were outlined in the individual journals section (e) above.

(e) Theses and dissertation repositories

Six theses and dissertation repositories will be targeted; Ethos [33], DART-Europe E thesis [34], National ETD Portal South Africa [35], Open Access Theses and Dissertations [36] and ProQuest Digital Dissertations and Theses [37]. Again, searches on these repositories will use the simple search terms that were outlined in the individual journals section (e) above.

(f) Stakeholder engagement

Through a process of stakeholder engagement, expert advice has been, and will continue to be, sought on relevant key papers from both the peer reviewed and grey literature. On the 18th December 2013, a short presentation on the project was provided to the UK Bushmeat Working Group, and feedback on the proposed approach and on relevant papers requested. The project has been summarised in a flyer to be disseminated via the project partners' websites and at relevant conferences and workshops attended by team members.

In addition, a call for case study examples of alternative livelihood projects and key papers that measure such intervention's effectiveness for achieving biodiversity conservation has been issued on relevant email listservs including: the Biodiversity-L Listserv, a peer to peer announcement list run by the International Institute for Sustainable Development [38]; and the Poverty and Conservation Learning Group (PCLG), an international network of conservation and development organisations coordinated by IIED [32]. A blog [39] has also been produced and promoted via social networks, to encourage the sharing of relevant case studies and key papers.

Further efforts will include requesting our Advisory Board of experts working within the field of conservation and development, who will guide the systematic review, to list five to ten key publications that are relevant to the research question.

The references of all documents recommended will be screened for relevance (see Study inclusion criteria), as well as the references of those documents subsequently included through this process, in a 'backward snowball' [40].

Estimating the comprehensiveness of the search

The iterative development of search terms - and their piloting against a test library of key papers - will ensure the search is as comprehensive as possible. In addition, searching in both French and Spanish languages will broaden the coverage of the literature captured. Potential limitations to the review's comprehensiveness include the lack of standard terminology to describe 'alternative livelihoods projects'. As a result, it could be that many programmes that fall into the definition used by this systematic review for 'alternative livelihoods projects' are not captured. The iterative development of search terms will help to minimise the potential biases caused by this limitation.

A further limitation is the extent to which grey literature can be accessed. It is anticipated that much of what is known about "alternative livelihood projects" may not be documented in scientific literature but rather in project records. This protocol outlines an extensive methodology that includes engaging with individuals and organisations to help in the process of identification and examination of as many repositories of information as possible. However, access to documents may be restricted in many cases.

Study inclusion criteria

After all the literature captured through searches is gathered, and duplicates are removed, the inclusion criteria detailed in Table 2 will be applied to each document. A Kappa analysis will be performed on a random (minimum of 100) sample of the titles and abstracts to measure the level of agreement between the five researchers in applying these criteria [41,42]. If the kappa is lower than 0.6, reviewers will discuss the discrepancies and clarify any differences in the interpretation of the inclusion criteria. The five reviewers involved in this process have a role in screening the titles, abstracts, and full texts.

When applying the inclusion criteria to the search results, documents will be assessed independently, with each reviewer assigned to a proportion of the literature. A reviewer will be able to include, exclude or query a document on the basis of its title, followed by its abstract. Where there is doubt, a reviewer will be encouraged to query the inclusion of a document with the other reviewers during weekly team meetings. Exclusion at this stage will be conservative, ensuring that if there is any doubt regarding a document's relevance it is retained until an assessment based on the full text.

Once documents have been included or excluded on the basis of their titles or abstracts, all reasonable effort will be made to obtain the full text electronic or paper copies of the remaining documents. This includes emailing corresponding authors where the full text is not available to download from web-based sources. At this point, any documents that the review team is uncertain about will be screened according to their full text.

Table 2 Primary inclusion criteria 1

Articles will be **included** in the review if they meet these criteria.	**1. Relevant population:** The study assesses an element of biodiversity (for example species, group of species, habitat).
	2. Relevant intervention: The study involves an intervention that aims to achieve conservation of biodiversity through a livelihoods intervention.

Potential effect modifiers and reasons for heterogeneity

There are a number of effect modifiers that the review team will consider including: ecosystem/habitat type, size and location; protection status including time since establishment; management type; presence of other conservation mechanisms or associated development projects; availability & access of other natural resources; country (including economic indices); population density; baseline poverty; livelihood strategies.

Study quality assessment

To minimize study bias and ensure validity of the studies with respect to the review question, we will use a checklist approach to assess the quality of included studies. The basic quality standard that we will apply is that studies must have included a measure of effectiveness in their analysis. Any studies that do not include a clearly articulated measure of effectiveness will be excluded from the synthesis. Beyond this basic standard we will also check for the following:

- Clear and repeatable methodology
- Methodology appropriate to the assessment of effectiveness
- Methodology allows for causal links to be established
- Reported findings consistent with methodology employed
- Sample size is specified and appropriate
- Confounding factors are taken into account

Studies which meet none of the checklist quality criteria will not be included in the synthesis. Our synthesis will then distinguish between studies which meet a high quality score (meets all the criteria on the checklist), those with a low score (2 or less criteria me) and those in between (medium quality score). The checklist criteria will be tested on key references and will be further refined during the data extraction process, including through consultation with the Advisory Group.

Data extraction strategy

We will use a short data extraction template, developed iteratively with stakeholders (modelled on Isasi-Catalá, 2009 [43]; itself adapted from Pullin & Knight 2001 [44]) to capture relevant data that will be analysed in the review. The data extraction strategy will be piloted using the 15 articles in the test library.

Data to be extracted will include:

1. Bibliographic information: author, year, title, publication, place published, publisher.
2. Basic information: location of study, alternative livelihood intervention (alternative resource/method/ occupation), date & duration of the intervention, stand-alone intervention or component of a wider project, project funder(s) & implementer(s).
3. Details of evidence type: source, location of primary author, study design, methodology, type of data, unit of analysis, duration of study.
4. Relevant detail considered by study: conservation target, target group of the alternative livelihood intervention, scale of the alternative livelihood intervention, and conceptual link between alternative livelihood intervention & conservation intention.

As part of the data extraction, articles will be assessed to determine if they have included some assessment of the effectiveness of the alternative livelihoods intervention. Those that have done so will be subjected to second phase of data extraction using a second similarly-derived template. Data to be extracted in this phase will include:

5. Details of outcomes: effectiveness measure used (attitudinal/behavioural/change in conservation status etc.), reported effect (positive/negative/no effect), duration of impacts, scale & sustainability of impacts, and the nature of any secondary impacts.

The data extraction templates will be designed so that wherever possible fixed answers will be selected from a coded, drop down list. Entering the answers into the spreadsheet in numerical code format will then allow for subsequent analysis with descriptive statistics. Where pre-determined, coded options are not possible, qualitative narrative data will be collected.

Data management

The final results from the searches outlined in 3.1.3 will be added to a Zotero library [45]. Microsoft Excel [46] will be used to collect the results of the data extraction.

Data synthesis and presentation

For the overall description of the state of the evidence base – the systematic map - we plan to provide a narrative mapping of all the articles that were reviewed in full. We will use descriptive statistics to summarise trends, to map the quantity and type of articles reviewed, and to highlight their relevance to the primary question. Table 3 outlines the main categories of the data that will be analysed through producing summary graphs and charts to provide an overview of the evidence base. The map will be provided in a narrative and graphic form (using charts and tables to illustrate the results found) and we will also explore the potential for recording data on a project-by-project basis in an Access - or custom-built – database that can subsequently be made available to other researchers and practitioners online.

For the more detailed analysis of the studies that meet our quality assessment criteria for the systematic review, we will again predominantly rely upon descriptive statistics and narrative synthesis. We will explore the possibility of meta-analysis in order to quantify and synthesise the overall findings. However, given the variation in types of interventions that the scoping study revealed we anticipate that meta-analysis is not likely to be feasible. Our

Table 3 Categories of data to be included in the data synthesis

Nature of evidence	• Sources of evidence (journal types and subjects, grey literature)
	• Types of evidence (study design, data sources, scale of analysis)
	• Producers of evidence (location of primary authors/institutions)
Representativeness and coverage of evidence	• Geographic coverage (countries, regions)
	• Ecological coverage (biomes, habitats)
	• Components/attributes of biodiversity studied (species, group of species)
	• Types of alternative livelihood projects (i.e. alternative resource, alterative occupation or alternative method)
	• Measure of effectiveness used
	• Nature of impacts reported (positive, negative, neutral) and relative proportions of each type
Measure of effectiveness	• Scale of impacts and numbers of studies that address scale
	• Nature of any secondary impacts (including social & economic) & evidence of any trade offs or synergies.
	• Sustainability of impacts - no's of studies that consider sustainability issues
Policy, institutions and governance	• Attention paid to underlying policy, institutional and governance issues

synthesis will include details of the comparative effectiveness of different alternative livelihood project approaches (alternative resource, alternative occupation, alternative method); the most common outcomes (changes in attitude compared to changes in behavior or changes in conservation status); and - if possible – the factors limiting effectiveness (access to markets; type and level of threat; implementation issues). In all cases, as mentioned above, we will distinguish between studies of different quality in our synthesis. The data extraction strategy and synthesis approach will be further refined during the review process, including through consultation with the Advisory Group.

Additional files

Additional file 1: A List of Search Terms used and the Test Library.

Additional file 2: The Search Strategy used for Bibliographic Databases.

Additional file 3: The Search Strategy used for Internet Search Engines.

Additional file 4 The Search Strategy to be used for Specialist Websites.

Competing interests
The authors declare that they have no competing interests.

Authors' contributions
DR, MD & FB co-drafted the manuscript, while SAW, WZ, NW, NH, JW, GP, TS, KR provided comments. FB, MD and WZ developed the search strategy, with input from GP. DR, MD, FB & GP jointly developed the data extraction strategy. DR will coordinate the review including the analysis and presentation of the results with help from MD, FB, WZ, SAW & GP. All authors read and approved of the final manuscript.

Acknowledgements
The authors would like to thank Professor Peter Kanowski and Dr Peter Holmgren of the Centre for International Forestry Research for their support for the evidence-based forestry initiative, which is funded by a grant to CIFOR from the UK's Department for International Development (DfID). We would also like to thank staff of the Bodleian library, University of Oxford for help locating sources of evidence.
This work has been conducted with the financial assistance of the UK Department for International Development, through its KNOW-FOR program grant to CIFOR, and funding partners who have contributed to the CGIAR Fund.

Author details
[1]International Institute for Environment and Development (IIED), 80-86 Grays Inn Road, London WC1X 8NH, UK. [2]Centre for International Forestry Research (CIFOR), Jalan CIFOR, Situ Gede, Sindang Barang, Bogor 16115, Indonesia. [3]Zoological Society of London (ZSL), Regent's Park, London NW1 4RY, UK. [4]Department of Life Sciences, Imperial College London, Silwood Park Campus, Buckhurst Road, Ascot SL5 7PY, UK. [5]Archipelago Consulting, P.O. Box 4750, Portland, Maine 04112-4750, USA. [6]Department of Zoology, University of Oxford, The Tinbergen Building, South Parks Road, Oxford OX1 3PS, UK. [7]Conservation Science Group, Department of Zoology, University of Cambridge, Cambridge CB2 3EJ, UK. [8]Durrell Institute of Conservation and Ecology, University of Kent, Canterbury, Kent CT2 7NZ, UK.

References
1. Secretariat of the Convention on Biological Diversity (SCBD): **Livelihood alternatives for the unsustainable use of bushmeat.** In *Report prepared for the CBD Bushmeat Liaison Group. Technical Series No. 60.* Montreal: SCBD; 2011:46.
2. Triet R: **Combining biodiversity conservation with poverty alleviation - a case study in the Mekong Delta Vietnam.** *Aquatic Ecosystem Health Manage* 2010, **13:**41–46.
3. Munthali S, Mughogho D: **Economic incentives for conservation: beekeeping and *saturniidae* caterpillar utilization by rural communities.** *Biodivers Conserv* 1992, **1:**142–154.
4. Gordan I, Ayiemba W: **Harnessing butterfly biodiversity for improving livelihoods and forest conservation: The Kipepeo Project.** *J Environ Dev* 2003, **12:**82–98.
5. Sievanen L, Crawford B, Pollnac R, Lowe C: **Weeding through assumptions of livelihood approaches in ICM: seaweed farming in the Philippines and Indonesia.** *Ocean Coast Manag* 2005, **48:**297–313.
6. Hill N, Rowcliffe M, Koldewy H, Milner-Gulland EJ: **The Interaction between Seaweed Farming as an Alternative Occupation and Fisher Numbers in the Central Philippines.** *Conserv Biol* 2011, **2:**324–334.
7. DeWan A, Green K, Li X, Hayden D: **Using social marketing tools to increase fuel-efficient adoption for conservation of the golden snub-nosed monkey, Gansu Province, China.** *Conserv Evid* 2013, **10:**32–36.
8. Lui M, Sadovy de Mitcheson Y: **Profile of a fishery collapse: why mariculture failed to save the large yellow croaker.** *Fish Fish* 2008, **2008:**219–242.
9. Mansfield D, Pain A: *Alternative Livelihoods: substance or Slogan?.* Kabul, Afghanistan: Afghanistan Research and Evaluation Unit (AREA) Briefing Paper; 2005. http://ageconsearch.umn.edu/bitstream/14650/1/bp05ma01.pdf.
10. Hilson G, Banchirigah SM: **Are alternative livelihood projects alleviating in poverty in mining communities? Experiences from Ghana.** *J Dev Stud* 2009, **45:**172–196.
11. U.S. Agency for International Development (US AID)/Central Africa Regional Program for the Environment (CARPE): **Regional Development Cooperation Strategy 2012–2020.** 1–38 [http://www.usaid.gov/sites/default/files/documents/1860/CARPE_RDCS_0.pdf]
12. Carney D: **Sustainable Rural Livelihoods: What contribution can we make?** In *Department for International Development's Natural Advisers' Conference, July 1998.* London: Department for International Development (DfID); 1998.
13. Department for International Development (DFID): **Sustainable livelihoods guidance sheets numbers 1–8.** 1999a, 2000d, 2001. [www.livelihoods.org]
14. Department for International Development (DFID): *Sustainable Livelihoods and Poverty Elimination.* London: Department for International Development (DfID); 1999b.
15. Brown K: **Three challenges for a real people-centred conservation.** *Glob Ecol Biogeogr* 2003, **12:**89–92.
16. International Union for the Conservation of Nature (IUCN): **Motion 145.** In *World Conservation Congress, 6-15th September 2012; Jeju, Korea.* 2012. Access online: [https://portals.iucn.org/docs/2012congress/motions/en/M-145-2012-EN.pdf]
17. **Scopus.** [http://www.scopus.com/]
18. **Web of knowledge.** [http://wok.mimas.ac.uk/]
19. **Cab abstracts.** [http://www.cabi.org/]
20. **OvidSP.** [https://ovidsp.ovid.com/]
21. **Google scholar.** [http://scholar.google.co.uk/]
22. **Cab thesaurus.** [http://www.cabi.org/cabthesaurus/]
23. **AGROVOC.** [http://aims.fao.org/standards/agrovoc/]
24. **NAL thesaurus.** [http://agclass.nal.usda.gov/]
25. **AGRIS.** [http://agris.fao.org/agris-search/index.do]
26. **AGRICOLA.** [http://agricola.nal.usda.gov/]
27. **Directory of Open Access Journals (DOAJ).** [http://www.doaj.org/]
28. **International Network for the Availability of Scientific Publications (INASP).** [http://www.inasp.info/]
29. **African Journals Online (AJOL).** [http://www.ajol.info/]
30. **Asian Journals Online.** [http://ajouronline.com/]
31. **Poverty and Conservation Learning Group (PCLG).** [http://povertyandconservation.info/]
32. **Google.** [https://www.google.com/]
33. **British Library Electronic Theses Online Service (EThOS).** [http://ethos.bl.uk/Home.do]
34. **DART Europe E- thesis Portal.** [http://www.dart-europe.eu/basic-search.php]

35. **National ETD Portal of South Africa.** [http://www.netd.ac.za/]
36. **Open Access Theses and Dissertations.** [http://oatd.org/]
37. **ProQuest.** [http://www.proquest.co.uk/en-UK/]
38. **Biodiversity-L Listserve.** [http://www.iisd.ca/email/biodiversity-L.htm]
39. Day M: **Are 'alternative livelihoods' projects effective?** 2014 [http://www.iied.org/are-alternative-livelihoods-projects-effective]
40. Hagen-Zanker J, Mallet R: *How to do a rigorous, evidence-focused literature review in international development, A Guidance Note.* Overseas Development Institute; 2013 [http://www.odi.org.uk/sites/odi.org.uk/files/odi-assets/publications-opinion-files/8572.pdf]
41. Cohen J: **A coefficient of agreement for nominal scales.** *Educ Psychol Meas* 1960, **20:**37–46.
42. Edwards P, Clarke M, DiGuiseppi C, Pratap S, Roberts I, Wentz R: **Identification of randomized controlled trials in systematic reviews: accuracy and reliability of screening records.** *Stat Med* 2002, **21:**1635–1640.
43. Isasi-Catalá E: **Is translocation of problematic jaguars (Panthera onca) an effective strategy to resolve human-predator conflicts?** 2009. CEE protocol 08–018 (SR55). Collaboration for Environmental Evidence. http://www.environmentalevidence.org/Documents/Final_protocols/Protocol55.pdf.
44. Pullin A, Knight T: **Effectiveness in conservation practice: pointers from Medicine and Public Health.** *Conserv Biol* 2001, **15:**50–54.
45. **Zotero.** [http://www.zotero.org/]
46. **Microsoft Excel.** [http://office.microsoft.com/en-gb/excel/]

Do mangrove forest restoration or rehabilitation activities return biodiversity to pre-impact levels?

Dominic A Andradi-Brown[1*], Caroline Howe[2], Georgina M Mace[2] and Andrew T Knight[1,3]

Abstract

Background: Mangrove forest restoration and rehabilitation programs are increasingly undertaken to re-establish ecosystem services in the context of community-based biodiversity conservation. Restoration is returning a habitat to the most natural condition, whereas rehabilitation often focuses on optimising ecosystem services alongside biodiversity. With many different restoration and rehabilitation objectives and techniques existing, it is difficult to assess the general effectiveness of restoration and rehabilitation on biodiversity and ecosystem services. This systematic review protocol presents a methodology that will be used to assess the impacts of mangrove forest restoration and rehabilitation on biodiversity and provisioning ecosystem services in a global context.

Methods: This review will assess studies that have undertaken biodiversity surveys of restored and rehabilitated mangrove forests by comparing them against suitable mature reference mangrove forests within the same region, or surveys prior to degradation of the forest. This review will investigate how the age and initial tree diversity of a restoration or rehabilitation activities determine the effectiveness of these initiatives. Taxa of commercial value to local communities will be assessed to identify whether rehabilitation for optimal ecosystem service provision is likely to conflict with the full restoration of mangrove forests.

Keywords: Mangroves, Tropical forest, Restoration, Rehabilitation, Biodiversity, Ecosystem services

Background

Mangrove forests are located in the tropical and subtropical areas at the boundary between the aquatic (normally marine or brackish) and the terrestrial environment [1]. Mangrove forests are biodiverse and provide a range of benefits (ecosystem services) mostly to local communities living in close proximity to the forests [2,3], as well as being closely coupled with other tropical marine environments such as coral reefs [4,5]. Despite their importance, mangrove forests have faced widespread destruction throughout the past century, with more than 25% of their global cover cleared between 1980 and 2000 [1] leading to fears they could be functionally extinct within 100 years [6]. In response to this many mangrove restoration and rehabilitation projects have been established [7].

Threats to mangrove forests vary regionally, but in many countries land conversion is causing the biggest losses [8]. Coastal land is often of high value creating large pressures for conversion for aquaculture and agriculture or infrastructure and development projects for facilities such as harbours and industries such as tourism [8,9]. In South-East Asia, mangrove clearance for shrimp aquaculture has become a major threat, increasing 20-30% annually throughout the early 1990s, and is responsible for half of all mangrove clearance in the Philippines between 1951 and 1988 [10]. Large ponds that can be filled from and drained into the sea are created in areas cleared of mangroves, fragmenting the remaining mangrove forests, leading to the loss of many traditional livelihoods [11]. Forest overharvesting is a major threat in some areas, mainly caused by timber demand for firewood, charcoal and construction materials, but also for mangrove fruit, leaves, bark and roots which have a traditional medicinal role [12]. The impacts on biodiversity can be severe: for example, prolonged human extractive use combined with land conversion of mangroves in the Sundarbans, Bangladesh mangrove ecosystem has led to local terrestrial species extinctions [13] and fisheries decline [14,15].

* Correspondence: dominic.andradi-brown08@imperial.ac.uk
[1]Department of Life Sciences, Imperial College London, Silwood Park Campus, Buckhurst Road, Ascot, Berkshire SL5 7PY, UK
Full list of author information is available at the end of the article

Biodiversity definitions, such as that adopted by the Parties to the Convention on Biological Diversity (CBD), tend to emphasise variability among and within species and the ecosystems they are part of. Biodiversity can therefore be considered at different levels, from genes through to whole ecosystems [16,17], though in the context of mangrove restoration and rehabilitation programs it is normally assessed using species richness and forest structural assessments [18]. Mangrove forests tend to have low tree species richness compared to other tropical forests, with around 70 tree taxa found in mangroves of which a small subset have the potential to become dominant [19]. Despite this, mangroves dominated by different tree species can have similar and rich invertebrate communities, believed to be more influenced by salinity, tidal inundation and sediment composition than tree species diversity [20]. Niche partitioning studies have also been conducted in mangroves and associated intertidal mud flats to identify how such diverse communities can be maintained with a relatively homogenous input of organic materials and nutrients [21]. Disturbance through lightning strikes, hurricanes, frost damage, plant pathogens, wood boring insects and forest fires generate forest gaps [22], helping maintain heterogeneity in tree age.

Mangroves provide many ecosystem services, which can broadly be grouped following the four categories proposed by the Millennium Ecosystem Assessment [23] including provisioning (timber, fisheries), regulating (coastal protection, carbon storage), supporting (sediment stabilisation, primary production, nutrient cycling) and cultural (religious values, tourism and recreation) services (Table 1) [2,24]. At a global level there is increased awareness of the role of mangroves in carbon storage [25], both through timber but more importantly as 'blue carbon'- carbon resources stored underground in associated stable mangrove sediment, which can be greater than that within the standing mangrove forest [26].

Quantifying the amount of an ecosystem service is difficult, so metrics often use underpinning biodiversity as a proxy. The definition of biodiversity emphasises variability, yet many ecosystem services depend on abundance and traits of several dominant species rather than overall biodiversity [17,38]. While additional species in a community often may increase ecosystem resilience [39], ecosystem properties can often be maintained despite species loss due to functional redundancy or little contribution to the process from the species [40]. The focus on abundance becomes clear when considering some metrics used to assess the amount of an ecosystem service (see Table 1). Assessing the abundance or biomass of key taxa can therefore give an indication of service levels. For example, consider a provisioning ecosystem service such as crab harvesting for food in South-East Asian and Pacific mangrove forests [28]. Of the many species of crab found within mangrove forests only a few species are suitable for harvesting (e.g. *Scylla serrata*), so the biomass of these few species give a better indication of food provision than the overall crab species richness within the forest.

Recent recognition of the importance of mangrove forests has led to widespread efforts to restore or rehabilitate deforested mangrove areas, with many of these projects hoping to combine re-establishing ecosystem services for local communities benefits with biodiversity conservation [7]. With a diverse range of techniques available

Table 1 Ecosystem services and example assessment metrics reported from mangrove forests

Ecosystem service type	Examples	Biodiversity or abiotic component responsible	Example assessment metrics	Example studies
Provisioning	Timber	Tree species	Stand density, stem diameter, tree height	[27]
	Fisheries	Fish and invertebrate species	Abundance and biomass of relevant species	[28,29]
Regulating	Coastal protection	Tree species	Mangrove area, tree density and structural complexity	[30,31]
	Carbon storage	Sediment, Tree species	Sediment depth, sediment carbon content, tree biomass	[32]
Supporting	Erosion control and sediment stabilisation (land expansion)	Tree species	Forest age, density, form and structural complexity of dominant tree species	[33,34]
	Primary production	Tree species	Leaf litter, tree biomass gain	[35]
	Nutrient cycling	Whole biotic community and sediment	Leaf litter biomass	[36]
Cultural	Religious values	Unclear	Unclear	[24]
	Tourism and recreation	Highly variable	Species richness and abundance of charismatic species	[37]

Ecosystem service examples grouped by type with example assessment metric identified from published literature. Ecosystem service types are based on the Millennium Ecosystem Assessment [23].

for restoration and rehabilitation, it is important to ensure that a clear objective exists [41]. Goals of mangrove restoration or rehabilitation often include: conservation, coastal protection, timber production forests or mixed use forests for high sustainable yields [18,42]. Restoration is defined by the Society for Ecological Restoration [43] as 'the process of assisting the recovery of an ecosystem that has been degraded, damaged, or destroyed', whereas rehabilitation focuses on 'reparation of ecosystem processes, productivity and services'. The goal of restoration is to make structural or functional changes to an ecosystem to return it to original condition (biotic community structure and species composition), while rehabilitation emphasises modifying ecosystem structural or functional characteristics to achieve a desired outcome. The difference between these two definitions highlights a potential conflict in objectives. Mangrove systems can be restored or rehabilitated at different scales through a suite of different methods, from natural regeneration through to direct planting of single or mixed species plantations [18]. Ensuring the correct hydrological regime by changing river and coastal water management regimes if required maybe essential for effectiveness, and often is enough for natural regeneration to occur [44].

Early mangrove rehabilitation programs focused on establishing forest cover often for timber value and coastal protection without directly considering further biodiversity components, such as tree or structural diversity [41]. Plantations in mangroves are often monocultures due to ease of planting or for silviculture, enabling harvesting at the same time [44]. While there has been a slight increase in multi-species plantations for restoration since the early 1980s, even when true ecosystem restoration is clearly the objective, fewer mangrove tree species are normally used in initial forest establishment and any subsequent supplementary planting than are found in natural forests [44]. Many mangrove restoration and rehabilitation projects are conducted through integrated conservation and development projects [45] or community based natural resource management programs [46], engaging local communities and focusing on ecosystem services provided by mangroves in part to try to establish protection of biodiversity through sustainable use. Often, projects conducted in this context labelled as mangrove restoration are actually rehabilitation, as they seek to establish mangrove forests that either maximise benefits to people or try to balance benefits to people and biodiversity conservation without directly aiming to recreate the most natural biotic community possible.

There is currently mixed evidence for the effectiveness of mangrove restoration and rehabilitation projects, in part due to a lack of objective goal-setting and a lack of will to fund full ecological restoration favouring other goals [41,47]. A previous study found rehabilitated mangrove forests had higher crab abundances than natural mangroves [48], but that the crab community was more similar to natural forests than the previous bare mud flat. In another study, planting technique during rehabilitation was found to affect epibiotic communities, through varying root surface area and so changing algal diversity and epibiotic community biomass and structure [49]. When considering ecosystem services more directly, a study found local people in Kenya used natural mangroves in preference and considered them more valuable than replanted forests for most ecosystem services [24]. The only exception was building materials, as replanted mangrove plantations were considered to contain more suitable timber for cutting wooden poles of high quality. This raises the concern that restored and rehabilitated forests may not contain biodiversity and associated ecosystem service benefits at the levels found in natural forests. This issue is not new, and a recent global study incorporating data from 89 studies across six different broadly defined biomes found restoration increased biodiversity by 44% and ecosystem services by 25% relative to levels prior to restoration, but despite this increase both biodiversity and ecosystem service provision remained lower than natural ecosystems [50].

Mangrove restoration and rehabilitation, however, could potentially be more effective than restoration or rehabilitation in other ecosystems, as natural mangrove forests have low tree species richness compared to other tropical forests and are often dominated by a single species or even exist as single species stands [51]. Mangrove leaf litter has an important role alongside imported phytodetritus from surrounding areas in decomposition and nutrient cycling between the mangrove biotic and abiotic components as well as linking mangrove vegetation to fauna [52]. Studies of crabs in isolation suggest different crab species feed preferentially on the same species of leaves [21], suggesting that while many rehabilitated forests may not have maximal tree species richness, they may still support many other dependent ecological communities.

Despite the proliferation of projects aiming to restore and rehabilitate mangroves there is currently very little evaluation of their effectiveness. A recent narrative review by Bosire *et al.* [7] investigated functionality of restored mangroves. While they identified some trends they also found contradictory studies documenting invertebrate species abundance and diveristy. However, the review only used a subset of the available literature on the topic, of which the majority was published in peer-review journals or FAO reports. With so much recent attention being given to mangrove loss [6], payment for ecosystem services schemes often with a focus on carbon [53], and community based conservation [46] it is important to systematically and quantitatively evaluate the effectiveness of mangrove restoration and rehabilitation schemes using all available evidence from a range of sources. This project seeks to understand the effectiveness of mangrove restoration and rehabilitation for biodiversity conservation and associated provisioning ecosystem services through a systematic review comparing restored and rehabilitated mangrove forests with reference mature mangrove systems globally across a range of taxonomic groups. Uniquely using this approach will minimise bias in the studies included and allow quantitative synthesis of results, giving indications of magnitude of differences between restored or rehabilitated mangroves and reference sites.

Objective of the review

Primary question

Do mangrove forest restoration and rehabilitation activities return biodiversity to pre-impact levels? This can be broken into components as shown in Table 2.

Mangrove restoration and rehabilitation sites are considered as any sites that have suffered large scale anthropogenic degradation resulting in the loss of mature tree canopy cover and subsequently through a change in management begun to establish new tree canopy cover. Reference mangrove forests are mature mangrove forests composed of native tree species which have never been subjected to major anthropogenic impacts and represent as closely as possible the biodiversity of mangrove forests in the region. Alternatively the reference could be the restored or rehabilitated mangrove surveyed prior to first impact.

Secondary questions

Are provisioning ecosystem services restored during mangrove restoration or rehabilitation?

Does the success of mangrove restoration and rehabilitation for biodiversity and ecosystem services vary in relation to taxa or service types?

Subgroups for taxa will be defined at family, genus and species level as well as by trophic group. Subsequent analysis will be conducted at the level most appropriate once data extraction is complete. This will be determined by the resolution of the taxonomic data provided within studies and will be the lowest taxonomic level at which there is a sufficient sample size for meta-analysis to be conducted.

Subgroups for ecosystem services will be grouped following the broad categories of ecosystem service: provisioning, regulating, supporting and cultural based on the Millennium Ecosystem Assessment definitions [23]. Additionally, if sufficient data allows, provisioning services will be subdivided into food, water, timber *etc.* as appropriate.

Methods

Searches

Searches will be limited to English. Non-English language articles providing a published title and abstract in English will be included in the review, with the full text being translated if required. Non-English language articles without an English title and abstract will be excluded from the review. Searches will not intentionally be restricted by year, however database access will be limited to certain years by institutional subscriptions (see below for details).

Databases

We aim to search the following databases:

- ISI Web of Knowledge platform - http://isiknowledge.com
  - Web of Science (1970-present)
  - BIOSIS Citation Index (1969–2008)
  - Current Contents Connect (1998- present)
  - CABI: CAB Abstracts (1973- present)
  - MEDLINE (1950- present)
  - Zoological Record (1864- present)
- Science Direct - http://www.sciencedirect.com (all years)
- JSTOR - http://www.jstor.org (all years)
- Index to Theses Online - http://www.theses.com (all years)
- Scopus - http://www.scopus.com
- AGRICOLA - http://agricola.nal.usda.gov (all years)
- Aquatic Sciences & Fisheries Abstracts - http://search.proquest.com/asfa (all years)
- Environmental Sciences & Pollution Management - http://search.proquest.com/espm (all years)

All relevant database records will be downloaded with abstracts (where possible) and stored in an EndNote library.

Table 2 Structure of the primary systematic review question

Subject (population)	Intervention	Comparator	Outcome
Mangroves	Restoration or rehabilitation	Restored or rehabilitated mangrove compared with adjacent reference mangroves. Restored or rehabilitated mangrove surveyed prior to impact and after restoration.	Difference in biodiversity or abundance measure. Difference in ecosystem service measure.

Specialist sources

The websites of the following organisations working on mangrove related projects will be searched. Relevant studies and references contained within will be identified. Links from these websites will be followed once, to any partner organisation that potentially may have studies available. The partner organisation website will then be searched for relevant studies or references.

- Bird Life International Mangrove Alliance http://www.birdlife.org/mangrove-alliance
- Center for International Forestry Research http://www.cifor.org/
- GLOMIS – Global Mangrove Database & Information System http://www.glomis.com
- International Society for Mangrove Ecosystems http://www.mangrove.or.jp
- MangroveRestoration.com http://www.mangroverestoration.com
- Mangrove Action Project http://www.mangroveactionproject.org
- RAMSAR http://www.ramsar.org
- Society for Ecological Restoration http://www.ser.org
- Society for Wetlands Scientists http://www.sws.org
- UN Food and Agriculture Organization http://www.fao.org
- USAID http://dec.usaid.gov
- Wetlands International http://www.wetlands.org
- World Fish Centre http://www.worldfishcenter.org

Search engines

An Internet search will be carried out using the following search engines:

- Google Scholar - http://scholar.google.co.uk
- Scirus - http://www.scirus.com

Search will be limited to Word and/or PDF documents when possible. The first 50 Word and PDF document hits will be examined and references contained within articles viewed at full text will be searched.

Search terms

Review scoping was conducted to guide the construction of this protocol. The search terms were developed following an iterative process. The search string was developed using ISI Web of Knowledge and trialled against a test library. The test library consisted of 12 studies containing relevant data to the systematic review and identified by the review team from existing publications reference lists (Additional file 1). The proportion of studies in the test library identified by each search string was recorded as the search terms were refined (Additional file 2).

The final search terms to be used are as follows (though these may be simplified when searching databases that do not allow Boolean operators or wildcards):

Mangrove* OR "Inter-tidal forest" OR "Intertidal forest" OR Rhizophora*

AND

Reforest* OR Restor* OR Replant* OR Secondary OR Plantation* OR Planted OR Regenerat* OR Rehabilitat* OR Stand

AND

divers OR Communit* OR "Yield table*"

Study inclusion criteria

In order to only select relevant studies the following inclusion criteria will be applied sequentially in three stages: 1) by title, 2) by abstract and 3) by full text. If there is uncertainty in relevance of an article it will be put through to the next stage. Citations will be stored in EndNote (version X5, http://www.endnote.com) and EPPI-Reviewer (version 4, http://eppi.ioe.ac.uk/cms/er4) at each stage.

Study inclusion criteria for title and abstract screening:

1. Relevant subject: Mangrove forest
2. Relevant intervention: Site where mangrove forests have been established either through direct plantations or protection to allow natural regeneration.
3. Relevant outcome: Indication that a component of species biodiversity, abundance, or directly biodiversity related ecosystem service measure is assessed/measured.
4. Relevant comparator: Indication that study attempts to evaluate the quality/success of the mangrove restoration. If comparators are specified then a reference mature mangrove forest for comparison or data on original mangrove forest structure pre-impact and post restoration.

During abstract screening it is not expected that comparators will be explicitly stated in abstracts, however, from preliminary search testing all relevant studies stated that they aimed to evaluate restoration success in their abstracts, with the majority specifically stating the existence of a comparator.

Study inclusion criteria for full text screening:

1. Relevant subject: Mangrove forest
2. Relevant intervention: Study must include a site where mangrove forest has been established (either

through: direct plantations, protection of an area allowing regeneration, natural regeneration combined with supplementary planting or altering environmental conditions to enable a mangrove forest to establish) where prior to the establishment/colonisation event there were not intact mangrove forests.

3. Relevant comparator: Restored mangrove compared with adjacent reference mangroves (mature mangrove forest(s) composed of native tree species which have never been subjected to major anthropogenic impacts and represent as closely as possible the biodiversity of mangrove forests in the region); or the restored mangrove was surveyed prior to first impact and after restoration and the surveys compared.
4. Relevant outcome: Empirical data contained for species biodiversity, abundance or biomass, or a difference in ecosystem service measure in both the intervention and the comparator. Studies must report mean, variance measure (variance, SD, SE, stated confidence intervals) and the sample size for both the intervention and the comparator.

Two researchers will assess a random 100 citations, applying the inclusion criteria at title and abstract level. Agreement between the two reviewers will be measured by kappa analysis, where a value of 0.6 or greater indicates good agreement [54]. If kappa is less than 0.6, differences will be discussed and clarified and the inclusion criteria modified if required. One reviewer will apply the agreed inclusion criteria to all remaining articles.

Potential effect modifiers and reasons for heterogeneity

Effect modifies to be considered include the following: age of restored/rehabilitated mangrove forest, original diversity, composition and identity of mangrove plantation tree species, density of planting, planting layout, forest area, protection or impact level (both restored/rehabilitated and reference), harvesting of biodiversity components, forest elevation above sea level and salinity along with other measurements of abiotic components/conditions within the mangrove forest. Additional possible effect modifiers reported in the studies will be recorded with and a decision on whether to include these made once it become clear how many studies provide that information.

Study quality assessment

While Before After Control Impact (BACI) studies are considered most robust, most studies are expected to contain control-impact (CI) quantitative assessments of biodiversity and/or ecosystem services in restored mangroves and a reference mangrove system. A detailed study critical appraisal methodology will be developed after the articles for inclusion in the review have been identified involving recording the spatial replication within and environmental similarity between the restoration and the comparator site. Studies will be classified as 'high', 'medium' or 'low' quality based these criteria to broadly define susceptibility to bias. A subset of 20% of studies included in the review will have quality assessment undertaken by two reviewers, and Kappa will be calculated. If Kappa is less than 0.6 the critical appraisal methodology will be clarified though consensus. One reviewer will then complete critical appraisal for all articles.

Data extraction strategy

Data will be extracted into a spreadsheet, recording outcomes, methodology, number of replicates and age of resorted mangrove forest along with all the above mentioned effect modifiers. The template spreadsheet for data extraction (Additional file 3) was developed during trial data extraction of studies identified during search term testing and provides a clear framework for extraction ensuring repeatability and efficiency. The template also provides space for any other relevant factors identified while reading an article at full text to be included. If any studies are found to be missing key relevant data an attempt to contact the authors will be made.

Data synthesis and presentation

From trial searches and data extraction we are confident effect sizes can be calculated and a meta-analysis conducted. For abundance and biomass data we plan to use the natural logarithm transformed response ratio (ln$R$) as an effect size, the natural logarithm of the ratio of mean abundance or biomass from the restoration or rehabilitation and the reference site [55]:

$$\ln R = \ln\left(\frac{\bar{X}_{REST}}{\bar{X}_{REF}}\right) \qquad (1)$$

Where $\bar{X}_{REST}$ and $\bar{X}_{REF}$ represent the mean abundance or biomass at the restoration or rehabilitation site and the mean abundance or biomass at the reference site respectively. Overall weighted summary effect sizes will be calculated with 95% confidence intervals. As many of the differences in observed effects may be caused by differences in restoration and rehabilitation implementation, a random effects model will be used during meta-analysis. For abundance and biomass this will use Eq. 2 [56], where LnR_i equals the effect size,

W_i equals the weighting and k equals the number of studies.

$$\overline{\ln R} = \frac{\sum_{i=1}^{k} W_i \ln R_i}{\sum_{i=1}^{k} W_i} \quad (2)$$

Abundance and biomass studies will be weighted by the inverse of the variance (Eq. 3) [55], where $\bar{X}$ equals the mean abundance or biomass, s equals the standard deviation of that mean and n equals the number of replicates.

$$V_{LnR} = \frac{s_{REST}^2}{n_{REST}(\bar{X}_{REST})^2} + \frac{s_{REF}^2}{n_{REF}(\bar{X}_{REF})^2} \quad (3)$$

Mixed effects models will be used to test the importance of effect modifiers. Sub-group analysis will also be run to investigate the effects of different taxonomic groups and ecosystem service types to address the proposed secondary questions.

It is expected that insufficient suitable studies will be identified to allow meta-analysis for all taxonomic groups or ecosystem services types. If this is the case, the outcomes for these taxonomic groups or ecosystem service types will be summarised in tables.

Additional files

Additional file 1: Test library of papers.

Additional file 2: Table of search terms trialled against number of hits from Web of Knowledge.

Additional file 3: Template data extraction spreadsheet.

Competing interests
The authors declare that they have no competing interests.

Authors' contributions
DAB, CH, GM and AK developed the review question. DAB conducted the pilot research. DAB wrote the protocol with support from CH, GM and AK. All authors read and approved the final manuscript.

Acknowledgments
ATK acknowledges the support of the ARC Centre of Excellence in Environmental Decisions at The University of Queensland, Australia. The authors would like to thank two anonymous reviewers and the journal editor for valuable comments which improved this protocol.

Author details
[1]Department of Life Sciences, Imperial College London, Silwood Park Campus, Buckhurst Road, Ascot, Berkshire SL5 7PY, UK. [2]Centre for Biodiversity and Environment Research, Department of Genetics, Evolution and Environment, University College London, London WC1E 6BT, UK. [3]Department of Botany, Nelson Mandela Metropolitan University, P.O. Box 77000, Port Elizabeth 6031, South Africa.

References

1. Hogarth P: *The biology of mangroves and seagrasses.* 2nd edition. New York: Oxford University Press; 2007.
2. Barbier EB, Hacker SD, Kennedy C, Koch EW, Stier AC, Silliman BR: **The value of estuarine and coastal ecosystem services.** *Ecol Monogr* 2011, **81:**169–193.
3. Vo QT, Kuenzer C, Vo QM, Moder F, Oppelt N: **Review of valuation methods for mangrove ecosystem services.** *Ecol Indic* 2012, **23:**431–446.
4. Demopoulos AWJ, Cormier N, Ewel KC, Fry B: **Use of multiple chemical tracers to define habitat use of Indo-Pacific mangrove crab, Scylla serrata (Decapoda: Portunidae).** *Estuar Coast Shelf Sci* 2008, **31:**371–381.
5. Mumby PJ, Hastings A: **The impact of ecosystem connectivity on coral reef resilience.** *J Appl Ecol* 2008, **45:**854–862.
6. Duke NC, Meynecke JO, Dittmann S, Ellison AM, Anger K, Berger U, Cannicci S, Diele K, Ewel KC, Field CD, Koedam N, Lee SY, Marchand C, Nordhaus I, Dahdouh-Guebas F: **A world without mangroves?** *Science* 2007, **317:**41–42.
7. Bosire JO, Dahdouh-Guebas F, Walton M, Crona BI, Lewis RR, Field C, Kairo JG, Koedam N: **Functionality of restored mangroves: a review.** *Aquat Bot* 2008, **89:**251–259.
8. Valiela I, Bowen JL, York JK: **Mangrove forests: One of the World's threatened major tropical environments.** *Bioscience* 2001, **51:**807.
9. Alongi DM: **Present state and future of the world's mangrove forests.** *Environ Conserv* 2002, **29:**331–349.
10. Primavera JH: **Socio-economic impacts of shrimp culture.** *Aquac Research* 1997, **28:**815–827.
11. Gunawardena M, Rowan JS: **Economic valuation of a mangrove ecosystem threatened by shrimp aquaculture in Sri Lanka.** *Environ Manage* 2005, **36:**535–550.
12. Warren-Rhodes K, Schwarz A-M, Boyle LN, Albert J, Agalo SS, Warren R, Bana A, Paul C, Kodosiku R, Bosma W, Yee D, Rönnbäck P, Crona B, Duke N: **Mangrove ecosystem services and the potential for carbon revenue programmes in Solomon Islands.** *Environ Conserv* 2011, **38:**485–496.
13. Seidenstlcker J, Hai A: *The sundarbans wildlife management plan: conservation in the Bangladesh coastal zone.* Gland, Switzerland: International Union for Conservation of Nature and Natural Resources; 1983.
14. Deb AK: **Fake blue revolution: environmental and socio-economic impacts of shrimp culture in the coastal areas of Bangladesh.** *Ocean & Coastal Management* 1998, **41:**63–88.
15. Islam MS, Haque M: **The mangrove-based coastal and nearshore fisheries of Bangladesh: ecology, exploitation and management.** *Rev Fish Biol Fish* 2004, **14:**153–180.
16. OTA: *Technologies to maintain biological diversity.* Washington, DC: U.S: Congress, Office of Technology Assessment, U.S. Government Printing Office; 1987.
17. Mace GM, Norris K, Fitter AH: **Biodiversity and ecosystem services: a multilayered relationship.** *Trends Ecol Evol* 2012, **27:**19–26.
18. Field C: **Rehabilitation of mangrove ecosystems: an overview.** *Mar Pollut Bull* 1999, **37:**383–392.
19. Duke NC, Ball MC, Ellison JC: **Factors influencing biodiversity and distributional gradients in mangroves.** *Glob Ecol Biogeogr Lett* 1998, **7:**27–47.
20. Michael Gee J, Somerfield PJ: **Do mangrove diversity and leaf litter decay promote meiofaunal diversity?** *J Exp Mar Biol Ecol* 1997, **218:**13–33.
21. Todd PA, Yeo DCJ, Ya BP: **Feeding ecology of two species of *Perisesarma* (Crustacea: Decapoda: Brachyura: Sesarmidae) in Mandai mangroves, Singapore.** *J Crustac Biol* 2008, **28:**480–484.
22. Duke NC: **Gap creation and regenerative processes driving diversity and structure of mangrove ecosystems.** *Wetlands Ecol Manage* 2001, **9:**267–279.
23. *Millennium ecosystem assessment. Ecosystems and human well-being: biodiversity synthesis.* [http://www.unep.org/maweb/documents/document.354.aspx.pdf]
24. Rönnbäck P, Crona B, Ingwall L: **The return of ecosystem goods and services in replanted mangrove forests: perspectives from local communities in Kenya.** *Environ Conserv* 2007, **34:**313–324.
25. Mcleod E, Chmura GL, Bouillon S, Salm R, Björk M, Duarte CM, Lovelock CE, Schlesinger WH, Silliman BR: **A blueprint for blue carbon: toward an improved understanding of the role of vegetated coastal habitats in sequestering CO 2.** *Front Ecol Environ* 2011, **9:**552–560.
26. Pendleton L, Donato DC, Murray BC, Crooks S, Jenkins WA, Sifleet S, Craft C, Fourqurean JW, Kauffman JB, Marbà N, Megonigal P, Pidgeon E, Herr D, Gordon D, Baldera A: **Estimating global "blue carbon" emissions from conversion and degradation of vegetated coastal ecosystems.** *PLoS ONE* 2012, **7:**e43542.
27. Kairo JG, Lang'at JKS, Dahdouh-Guebas F, Bosire J, Karachi M: **Structural development and productivity of replanted mangrove plantations in Kenya.** *Forest Ecol Manag* 2008, **255:**2670–2677.

28. Bonine KM, Bjorkstedt EP, Ewel KC, Palik M: **Population characteristics of the mangrove crab *Scylla serrata* (Decapoda: Portunidae) in Kosrae, Federated States of Micronesia: effects of harvest and implications for management.** *Pac Sci* 2008, **62:**1–19.
29. Rönnbäck P: **The ecological basis for economic value of seafood production supported by mangrove ecosystems.** *Ecol Econ* 1999, **29:**235–252.
30. Das S, Vincent JR: **Mangroves protected villages and reduced death toll during Indian super cyclone.** *Proc Natl Acad Sci* 2009, **106:**7357–7360.
31. Mazda Y, Magi M, Kogo M, Hong PN: **Mangroves as a coastal protection from waves in the tong king delta, Vietnam.** *Mangrove Salt Marshes* 1997, **1:**127–135.
32. Donato DC, Kauffman JB, Murdiyarso D, Kurnianto S, Stidham M, Kanninen M: **Mangroves among the most carbon-rich forests in the tropics.** *Nat Geosci* 2011, **4:**293–297.
33. Carlton JM: **Land-building and stabilization by mangroves.** *Environ Conserv* 1974, **1:**285–294.
34. Alongi DM, Pfitzner J, Trott LA, Tirendi F, Dixon P, Klumpp DW: **Rapid sediment accumulation and microbial mineralization in forests of the mangrove Kandelia candel in the Jiulongjiang Estuary, China.** *Estuar Coast Shelf Sci* 2005, **63:**605–618.
35. Amarasinghe MD, Balasubramaniam S: **Net primary productivity of two mangrove forest stands on the northwestern coast of Sri Lanka.** *Hydrobiologia* 1992, **247:**37–47.
36. Twilley RW, Lugo AE, Patterson-Zucca C: **Litter Production and Turnover in Basin Mangrove Forests in Southwest Florida.** *Ecology* 1986, **67:**670.
37. Bacon PR: **Use of wetlands for tourism in the insular Caribbean.** *Ann Tour Res* 1987, **14:**104–117.
38. Cardinale BJ, Duffy JE, Gonzalez A, Hooper DU, Perrings C, Venail P, Narwani A, Mace GM, Tilman D, Wardle DA, Kinzig AP, Daily GC, Loreau M, Grace JB, Larigauderie A, Srivastava DS, Naeem S: **Biodiversity loss and its impact on humanity.** *Nature* 2012, **486:**59–67.
39. Elmqvist T, Folke C, Nystrom M, Peterson G, Bengtsson J, Walker B, Norberg J: **Response diversity, ecosystem change, and resilience.** *Front Ecol Environ* 2003, **1:**488–494.
40. Hooper DU, Chapin FS III, Ewel JJ, Hector A, Inchausti P, Lavorel S, Lawton JH, Lodge DM, Loreau M, Naeem S, Schmid B, Setälä H, Symstad AJ, Vandermeer J, Wardle DA: **Effects of biodiversity on ecosystem functioning: a consensus of current knowledge.** *Ecol Monogr* 2005, **75:**3–35.
41. Lewis RR III: **Ecologically based goal setting in mangrove forest and tidal marsh restoration.** *Ecol Eng* 2000, **15:**191–198.
42. Lewis RR III: **Methods and Criteria for Successful Mangrove Forest Restoration.** In *Coastal Wetlands: An integrated ecosystem approach.* Edited by Perillo GME, Wolanski E, Cahoon DR, Brinson MM. Oxford: Elsevier; 2009:787–800.
43. Society for Ecological Restoration International Science Policy Working Group: *Society for ecological restoration international science & policy working group.* 2nd edition. The SER International Primer on Ecological Restoration; 2004. www.ser.org & Tucson: Society for Ecological Restoration International.
44. Ellison AM: **Mangrove restoration: do we know enough?** *Restor Ecology* 2000, **8:**219–229.
45. Smith AH, Berkes F: **Community-based use of mangrove resources in St. Lucia.** *Int J Environ Stud* 1993, **43:**123–131.
46. Datta D, Chattopadhyay RN, Guha P: **Community based mangrove management: a review on status and sustainability.** *J Environ Manage* 2012, **107:**84–95.
47. Lewis RR III: **Ecological engineering for successful management and restoration of mangrove forests.** *Ecol Eng* 2005, **24:**403–418.
48. Bosire JO, Dahdouh-Guebas F, Kairo JG, Cannicci S, Koedam N: **Spatial variations in macrobenthic fauna recolonisation in a tropical mangrove bay.** *Biodivers Conserv* 2004, **13:**1059–1074.
49. Crona BI, Holmgren S, Rönnbäck P: **Re-establishment of epibiotic communities in reforested mangroves of Gazi Bay, Kenya.** *Wetlands Ecol Manage* 2006, **14:**527–538.
50. Benayas JMR, Newton AC, Diaz A, Bullock JM: **Enhancement of biodiversity and ecosystem services by ecological restoration: a meta-analysis.** *Science* 2009, **325:**1121–1124.
51. Ashton EC, Hogarth PJ, Macintosh DJ: **A comparison of brachyuran crab community structure at four mangrove locations under different management systems along the Melaka Straits-Andaman Sea Coast of Malaysia and Thailand.** *Estuaries* 2003, **26:**1461–1471.
52. Bouillon S, Koedam N, Raman A, Dehairs F: **Primary producers sustaining macro-invertebrate communities in intertidal mangrove forests.** *Oecologia* 2002, **130:**441–448.
53. Lau WWY: **Beyond carbon: conceptualizing payments for ecosystem services in blue forests on carbon and other marine and coastal ecosystem services.** *Ocean & Coastal Management* 2013, **83:**5–14.
54. Landis JR, Koch GG: **The measurement of observer agreement for categorical data.** *Biometrics* 1977, **33:**159–174.
55. Hedges LV, Gurevitch J, Curtis PS: **The meta-analysis of response ratios in experimental ecology.** *Ecology* 1999, **80:**1150–1156.
56. Borenstein M, Hedges LV, Higgins JPT, Rothstein HR: *Introduction to meta-analysis.* John Wiley & Sons, Ltd: Chichester, UK; 2009.

7

Human well-being impacts of terrestrial protected areas

Andrew S Pullin[1*], Mukdarut Bangpan[2], Sarah Dalrymple[1], Kelly Dickson[2], Neal R Haddaway[1], John R Healey[3], Hanan Hauari[2], Neal Hockley[3], Julia P G Jones[3], Teri Knight[1], Carol Vigurs[2] and Sandy Oliver[2]

Abstract

Background: Establishing Protected Areas (PAs) is among the most common conservation interventions. Protecting areas from the threats posed by human activity will by definition inhibit some human actions. However, adverse impacts could be balanced by maintaining ecosystem services or introducing new livelihood options. Consequently there is an ongoing debate on whether the net impact of PAs on human well-being at local or regional scales is positive or negative. We report here on a systematic review of evidence for impacts on human well-being arising from the establishment and maintenance of terrestrial PAs.

Methods: Following an *a priori* protocol, systematic searches were conducted for evidence of impacts of PAs post 1992. After article title screening, the review was divided into two separate processes; a qualitative synthesis of explanations and meaning of impact and a review of quantitative evidence of impact. Abstracts and full texts were assessed using inclusion criteria and conceptual models of potential impacts. Relevant studies were critically appraised and data extracted and sorted according to type of impact reported. No quantitative synthesis was possible with the evidence available. Two narrative syntheses were produced and their outputs compared in a metasynthesis.

Results: The qualitative evidence review mapped 306 articles and synthesised 34 that were scored as high quality. The quantitative evidence review critically appraised 79 studies and included 14 of low/medium susceptibility to bias. The meta-synthesis reveals that a range of factors can lead to reports of positive and negative impacts of PA establishment, and therefore might enable hypothesis generation regarding cause and effect relationships, but resulting hypotheses cannot be tested with the current available evidence.

Conclusions: The evidence base provides a range of possible pathways of impact, both positive and negative, of PAs on human well-being but provides very little support for decision making on how to maximise positive impacts. The nature of the research reported to date forms a diverse and fragmented body of evidence unsuitable for the purpose of informing policy formation on how to achieve win-win outcomes for biodiversity and human well-being. To better assess the impacts of PAs on human well-being we make recommendations for improving research study design and reporting.

Keywords: National Park, Reserve, Community, Governance, Conservation, Poverty, Development, Biodiversity, Systematic review

* Correspondence: a.s.pullin@bangor.ac.uk
[1]Centre for Evidence-Based Conservation, School of Environment, Natural Resources and Geography, Bangor University, LL57 2UW Bangor, Gwynedd, UK
Full list of author information is available at the end of the article

Background

The concept and practice of protecting areas for the purposes of conservation has been at the heart of conservation policy since its inception in the 19th Century. The idea that intervening to protect areas from human activity is an effective way of conserving species and habitats and preventing habitat loss and species extinction is arguably as pervasive today as it was when the first protected areas (PAs) were established (MEA 2005). The central place of PAs in the conservation movement has been reflected in the increase in both the number of PAs and the area of land and sea placed under protection. The proportion of total area of land under some form of protection has now reached nearly 13% [1,2].

The process of protecting areas from the threats posed by human activity will by definition inhibit some human actions and therefore has the potential to negatively impact human well-being. There are many historical records to suggest that few PAs were uninhabited wildernesses before designation, and the early history of PAs, for example in the USA and East Africa, is of forced eviction and persecution of local communities by colonial powers [3]. This scenario has continued in some countries with, in some cases, the colonial powers being replaced by multinational corporations or international conservation Non-Governmental Organisations (NGOs) [4]. The problem of negative impacts of PAs on human well-being gained official recognition in the Convention on Biological Diversity (CBD) arising from the Rio Summit in 1992. The principle that PAs should do no harm to local people was only established at the World's Park Congress in 2003 (at which, the Durban Accord was proclaimed). Prior to 1992, the establishment plans of PAs did not normally have objectives concerning human well-being.

However, negative impacts could be balanced by positive impacts as PAs may also improve human well-being and alleviate poverty [5]. By preventing conversion of natural habitats, PAs may improve the provision of some valued ecosystem services to some users. For example, downstream farmers may benefit from conservation of forested watersheds [6]. PAs may also directly introduce new livelihood options into a region through the expansion of tourism or research, or improvements to infrastructure may indirectly result in economic development.

Recently there has been considerable debate on whether, apart from their effects on global environmental benefits, the net impact of PAs on human well-being at local or regional scales are positive or negative [7-10]. There is concern that continuing with a policy of PA establishment could conflict with goals of poverty alleviation [11]. The CBD Aichi targets include a target of 17% of terrestrial and inland areas covered by well-managed PAs by 2020. Future policy decisions on support for PA establishment and future management need to be informed by the best available evidence on their human well-being impacts. That there have been and continue to be, some major negative impacts on local communities caused by the establishment of some protected areas, is not in dispute. However, of particular interest to policy makers is the balance of positive and negative impacts on human livelihoods that arise from the PA establishment, the distribution of these benefits and costs, and the factors that might cause this to vary.

We report here on the conduct and outcome of a systematic review of evidence for impacts on human well-being arising from the establishment and maintenance of terrestrial PAs. Establishing the state of the evidence base through systematic review will inform decision making concerning future investment in PAs and future research needs. The review question was formulated by the Scientific and Technical Advisory Panel (STAP) of the Global Environment Facility (GEF).

At the outset of the review the following broad categories were posed by STAP as a starting point for a conceptual model concerning the potential impact of PA establishment (or change in PA status) on people and/or the communities of which they are part. These categories and associated questions were used to guide development of specific evidence inclusion criteria (categories were subsequently modified based on an in-depth understanding of the literature in order to code and present the available evidence, see methods).

Livelihood strategies

Did the establishment or change in status of the PA or management activities within the PA generate or decrease specific production opportunities (e.g. more demand for labour, herding activities and associated products no longer viable, new demand for particular food, handicraft, services or products etc.)? Did the PA influence (i.e. increase or decrease) migration generally, and of particular social groups? Has this differentially impacted (positively or negatively) the most vulnerable groups in local communities (e.g. women, children, poorest sectors of the community)?

Social capital

Did the establishment and management of the PA affect the development of social networks? Did it positively or negatively impact education and capacity building, e.g. by generating or decreasing opportunities for formal and/or informal education? Has PA establishment differentially affected more vulnerable groups (e.g. women, children, poorest sectors within local communities) in a positive or negative way?

Empowerment

Did the PA empower or disempower local communities and any particular social groups? Were new organizations/

institutional arrangements that represent the interests of communities and any particular social groups created or existing ones undermined? Have these organizations developed activities aimed at improving their livelihoods (e.g., legislation to support local livelihoods, land tenure, co-management of local resources, other social benefits) or have existing activities been negatively affected?

Human rights

Whilst recognising that the scope of human rights is very broad, in this review we focus on the following question; were the rights of any local stakeholders, affected either positively or negatively by the PA (considering, e.g., rights to education, adequate access to food, clothing, health, choices)?

Access to ecosystem goods and services and natural resources essential for well-being

Did the PA have any positive or negative impact on access to ecosystem services and natural resources? For example were there changes in the cost (in terms of money, level of effort, or time) in obtaining firewood, clean water, and other resources/services? Was access to culturally significant places (e.g. sacred grounds) affected? Did self-sufficiency in food (by locally cultivating, hunting, raising animals or gathering) or access to medicinal plants change? Has this been a consequence of the direct impact of the PA through legal prohibition of access or indirect as a consequence of changes in infrastructure and/or institutions? Have any of these positive or negative impacts been disproportionably high or low on particular sectors of society?

Measuring impacts needs to take into account that terrestrial PAs are varied in their status, management and objectives. For example, many protected areas will have been established with a primary aim of landscape or biodiversity conservation, possibly at a time when potential impacts on local human well-being were not widely considered. Types of PAs have been categorised by the World Commission on Protected Areas (WCPA) of the International Union for the Conservation of Nature (IUCN) and are used to classify entries in the World Database of Protected Areas (http://www.protectedplanet.net/). The PA categories reflect the range of management objectives of different PAs, from strict protection of nature or wilderness, conservation of ecosystems, protection of national monuments or management of important habitat or species (categories I-IV) through to objectives which place human use of the landscape much more centrally (categories V-VI). Progressively, since 1992 the requirement of PAs to 'Deliver benefits to resident and local communities consistent with the other objectives of management' has become a common objective [12] with the categorisation representing a gradation of increasing human intervention.

In this review we also recognise that the impacts of PAs on human well-being are likely to be highly context dependent and vary over the lifetime of the PA. For example, initial establishment may have significant and immediate negative impacts on the communities within the PA but in the longer term there may be positive impacts on well-being. Similarly, spatial context of the PA and communities within and around it will influence impact. Some communities will be located in a PA, some in buffer zones around the edge and others more distant. Other PAs may be close by and exert an influence. Presence or absence of infrastructure may influence the spatial scale over which impacts occur. Impacts that are negative 'locally' may have to be balanced against impacts that are positive but experienced more widely or more remotely.

Given the diversity of PAs, contexts of establishment and of communities affected, this review did not aim to estimate a mean effect of PA establishment on local communities but to critically appraise the range of effects that have been reported. The nature and scope of the question posed in this review (including as it does issues of empowerment and social capital) suggested to the Stakeholders and the Review Team that a combination of quantitative and qualitative evidence might provide the most reliable base with which to inform future decision making.

Objective of the review

The primary research question was:

What are the human well-being impacts of terrestrial protected areas?

We aimed to synthesise the empirical evidence of positive, negative or neutral impacts of PAs on human well-being at the local to regional scales, with emphasis on local communities and contemplating as broad a definition of well-being as possible (see categories to be considered below).

We were also interested in two secondary questions;

1. How are costs and benefits distributed among and within local communities living inside and in the buffer zones of PAs (by socio-economic status, gender, age etc.)?
2. How do costs and benefits vary with governance, resource tenure arrangements, and site characteristics? As the aim is to assess the human well-being impacts of PAs in their current form and capture lessons for future interventions, relevant evidence may be provided by studies that generate hypotheses about impact, as well as studies that test hypotheses about impact.

Methods

Design of review

An *a priori* protocol was established, peer reviewed and posted on the CEE website by CEE [13]. This review was divided into two separate processes; a qualitative synthesis of people's views, observations and related documentary evidence (led by the EPPI-Centre) and a synthesis of quantitative evidence of impact, including people's attitudes and views (led by Bangor University). Henceforth these will be referred to as 'qualitative' and 'quantitative' as reflected in the protocol and the methods below. A single systematic search was conducted to identify potentially relevant studies to answer the review questions. Initial screening of title was undertaken by Bangor University. Further supplementary searching, screening, coding and synthesis were conducted independently by each of the review teams. Thus, the searching phase and initial screening by title were not separate, but thereafter the report sections reflect the different approaches.

Searches

Search terms were selected to capture all information pertaining to PAs (the intervention) and relevant outcomes associated with human well-being. No search terms for the study population, in this case local communities, were used as these were not likely to be included in the title or abstract and therefore their incorporation in the search might have risked excluding relevant studies. The use of the asterisk denotes a wildcard character that prevents plurals or alternative word endings being excluded. The search terms are as follows:

> *Exposure:* protected area*, nature reserve*, wilderness area*, national park*, natural monument*, natural feature*, management area*, world heritage site*, biosphere reserve*, biodiversity conservation.
> *Outcomes*: poverty, human well*, socio-econom*, econom*, human health, livelihood*, social capital, social welfare, empowerment, equity, ecosystem service*, perception*, attitude*.

Where the databases searched could accommodate all search terms simultaneously, they were separated using Boolean operators. The search terms are grouped by intervention and outcome and we refer to these groups as 'sets'. The Boolean operator 'OR' was used to separate search terms within a set and the operator 'AND' used to separate the two sets thus ensuring that the search returned references including at least one term from each set.

In many cases, the literature databases could not accommodate all the search terms above and therefore search strings had to be adapted as necessary. In some cases, this meant using pairs of search terms, one from each set, separated by 'AND'. Occasionally, search terms had to be entered individually and/or the database did not accommodate Boolean operators, in these cases only the search terms pertaining to the intervention were used. Full details of the search are recorded in Additional file 1.

Where studies are reported in other languages, relevance was assessed initially from their titles and abstracts (translated if necessary). Non-English language articles that could not be considered in this review were recorded for future assessment (see Additional file 2).

For the qualitative synthesis, non-English papers were translated using Google Translate online translation tool. Google Translate was tested for accuracy using a sample report that included both and English and non-English abstracts. By translating the non-English abstract of the paper and comparing with the English abstract given in the same paper we were able to see how well the tool 'understood' and conveyed the same meanings in the text. Google Translate was found to be a fairly accurate tool, although more accurate on some languages than others. Using Google Translate, non-English articles were translated then screened on full text and included or excluded.

Estimating the comprehensiveness of the database search

All search terms were included in a search string formatted according to requirements for searching in the Web of Knowledge database. A set of 18 references were identified by the review team as being relevant to the review question and used as a 'test library' to check whether the search strings captured the expected studies. The test library did not consist entirely of studies that would be included at every stage of the systematic review and data extraction. Instead, there were some studies which were expected to be captured in the search but did not include a suitable intervention, or were not primary literature. Whilst it was not required for the database search to capture these articles, it was a valuable test of how specific the search strings were to the subject area whilst retaining the sensitivity required to capture all available literature. The balance between specificity and sensitivity is key to achieving a comprehensive search whilst avoiding capturing too many spurious hits. The search string employed (see Additional file 1), consisting of the terms listed above, was the last of 16 iterations and was arrived at using the test library to evaluate the search results returned by Web of Knowledge.

Publication databases searched

The search included the following computerized databases:

- Web of Knowledge
- Scopus
- Agricola
- CAB Abstracts
- PubMed
- Econlit
- Directory of Open Access Journals (DOAJ)
- LILACS (Latin American & Caribbean Health Sciences Literature – Spanish language)

A supplemental search was undertaken using an expanded set of intervention and outcome terms to test how many extra relevant articles they might provide. The following intervention and outcome terms were entered into Web of Knowledge, and articles obtained were assessed as above.

Exposure: ecotourism, eco-tourism, eco tourism, wildlife tourism, trophy hunting, conservation corridor*, community conservanc*
Outcomes: natural resource, farm, enterprise*, human AND migration, gender, discriminat*

Internet searches conducted
Internet searches using online search engines were not performed as part of this review. It was felt that the non-transparent and transient nature of search engine functionality would result in an unacceptable lack of replicability. We therefore focussed on an extensive search of specialist sources (see next section) to identify grey literature and reduce possible publication bias.

Specialist searches
A list of thirty-three relevant stakeholder organisations was identified and their websites searched by members of the qualitative and quantitative review teams and/or contacted by email with a request for relevant literature (Table 1).

Websites were searched using a hierarchical approach, from the original search string down to individual words. Where this was not possible, the following individual terms were searched; national park*, protected area*, and reserve*. Boolean operators and wildcards were used where possible. All returns were assessed except where searches resulted in large numbers of results (i.e. > 100) when the first 50 returns were scanned for relevance. In each case, results were assessed at title and then full text, for relevance. Full texts were visually scanned for relevant data, along with within-document searches for the following terms; park, protect, and reserve. Due to the disparate modes of operation of the website search engines, a wide range of approaches was necessary and is documented in Additional file 3.

Bibliographic searches
Where studies identified in the search used data reported in earlier primary literature, the original reference was sought and included in the data extraction process. These references were catalogued in a separate Endnote library. Five key reviews were identified from the above searches. The references within the five relevant reviews were examined to identify pertinent articles for inclusion in the review (for details, see Additional file 4). These references were assessed at title, abstract, and full text level where not already included in the above database and website searches.

Search update
In May 2013 an update to the searches was undertaken to supplement the review with all relevant evidence published since the original search was carried out in October/November 2011.

The original search string (October 2011) was combined with the supplemental search string (March 2012) and entered into the database *Web of Knowledge*. Details of this search string and the number of hits returned can be found in Additional file 1.

Screening
Inclusion criteria
The following inclusion criteria were applied to those articles captured by the search.

Populations Human populations/communities currently or previously living in or near terrestrial PAs. In order to avoid subjectivity in deciding what is 'local', studies were included if the subjects were any type of community within the same country as the PA(s) that formed the focus of the paper.

Exposure Establishment/implementation, presence or change in status of terrestrial PAs areas with IUCN classifications I-VI as defined in the World Database of Protected Areas. Specifically, studies were included only where the collection of data on the impacts of PAs was undertaken during or after 1992. This included changes in status of PAs. The impacts of individual PAs were treated independently where possible but in some cases the studies looked at multiple PAs in such a way as to prevent the separation of impact. Projects that were established within or surrounding PAs as a result of the existence of that PA were also included.

Types of study Studies were considered for inclusion in this review if they satisfied one of the following criteria:

a) Evaluate the impact of PAs on human well-being (outcome evaluations which use the following study designs with appropriate comparators (i.e. a comparable state with which the intervention or exposure can be compared): Randomised Controlled Trials (RCTs) / Controlled Trials (CTs), control-intervention site comparisons, interrupted time series, Before-After/Control-Intervention (BACI) designs);
b) Reported economic valuation of welfare changes in monetary terms: based on stated or revealed preferences or production function approaches, in

Table 1 Relevant stakeholder organisations and their websites searched for relevant literature and contacted with a request for information

Organisation	Web Site	Email
Asociación Campesina e Indígena de Agroforestería Comunitaria Centroamericana	http://www.acicafoc.org/	Yes
Conservation International	http://www.conservation.org	Yes
Cultural Survival	www.culturalsurvival.org/	Yes
Department for International Development	http://www.dfid.gov.uk/	Yes
Development Experience Clearinghouse (DEC)	http://dec.usaid.gov/index.cfm	Yes
EcoTerra	www.ecoterra.net/	Yes
Eldis (Livelihoods Connect)	http://www.livelihoods.org	Yes
Environment Knowledge Hub	http://ekh.unep.org/	No
European Tropical Forest Research Network	http://www.etfrn.org/etfrn/index.html	Yes
First Peoples Worldwide	www.firstpeoplesworldwide.org/	Yes
Food and Agriculture Organisation	http://fao.org	Yes
Forest Peoples Programme	http://www.forestpeoples.org/	Yes
GEF Evaluation Office	http://www.thegef.org/gef/gef_Documents_Publications	Yes
GEF Small Grants Programme	http://sgp.undp.org/	No
Indigenous Knowledge	http://indigenousknowledge.org/	Yes
International Fund for Agricultural Development	http://www.ifad.org/	Yes
International Institute for Environmental Development	http://www.iied.org	Yes
International Union for Conservation of Nature	http://www.iucn.org	Yes
Nature Valuation and Financing Network	http://www.naturevaluation.org	Yes
Overseas Development Institute	http://www.odi.org.uk/	Yes
Pacific Forestry Centre	http://www.pfc.cfs.nrcan.gc.ca/	Yes
Poverty and Conservation	http://povertyandconservation.info	Yes
Poverty-Well-being Platform	www.poverty-well-being.net	Yes
RECOFTC – The Center for People and Forests	http://www.recoftc.org/site/	Yes
Rights and Resources Initiative	http://www.rightsandresources.org/	Yes
Survival International	www.survivalinternational.org/	Yes
Tropenbos International	http://www.tropenbos.org/	Yes
United Nations Development Programme	http://www.undp.org/	Yes
United Nations Environment Programme	http://www.unep.org/	Yes
United States Agency International Development	http://www.usaid.gov/	Yes
Waldbau-Institut, University of Freiburg	http://www.waldbau.uni-freiburg.de/	Yes
World Bank	http://web.worldbank.org (http://documents.worldbank.org/curated/en/home)	Yes
World Conservation Monitoring Centre	http://www.unep-wcmc.org/	Yes

which the comparator may be modelled and/or implicit in survey responses (hypothetical);

c) Sought to identify PA factors that influence human well-being or
d) Sought evidence to support explanations or meaning of impact from people's views about PAs and human well-being, and observations or related documentary analysis

We included studies that obtained data through direct measurement, self-reported measures by respondents, and other data sources that were not accessible, but the methods for which were adequately reported (such as national census data).

Additional inclusion criteria for quantitative evidence

Comparators Comparators were classified as temporal, spatial or modelled/hypothetical. Temporal comparators included time series, before and after, change over time (i.e. single time point), and reported/perceived changes. Spatial comparators included dichotomous (i.e. inside/outside and near/far) and continuous (i.e. linear distance) variables. BACI comparators included both spatial and temporal comparisons. Studies may account for confounding variability between comparator and exposure populations by using matching techniques, testing for differences in confounding variables, or including these variables in statistical models.

Whilst economic valuation studies may lack before-after or matched site comparators, economic valuations are inherently relative to a counter-factual (which may be modelled or subjective). Not all of these inherent comparators are suitable for this review, however. Many studies purport to value a PA. However, on closer inspection, they are often valuing the benefits of the ecosystem-protected area complex as a single entity, and may present no information with which to estimate the value of the ecosystem in the absence of the PA as an appropriate comparator. These studies were rejected on the grounds that they lack an appropriate comparator (the comparator is usually the complete and sudden non-existence of the ecosystem, which is deemed inappropriate for the purposes of this review). Our minimum inclusion criteria are that the study makes some attempt to model what would happen to the ecosystem in the absence of the PA, for example, estimating rates of degradation before the PA was established and extrapolating these into the future to calculate that portion of the total value of the site which is attributable to the PA. Other economic valuation studies may value opportunity costs of the PA, and in these instances the instantaneous loss of access to resources following establishment of the PA constitutes an appropriate comparator.

Outcomes Specific human-well-being indicators linked to those broad questions set out in the Background section (e.g. Livelihood strategies, social capital, empowerment, human rights, access to ecosystem services and natural resources). Examples are measures of income, education, health and other mainstream socio-economic indicators; World Bank and United Nations Development Programme (UNDP) human development index measures and measures that consider health, longevity, education, gender equity, food security, livelihood diversity, subjective/reported measures of well-being, resilience, measures of social capital, and indicators of human rights.

Additional exclusion criteria for qualitative evidence

Studies of people's views were excluded if they: focused solely on the development or validation of a measurement tool without also presenting views separately from the validation of the tool, or reported trials or other outcome evaluations, unless it was clear from the abstract that they collected data about views as part of a process evaluation.

Preliminary screening process

Articles captured by the searches were stored in an Endnote library. Replicates were removed and titles examined for relevance to the inclusion criteria. Potentially relevant titles were then separately screened as to whether they contained evidence of people's views and/or quantitative evidence of impact as described in the following sections.

Screening articles for qualitative evidence

Following preliminary screening by title (3.3.2) articles were screened for evidence of people's views of PAs by abstract and then full text. Any articles missing an abstract were accepted for full text screening. The inclusion and exclusion criteria were re-applied to the full texts (70.3% were obtainable) and those that did not meet inclusion criteria were excluded. A coding tool was developed in order to ascertain what relevant information was held within articles (see Additional file 5). 'EPPI-Reviewer' (Thomas et al. 2010) software was used for screening using a single web location to house the documents and monitor progress of the review. Rejected articles and their corresponding reasons are listed in Additional file 6.

The systematic search and screen identified many studies relevant to the review question. The studies identified covered a broad range of geographical areas, types of PAs, and sampled a diverse population group using a variety of qualitative research methods. The characterisation of studies by methodological and contextual features provided the starting point for deciding which studies to include in the qualitative synthesis (see section Studies included in the qualitative synthesis for further details).

Descriptive information about included studies was collected and presented as a 'systematic map' of research into explanations and meaning of impact to people living in or near PAs. The 'map' provided a basis for informed discussion and decision making between both review teams about the focus of the qualitative synthesis which provides a detailed investigation of a more focused subset of this wider literature. As the synthesis was focused and narrowed down, a second set of inclusion criteria was developed and applied to the studies initially identified.

Understanding perceptions of the impact of protected areas requires appropriate research methods and full reporting of context. Further, understanding the meaning of protected areas for well-being requires qualitative data collection methods that allow people to express their views freely rather than merely responding to predetermined categories, and where the analysis provides 'thick' or 'rich' findings. 'Thick' findings have been defined as a rich, detailed description of specifics (as opposed to summary, standardization, generalization, of variables); "it captures the sense of what occurred and the drama of events, thereby permitting multiple interpretations" [11]. We therefore selected for in-depth review those studies using qualitative research methods which elicited the views of people living in or near protected areas where:

- The IUCN category, and the date this was assigned, was known

- The methods and date of data collection were clearly reported
- The analysis produced 'thick' or 'rich' findings
- Findings were linked to specific people (e.g. distinguishing ethnicities, employments or locations)

Screening articles for quantitative evidence

Following preliminary screening by title (3.3.2) articles were screened for quantitative evidence of impact by abstract and then full text. Any articles missing an abstract were accepted for full text screening. Where information in titles or abstracts was insufficient, articles were accepted for screening at full text.

A kappa test for consistency of decision regarding inclusion/exclusion at abstract level returned a moderate level of agreement (n = 100, kappa = 0.442, 95% C.I. 0.270 to 0.614) between two independent reviewers (SED and ASP). Twenty-eight studies were treated differently by the reviewers and these were re-read and the reasons for inclusion or exclusion by each reviewer discussed. These discussions were used to come to consensus on the interpretation of the titles and abstracts and informed the subsequent inclusion process.

Full text screening resulted in the rejection of articles that were not pertinent to the review in hand. Reasons for exclusion were recorded (Additional file 6). Relevant reviews were identified for use in the bibliographic searching phase described in Section 3.2.

We attempted to obtain all articles deemed relevant during abstract-level screening, for assessment at full text. Those articles that we failed to obtain within the limitations of time and resources of the review are listed in Additional file 2.

Qualitative synthesis of explanations and meaning of impact

There are three distinctive approaches to synthesising findings from research: aggregating the findings of very similar studies where the key concepts are clearly defined in advance; configuring the findings of dissimilar studies by investigating the implications of the differences in their methods, context and findings in order to define key concepts and develop theoretical understanding; and a combination of the two which can be applied where studies have important differences, but nevertheless share some important key concepts of well-being [15].

We chose this third approach because some (but not all) key concepts were decided and defined in advance: PAs with their clearly defined categories; populations and outcomes of interest. Despite these concepts being identified by the funders, a synthesis method was required that also allows new concepts to emerge from the data in order for the impact on human well-being to be understood from the perspective of people living in or near PAs. Framework synthesis does this because dimensions of a framework for structuring the synthesis can be identified by policy interests, research interests and concepts emerging from the data [16]. Framework synthesis also takes into account differences in context such as the different countries hosting PAs, the different categories of PA and the different times and stages in their history that each area was studied. Based on framework analysis of primary research data [17], an initial conceptual framework, either built from assumptions held by stakeholders or borrowed from related bodies of knowledge, evolves during the synthesis as the reviewers become more familiar with the literature being reviewed [16]. The five neat stages of framework analysis described for primary research [18] are in practice a more iterative process for making sense of a pile of studies [19]. The process starts by delving into the abstracts then full texts to see the key issues and recurrent themes, some of which emerge from the data while others are purposely sought in response to the review question or prior knowledge. Once most of the key themes are identified, each study is coded with themes and the literature is sub-divided into sections for in-depth analysis. The conceptual framework is refined in light of these themes, and coding continues iteratively but systematically as the framework becomes increasingly coherent and simultaneously accommodating of the available data. Subsections of the framework shape data tables under key themes in order to develop concise summaries. Conclusions are drawn from the themes and associations between them.

Data extraction strategy

Included studies were inspected in detail for the focus of their findings. The coding framework described above (Additional file 5) was applied to each study. This framework covered: health (mental health, physical health, safety and wildlife conflict); socio-economic position (livelihood strategies, social capital, human rights, empowerment, environmental capital, governance); inequalities; and context. Study text was extracted in the form of: participant data (e.g. direct quotes); authors' descriptions of findings; and authors' conclusions, implementations and recommendations. The coding framework was tested by two or more reviewers independently coding successive studies, comparing and discussing how they applied the codes, and refining the framework and definitions of codes until a shared understanding was achieved across the team.

Synthesis

Synthesis began by developing the initial key concepts into a coherent framework that could accommodate the main approaches to establishing and maintaining PAs. Figure 1 illustrates two contrasting approaches to terrestrial PAs. The first is governance models that are imposed

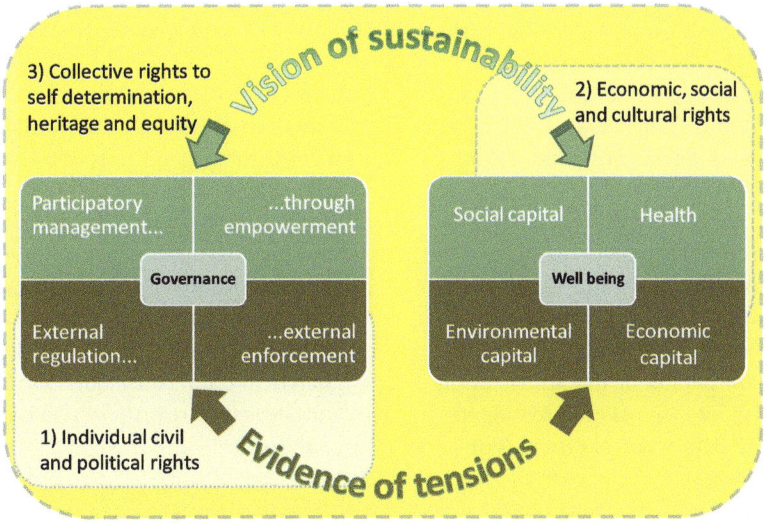

Figure 1 Conceptual framework for structuring the qualitative synthesis.

and enforced by external authorities and the second is participatory approaches such as community-based natural resource management. The latter has been developed in response to tensions raised by restrictions and resettlement having a deleterious impact on economic and environmental capital (agriculture, logging, tourism) and social capital (family and community relationships). In contrast, participatory approaches, within a regulatory framework, seek a vision of sustainability through building on social capital and good health. That vision of sustainability may be on a small scale, such as with IUCN category VI PAs, where the aim is sustainability within the boundaries. Alternatively it may be on a wider scale, across IUCN categories I or II and the wider area. These models are set against a backdrop of an evolving consensus about human rights that began with protecting individual civil and political rights, followed by the support for economic, social and cultural rights and then setting these rights within a broader framework to harness the combined efforts of individuals, states and other bodies to build collective rights to self determination, heritage and equity [20]. This synthesis aims to assess the extent to which these models and their anticipated impacts are supported by perceptions of impact on human well-being held by people living in or near protected areas and others working alongside them.

The framework was populated with research findings by dividing the studies into three overlapping subsets whose findings clearly addressed: participation (human rights, empowerment and governance), health (physical, mental and safety) and socio-economic position (wildlife conflict, livelihood strategies, social capital and environmental capital). Each set was taken by a different member of the team to seed the synthesis with studies focussing on findings in one of these three areas. Understanding the context of each study and the interplay between different concepts required reviewers to return to the full report of each study, alongside the text extracted for each code, and distil the key study characteristics and findings. The resulting coherent text was placed appropriately within the emerging framework (Figure 1). Where studies contributed substantial findings to different elements of the framework, these findings were distributed across the framework to allow each element to draw on all relevant studies. As the framework became more coherent with growing understanding of the relationships between its different elements, the elements were reordered to present governance issues first, and then well-being issues. The findings were summarised for each element of the framework, taking into account the governance and time context of the studies. The contexts of the studies contributing to the synthesis were recorded noting: the location (name and country of the protected area); IUCN category (present category, years assigned and changed); governance (national government, local government, private or community); and timing of study (year data was collected and the time lapse since the last assignment of IUCN category). Finally, all studies were revisited for additional evidence that appeared significant in light of emerging findings.

Synthesis of quantitative evidence

Data extraction strategy

A preliminary data extraction phase was carried out during full text screening, where each reported study was categorised in terms of which broad outcome types were represented in the quantitative data. These outcomes were categorised according to the coding framework described in (and in detail in Additional file 5). Furthermore, details about the following were ascertained; the broad study methodology (i.e. self-reported data from close-ended questioning in interviews and questionnaires, direct data collection, valuation or data taken from other sources such as national census documents and similar), and whether there was a valid comparator.

Comparator categories (i.e. no comparator, before-after, inside-outside, before-after-control-impacts, econometric) were used to group the studies for a second stage of detailed data extraction. Only those studies that possessed an appropriate comparator were taken on to this second stage of data extraction, where details of the study methodology, potential effect modifiers or reasons for heterogeneity and relevant outcome data were obtained. For all outcomes, averages (means/medians) and variability measures (standard deviation, standard error and confidence intervals) were obtained where possible, along with statistical test results and sums of counts (i.e. percentages of respondents in agreement).

Study quality assessment

Details relating to study quality were extracted for each study to allow critical appraisal of relevance (external validity) with respect to the review question and reliability (internal validity). Critical appraisal was conducted in two stages. Initially assessment was made as to whether a study possessed one or more of the following: confounding variables, within-study contradictory results, inappropriate implicit comparator, inappropriate population, inappropriate spatial comparator, inappropriate statistics, inappropriate temporal comparator, inappropriate time frame, insufficient detail in methods, data linked with marine ecosystem protection, non-standardised data (incomparable), protected area establishment too recent/not yet established, data collection prior to 1992, extremely small sample size (e.g. one study, where wood extraction was quantified in only nine households with one intervention and one control village), superseded by more recent study, unbalanced questioning/questioning bias, and unmatched methods (see Additional file 6 for further detail). Studies with one or more of these characteristics were excluded from the review.

Studies accepted following this first critical appraisal stage were then scored for reliability with respect to the question using the concept of susceptibility to bias (the extent to which a study is able to yield an unbiased estimate of effect; [21]). Susceptibility to bias in study design and reporting can affect both internal validity (study quality) and external validity (study generalisability).

Susceptibility to bias scores were assigned using a repeatable, but partially subjective, set of four criteria, each assessed on a scale of 0 to 2, and thus ranging from a minimum possible value of 0 and a maximum of 8 (Table 2). The four criteria were as follows:

i) Clarity of methods; the extent of detail given in the study's methodology regarding questionnaire design, survey implementation, replicate number and selection, and data analysis
ii) Study design; the rigour of study design in terms of replicate number and sampling, location and choice of replicates and controls, questionnaire design, and survey implementation
iii) Appropriateness of analysis; the suitability of analytical techniques such as the implementation and choice of statistics, and comprehensiveness of analysis
iv) Implementation bias; the presence of one of the following biases in measurement and analysis; Neyman bias (the intervention in question causes systematic drop-out within the population, e.g. the most affected families move away), questioning bias (unbalanced/leading questions), potentially influential variables, recall bias (response affected by memory).

Susceptibility to bias scores were then combined with an objective weighting by basic study design in order to provide a categorical assessment (high, medium or low) of susceptibility to bias (Table 3).

Potential effect modifiers and reasons for heterogeneity

Data on potential effect modifiers that were extracted from articles, included the following variables; protected area studied, country, IUCN category, protected area size, date of establishment, and history of protection prior to formal establishment, and residence history and ethnicity of the study population. Where sufficient information about IUCN category, protected area size and year of establishment was not presented in the article itself, data were supplemented from the World Database on Protected Areas (WDPA) (http://www.protectedplanet.net/).

Data synthesis and presentation

Extracted data were presented in narrative synthesis tables that summarise the studies' aims, methodology, study population, and results (Additional file 7). No further quantitative synthesis of data on outcomes was possible with the available data.

Results

Review descriptive statistics

The main literature search was conducted between 11 October and 14 November 2011 with an update carried out in May 2013.

Table 2 Scoring criteria for subjective assessment of susceptibility to bias

Assessment criteria	0	1	2
Clarity of methods	All sample sizes provided, selection method described, questionnaire design fully disclosed	Some information regarding sample size, selection method, or questionnaire design provided	Sample size not given in full, sample selection not stated, questionnaire design not discussed
Experimental design	Good sample size, appropriate sampling regime, control and intervention well matched, survey appropriately implemented	Low effective sample size, poorly randomised design, control and intervention choice not ideal	Very small sample size, pseudoreplication, non-random sample selection, control and intervention poorly matched/low consistency in sampling over time
Appropriateness of analysis	Confounding variables accounted for, appropriate metrics reported	Confounders only partly accounted for/only some low risk confounders ignored	Significant confounding variables unaccounted for, inappropriate metrics reported, incorrect statistical analysis
Implementation bias	No identifiable bias reported/evident	Low level bias present but ignored/ strong bias accounted for	Strong bias present and unaccounted for

The main search returned 13,248 articles (following removal of duplicate and marine-oriented articles) and title-level screening left a total of 1,710 articles. The search update returned 3,122 articles (following removal of duplicates) and title-level screening left a total of 204 additional articles. From this point forwards, screening at abstract was independent for the two review sections.

Qualitative evidence

Abstract screening for the qualitative review resulted in the final inclusion of 30 studies as illustrated in Figure 2. Four additional studies were included in an identical process following the search update in May 2013.

Quantitative evidence

Abstract screening for the quantitative review resulted in the final inclusion of 79 studies as illustrated in Figure 3.

In addition to the above searches 2,345 articles were identified by the supplemental search conducted on 23 March 2012: 535 remained following title-level screening, and 171 following abstract-level screening. Twelve studies from this supplemental search were included following critical appraisal.

Relevant organisation website searches conducted between 6 and 29 March 2012 yielded 94 relevant articles for the quantitative review. In addition to these searches, 50 relevant articles were identified through bibliographic checks and secondary sources, yielding a total of 1,164 potentially relevant articles. Of these articles, 76% were retrievable for full text assessment, whilst 275 were unobtainable within the timeline and resource limitations of the review. These numbers are represented visually in Figure 3. One hundred and fifty-seven of the articles identified in the May 2013 search update were obtainable, although five of these were in Chinese and could not be assessed (see Additional file 2).

Table 3 Categorization of susceptibility to bias

	Susceptibility to bias score		
Comparator	**0-2**	**3-5**	**6-8**
RCT* (randomised control trials)	Low	Low	Medium
BACI (before-after-control-intervention)	Low	Medium	High
Control-intervention/Before-after/ Econometric	Medium	High	High

*There were no RCT studies and this line is only provided for a theoretical perspective

Following full text assessment 177 articles from the original search, 16 from the supplemental search and 17 from the update met the inclusion criteria and were subsequently critically appraised. Following critical appraisal, articles/studies were rejected at full text for a wide range of reasons (see section Study quality assessment). A list of these articles and associated reasons is provided in Additional file 6. A final set of 70 articles was included, reporting on 79 studies and these are listed in Additional file 7.

Comparison of included studies from the two review processes

The following statistics describe the 30 studies on qualitative evidence of people's views and 79 on quantitative evidence of impact included in the synthesis.

Study location Figure 4 displays the countries from which data were collected for the included studies from the two processes. Frequently studied countries include Cameroon, China, India, and Uganda for the qualitative synthesis and India, Nepal, South Africa and Uganda for the 'quantitative impacts' review.

Category of PA studied The IUCN categories of the PAs (see below) examined in the included studies are shown in Figure 5. PAs predominantly fell under category II in both reviews, with 17 percent either unreported by WDPA or not present in the database for the 'quantitative impacts' review (NB this group includes only 'not applicable' for the qualitative synthesis, since studies were only included in this review if the protected area IUCN category was stated). A post-hoc decision to exclude category III PAs was made as they are small in number and somewhat atypical of the set of categories

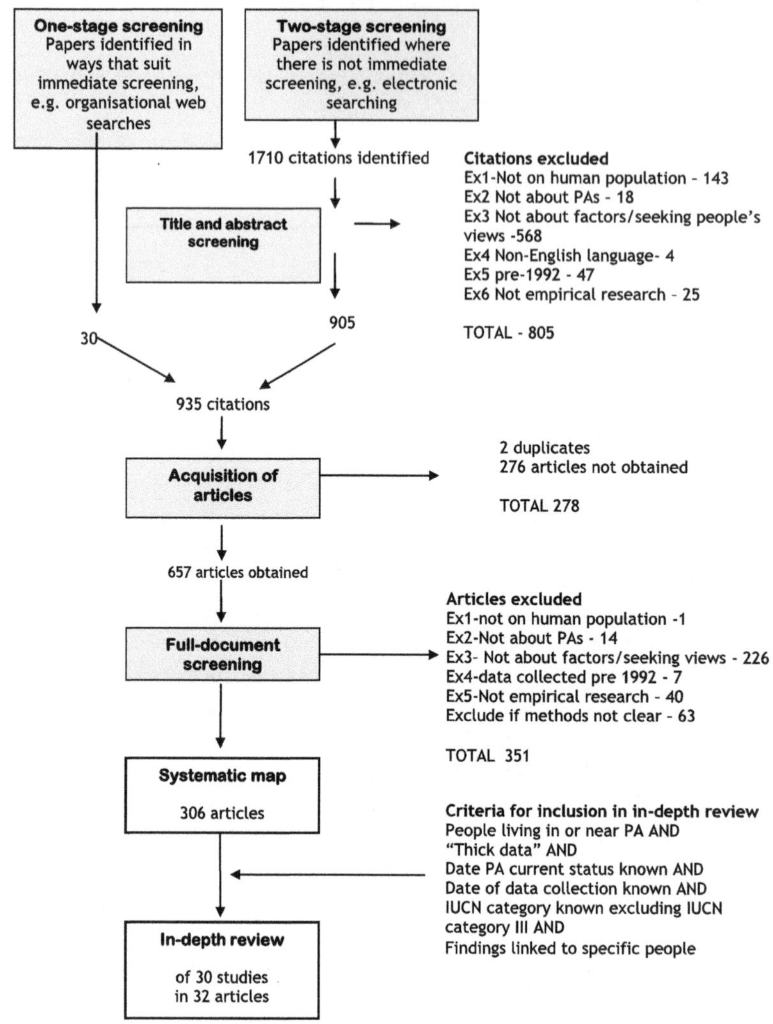

Figure 2 **Number of articles and studies at progressive stages of inclusion for qualitative synthesis.**

(specifically protecting natural monuments). Similar patterns were observed in both review processes Figure 6 displays PA establishment year, showing a peak in new PAs centered around the 1980s.

IUCN Protected Area Categories are described as follows:

Ia **Strict Nature Reserve** - protected from all but light human use in order to preserve all geological and geomorphological features of the region and their biodiversity, which is often dense and restricted to exclusively to scientific monitoring, study or education. Occasionally Strict Nature Reserves are of spiritual significance to surrounding communities in which case the people are generally allowed to continue the practice of their faith and may be directly involved in the area's conservation and management objectives, though perpetual human intervention would more suitably be allocated to categories IV or V.

Ib **Wilderness Area** - generally larger than Strict Nature Reserves, the main objectives of these areas is to provide an environment in which biodiversity and ecosystem processes (including evolution) are allowed to flourish or experience restoration if previously disturbed by human activity. Human use is limited, often allowing only those who are willing to travel of their own accord rather than via established touristic activities. Wilderness areas can be classified as such only if they are devoid of modern infrastructure, although they allow human activity to the level of sustaining indigenous groups living wilderness-based lifestyles.

II **National Park** - provide protection for functioning ecosystems (similar to the objectives of Wilderness Areas), but tend to be more lenient with human visitation and the supporting infrastructure. National Parks are managed in a way that may contribute to local economies through promoting educational and

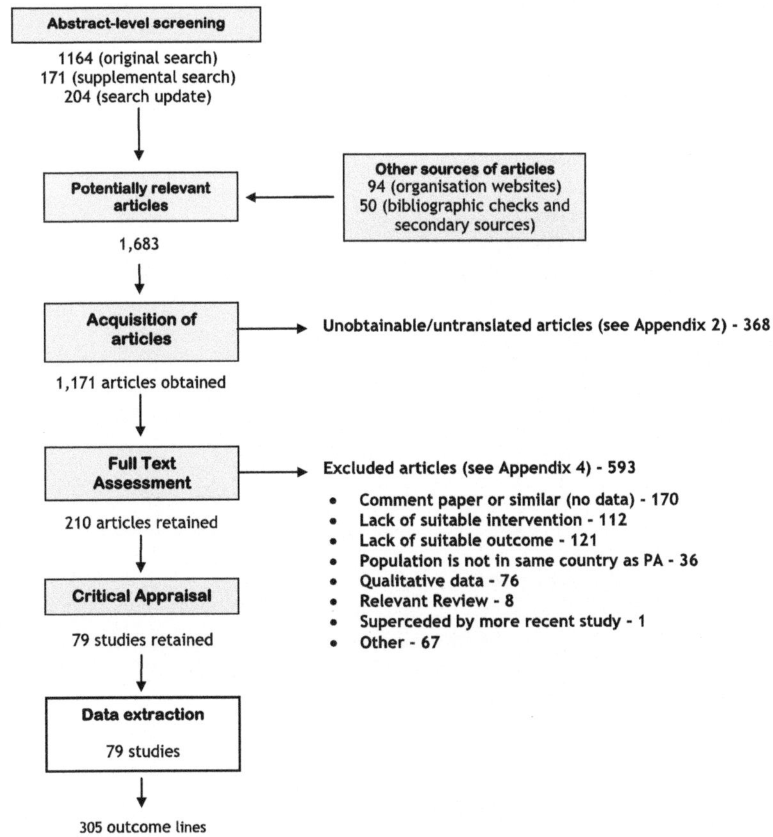

Figure 3 Number of articles and studies at progressive stages of inclusion for quantitative review.

recreational tourism on a scale that will not reduce the effectiveness of conservation efforts. The surrounding areas of a National Park may be for consumptive or non-consumptive use, but should nevertheless act as a barrier for the defence of the protected area's native species and communities to enable them to remain sustainable in the long term.

III **Natural Monument or Feature** - comparatively smaller areas, specifically allocated to protect a natural monument and its surrounding habitats. Natural Monuments or Features can be natural in the wholest sense, or include elements that have been influenced or introduced by humans. The latter should hold biodiversity associations or could otherwise be classified as a historical or spiritual site, though this distinction can be quite difficult to ascertain. As such, the classification then falls into two subcategories, those in which the biodiversity in uniquely related to the conditions of the natural feature, and those in which the current levels of biodiversity are dependent on the presence of the sacred sites that have created an essentially modified ecosystem.

IV **Habitat/Species Management Area** - focus on more specific areas of conservation in correlation to an identifiable species or habitat that requires continuous protection. These protected areas will be sufficiently controlled to ensure the maintenance, conservation and restoration of particular species and habitats - possibly through traditional means - and public education of such areas is widely encouraged as part of the management objectives. Habitat or Species Management Areas may exist as a fraction of a wider ecosystem or protected area and may require varying levels of active intervention including - but not limited to - the prevention of poaching, creation of artificial habitats, halting natural succession and supplementary feeding practices.

VI **Protected Landscape/Seascape** - cover entire bodies of land or ocean which engage a range of for-profit activities within the management plan. The main objective is to safeguard regions that have built up a 'distinct character' in regards to their ecological, biological, cultural or scenic value. Protected Landscapes and Seascapes allow a higher level of sustainable interaction with -surrounding communities (such as traditional agricultural and forestry systems) and should represent an integral balance between people and nature.

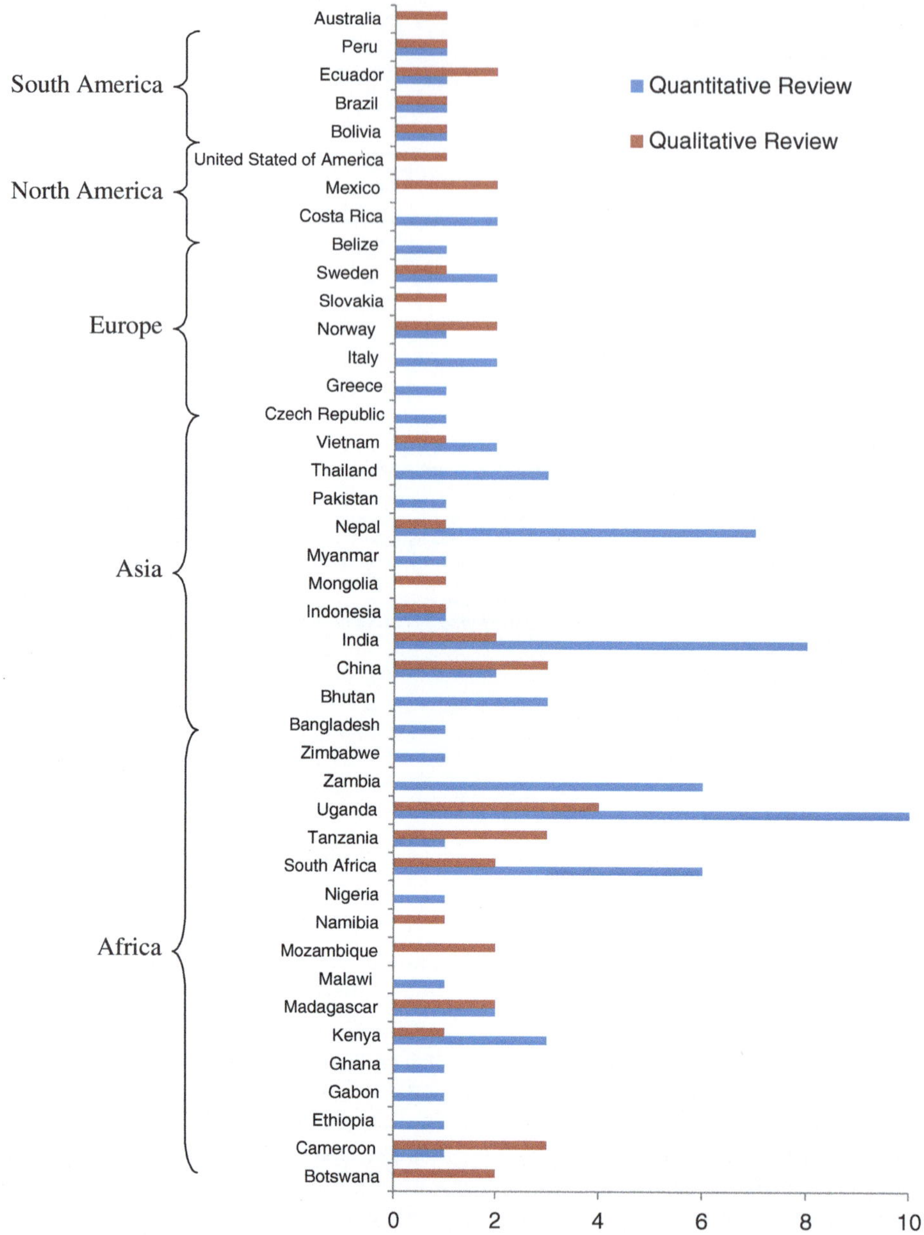

Figure 4 **Frequency of countries, arranged by continent, hosting the protected areas within included studies for the two reviews.**

VI **Protected area with sustainable use of natural resources** - based on a mutually beneficial relationship between nature conservation and the sustainable management of natural resources in correspondence the livelihoods of surrounding communities. A wide range of socio-economic factors are taken into consideration in creating local, regional and national approaches to the use of natural resources. Though human involvement is a large factor in the management of these protected areas, developments are not intended to allow for widescale industrial production. Governance has to be developed to adapt to the diverse and possibly growing range of interests that arise from the production of sustainable natural resources.

(Text adapted from United Nations Environment Programme (UNEP) World Conservation Monitoring Centre (WCMC) website (http://www.unep-wcmc.org/iucn-protected-area-management-categories_591.html; accessed 03/10/2012)

Study timescale Figure 7 displays the survey years for the included studies for both reviews. A significant number of studies failed to report the survey year (applicable

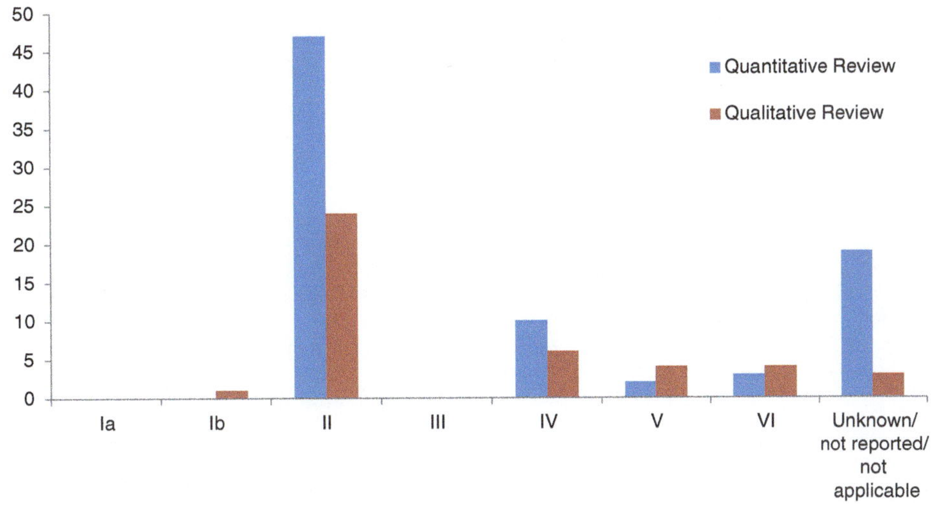

Figure 5 **Frequency of all named PAs by IUCN category within included studies.**

only to the 'quantitative impacts' review, since this was an exclusion criterion for the qualitative synthesis). The reduction in the number of surveys from 2005–2006 for both reviews likely relates to a publishing delay.

Further descriptive statistics for the review of quantitative evidence of impacts

Study comparator Figure 8 displays the range and frequency of comparators used in included studies. Spatial comparators (site comparators and linear distance) were common, along with reported change over time. 'Before and after, control intervention' (BACI) studies were very rare, with only one instance of a full BACI study.

Measured outcome Within the five key themes described in Section 1 (Livelihood Strategies, Human Rights, Social Capital, Empowerment, and Access to Ecosystem Goods and Services), 24 individual outcomes were identified in an iterative process during screening. The reported outcomes were separated in order to facilitate the description and analysis of similar, comparable data. The frequency of these broad outcome types is shown in Figure 9. Attitude, income, and wildlife conflict outcomes were most common.

Exclusion of studies following critical appraisal resulted in several outcomes identified during full title assessment being unrepresented by data namely, 'empowerment' and 'religion and spirituality'. Other outcomes are represented in

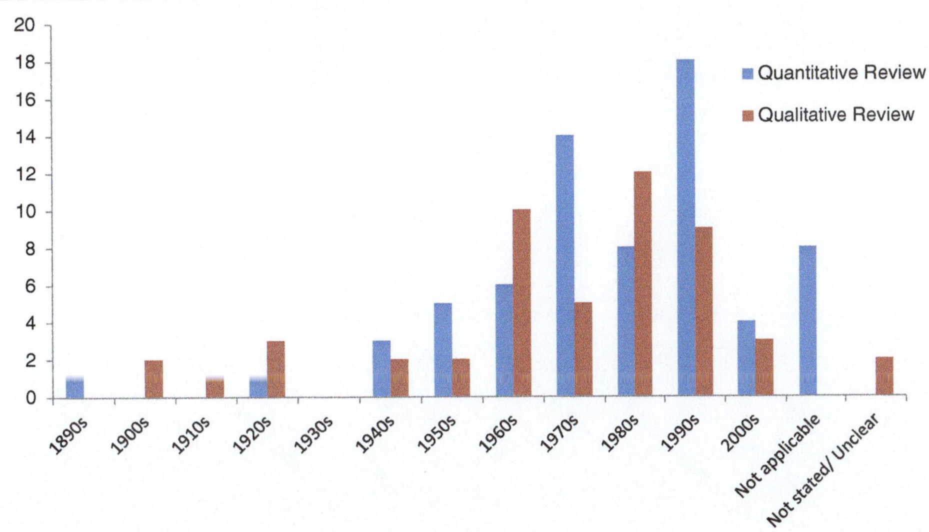

Figure 6 **Frequency of year of establishment of PAs within included studies.**

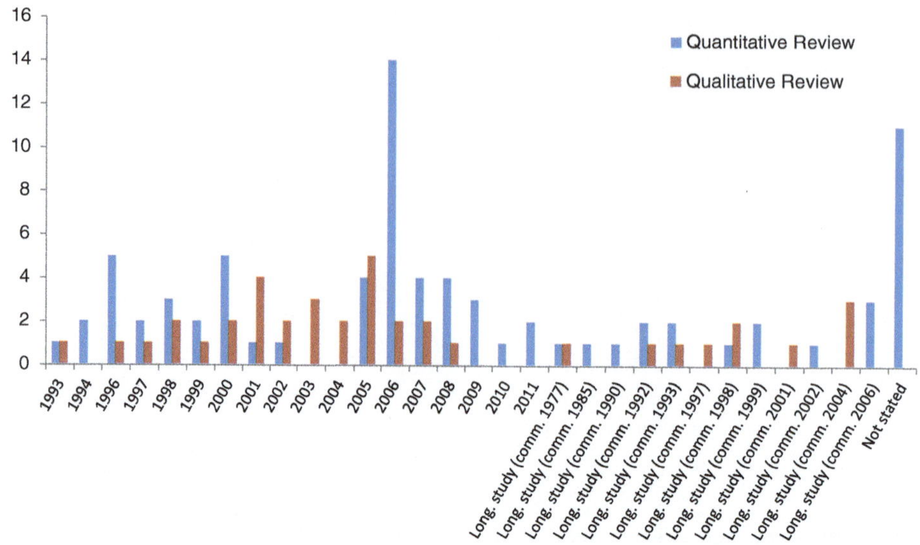

Figure 7 Frequency of year of PA survey for included studies. Surveys commence on stated year and are classified as 'longitudinal study' if survey spans more than two years.

two studies or fewer; 'gender equity', 'interactions with PA authorities', 'livelihood diversity', 'medicinal plants/animals', 'protect for future', and 'resettlement and displacement'.

Methodology of data collection Figure 10 displays the frequency of different data sources in included studies, showing that self reported measures formed the majority of data reported (63 of the 79 included studies).

Qualitative synthesis of explanations and meaning of impact

Studies included in the qualitative synthesis

The criteria for selecting studies to include in the in-depth qualitative synthesis is provided in section Screening articles for qualitative evidence. Although not statistically representative of the 306 studies of people's views identified (characterised in the map see Additional file 8), studies with 'thick' data were chosen for their ability to provide explanations and meaning of impact for people living in or

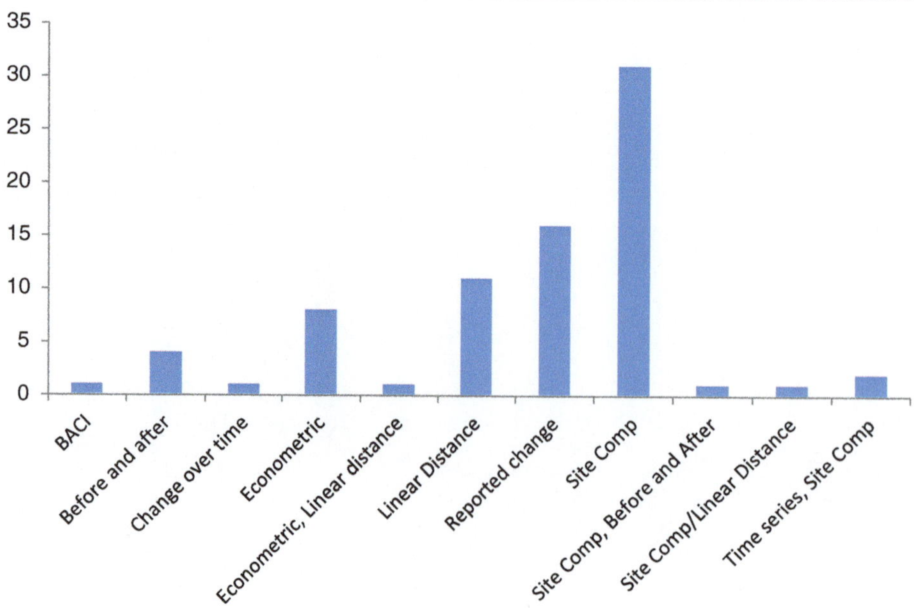

Figure 8 Frequency of comparators used in included studies.

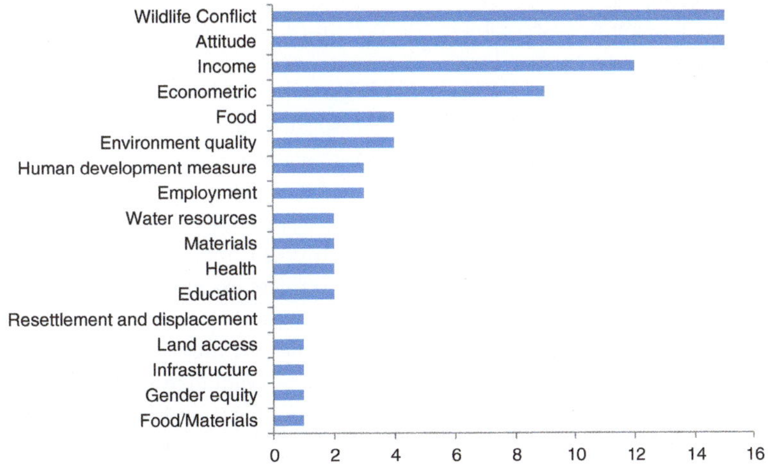

Figure 9 Frequency of broad outcome types in included studies.

near the PA. They also spanned the range of the broader literature, in terms of countries frequently studied, as described in the map. However, poor reporting makes it difficult to conduct further comparisons between the studies reviewed in-depth with the wider map in terms of IUCN categories and key dates.

The vast majority of included studies were by academic authors, with a small minority from NGOs, commercial organisations and protected area authorities. These different perspectives are likely to shape the focus of studies and, consequently, their findings.

Residents' views were most often sought for IUCN category II, where there are strict restrictions on settlements, but rarely for IUCN category VI, where the aim is to balance the aims of conservation and the needs of the local population. Leaders' views were most often sought about IUCN category I protected areas, which does not allow human settlements. There was also a noticeable absence of studies addressing the views of protected area authorities or employees in category IV (similar to the map).

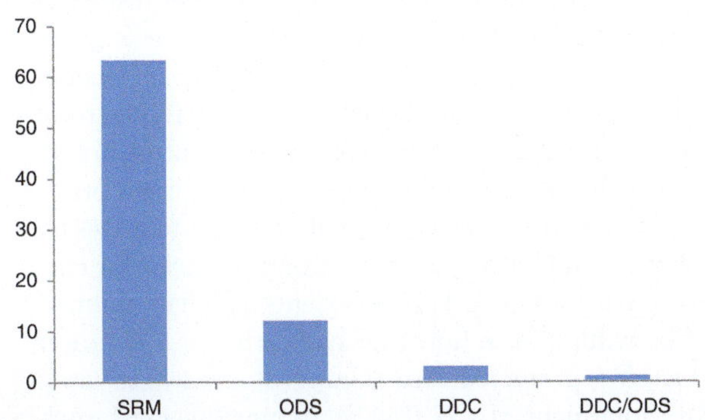

Figure 10 Frequency of data sources in included studies. DDC, direct data collection by study authors; ODS, other data sources; SRM, self-reported measure via survey responses.

The other stakeholders not well represented in the studies presenting 'thick' findings are visitors to protected areas. Studies seeking visitors' views often did so using highly structured questionnaires that provided little understanding of what protected areas mean to them. The one study of visitors included in the synthesis was set in a category IV protected area, so was not typical of most visitor surveys which, in the map, were most often used in category II.

This literature is thus a likely source for understanding people's views, and setting those views in context, across the IUCN categories, although less so for category IV.

Summary findings of qualitative synthesis of explanations and meaning of impact

This section presents a summary of a full synthesis of findings from studies that provide explanations or explore the meaning of the impact of protected areas on people living in or near PAs. The full synthesis can be found in Additional file 9. It also draws on the perspectives of other stakeholders relevant to those experiences and relevant policy and historical documents. The synthesis is presented within a conceptual framework that was informed by conservation policies and refined by the emerging research literature (Figure 1).

The synthesis process identified the implementation issues and characteristics of well-being that were associated by research participants with either (a) tensions arising from governance models imposed and enforced by external authorities or (b) a vision of sustainability sought through participatory management and empowerment commonly known as community-based natural resource management.

The latter approach has been developed in response to tensions raised by restrictions and resettlement having a deleterious impact on economic and environmental (natural) capital and social capital. Participatory approaches, within a regulatory framework, seek a vision of sustainability through building on social capital and good health. That vision of sustainability may be on a small scale, such as with IUCN category VI PAs, where the aim is sustainability within the boundaries. Alternatively it may be on a wider scale, across IUCN categories I or II and the wider area. These models are set against a backdrop of an evolving consensus about human rights that began with protecting individual civil and political rights, followed by the support for economic, social and cultural rights and then setting these rights within a broader framework to harness the combined efforts of individuals, states and other bodies to build collective rights to self determination, heritage and equity [20]. This synthesis aims to assess the extent to which these models and their anticipated impacts are supported by the perceptions of people living in or near protected areas and others working alongside them.

The characteristics of the included studies are summarised in Table 4, and their findings are subsequently synthesised in sections that match the conceptual framework (see Figure 11) to present a coherent narrative which explores key themes within governance (source of authority and nature of implementation) and then considers the impact on health and well-being in terms of environmental (natural), economic and social capital. There is necessarily some overlap between the governance issues and the impact on health and well-being. The key messages for both are presented in boxes to help readers navigate the breadth and depth of the available evidence. The contextual details of each study described in terms of IUCN Category, the date the PA was established, date of current PA status being assigned, time between current status being assigned and data collection are reported in Additional file 9. As the complexity of the links between governance and well-being precludes simple distinctions between PA approaches and their impacts, we focus first on governance and then on well-being, acknowledging that there is necessarily some overlap. The key messages for both are presented in boxes to help readers navigate the breadth and depth of the available evidence.

Governance

Matching study findings to the conceptual framework facilitated their comparison and interpretation to reveal problems and potential solutions during implementation and instances of both benefit and harm.

- Portraying longstanding residents as an environmental risk to pristine protected areas is controversial
- PA regulations can flounder as a result of inaccurate maps and poorly drafted legislation, and further confusion arises from discrepancies between state rules and local institutions and typically poor communication
- Memories of forced or induced migrations negatively influence subsequent community responses to authorities
- Inadequate or non-existent compensation is a widely held concern
- Externally imposed regulations were incompatible with traditional regulations and did not take into account cultural and social diversity; respect for the rules was greater where they were locally adapted and allowed income-generating possibilities.
- A major challenge is distinguishing subsistence activities for a sustainable environment from larger scale industrial activities.
- Success came from PAs staff having: prior experience of working with locals; clear guidelines; extensive training in community development, gender issues and a variety of participatory methodologies; meeting locals informally and working with existing kinship networks.
- All studies contributed some evidence related to governance.

Implementation problems The earliest source of tension in the life course of protected areas can be controversy about environmental risks posed by local inhabitants. Studies in Australia [22], Nepal [23], Norway [24], Mexico [25], Indonesia [26] and USA [27] revealed local inhabitants appreciating areas for their aesthetic and spiritual values as well as environmental products and economic and leisure opportunities. Mexican farmers in particular value the land for its provision of food, water, wood and other products, and they have developed farming styles along a spectrum of reciprocal relationships between man and nature between wilderness and urbanisation [28]. Diaw [29] claims that a resettlement policy to establish a IUCN II category park in Cameroon in 1961 was driven by scientific myths of a pristine forest whose protection was incompatible with indigenous residents despite historical analysis showing that the current forest structure was the result of sustained use over centuries. Economic arguments favouring resettlement were flawed, with excessively strong assumptions about tourism benefits, flood control, forest use, research discoveries, soil fertility and agricultural productivity. Similarly, in Norway residents pointed to a lack of convincing scientific evidence supporting the need for certain protective measures [24]. Residents of Utah claimed the land benefitted from how they had cared for it before it was declared a protected area (Grand Staircase Escalante National Monument, IUCN II [27]). In Norway, local tourist firms were frustrated by the lack of opportunities to play a part in the protected area management (IUCN II). They claimed that local expertise, based on generations of

Table 4 Summary table of included studies in the qualitative synthesis

Citation	Protected area	IUCN Category	Date PA established	Date of current IUCN category	Years since IUCN category change/ establishment	Year of data collection	Governance model	Country
Allendorf et al. (2007)	Royal Bardia NP	II	1969	1976	14	1990	Government: federal/ national	Nepal
Almudi and Berkes (2010)	Peixe Lagoon National Park	V	1986	2001-2010	3	2005 & 2007	Cooperatively managed: collaboratively	Brazil
Bedunah and Schmidt (2004)	Gobi Gurvansaikhan	II	1993	1993	5-7	1998 – 2000	Information not found	Mongolia
Bizikova et al. (2012)	Slovensky Raj Park	II	1998	1998	8	2006	Government: federal/ national	Slovakia
Bolaane (2004)	Moremi Game Reserve	IV	1965	1965	32	1997 – 2001	Private	Botwana
Bruyere et al. (2009)	Buffalo Springs Samburu	II	1985	1985	20	2005	Government: local	Kenya
Castillo et al. (2005)	El Vizcaíno	IV	1988	1988	16	2004–2007	Government: federal/ national	Mexico
	Sian Ka'an		2006	2008	3			
	Monarch Butterfly		2008		< 1			
Davis (2011)	Tarangire National Park	II	1970	1970	35	2005 – 2007	Government: local	Tanzania
Diaw (2010)	Korup National Park	II	1961	1985	18	2003	Government: federal/ national	Cameroon
First Peoples (2006)	Mgahinga National Park	II	1930	1930	Not stated		Government: federal/ national	Uganda
	Bwindi National Park		1991	1991	14	2005	Government: federal/ national	Uganda
	Awa Forest Reserve Zone	VI	1988	1988	10-20	1998 – 2008	Community: indigenous	Ecuador
Gerritsen (2002)	Sierra De Manantlan Biosphere Reserve	VI	1987	1987	6 – 11	1993 – 1998	Government: federal/ national	Mexico
Hartter (2009)	Kibale national park	II	1932	1993	16			Uganda
Haukeland (2011)	Rondane National Park	II	1962	1962	47	2009		Norway
	Jotunheimen National Park	II	1980	1980	18		Government: national	
Herrold-Menzies (2011)	Caohai Nature Reserve	V Protected landscape/ Seascape	1985	1985	13-16	1998-1999, 2000, 2000-2001	Government managed: local	China
Hoole and Berkes (2010)	Etosha National park	II	1975	1975	31	2006 - 2007	Government	Namibia
Keskitalo and Lundmark (2010)	Sarek National Park;	Ib	1909	1909	95 years	2001	Government: federal/ national	Sweden
	Stora Sjöfallet		1909	1909	95 years			
	National Park; Abisko	II	1920	1920	83 years			
	National Park; Pieljekaise National Park; Vadvetjåkka		1962	1982	22 years			

Table 4 Summary table of included studies in the qualitative synthesis *(Continued)*

	National Park; Padjelanta National Park							
	Haparanda-Sandskär Nature Reserve							
Lunstrum (2008)	Limpopo National Park	II	1979	1979	25	2004 – 2005	Cooperative: transboundary/ collaborative	Mozabique
Mbaiwa (2005)	Moremi Game Reserve	IV	1965	1965	38	1998, 2001, 2003	Private: for profit	Botswana
Mehring et al. (2011)	Lore Lindu National Park	II	1982	1982	24	2006 – 2008	Government: federal/ national	Indonesia
Milgroom and Sperienburg (2008)	Limpopo National Park	II	2001	2001	6 – 7	2007 – 2008	Cooperative: transboundary/ collaborative	Mozabique
Nguiffo (2001)	Dja Wildlife Reserve	IV	1950	1950	51	2001	Government: federal/ national	Cameroon
Ogra (2008)	Rajaji National Park	II	1983	1983	20	2003 – 2004	Information not found	India
Ormsby and Kaplan (2005)	Masoala National park	II	1997	1997	4	2001	Private/ non-profit	Madagascar
Petrzelka and Marquart-Pyatt (2013)	Grand Staircase	V	1996	1996	0	1996	Government: federal/ national	USA
	Escalante National Monument				10	2006		
Slater (2002)	Qwaqwa National Park	IV	1992	1992	6 – 7	1998 – 1999	Government: local	South Africa
Sletten et al. (2008)	Mount Elgon NP	II	1951	1951	47	2002	Government: federal/ national + collaborative	Uganda
Spenceley and Goodwin (2005)	Kruger National Park	II	1926	1930	74 – 75	2000 - 2001	Information not found	South Africa
Songorwa (1999)	Selous Game Reserve	IV	1905	1905	91	1996	Community: local	Tanzania
Stone and Wall (2004)	Jianfengling National Forest Park and Diaoluoshan National Forest Park in Hainan Province, China	V	1986	1986	16 years	2001	Government: local	China
Strickland-Munro and Moore (2012)	Purnululu National Park	II	2003	2003	5 years	2008	Government: federal/ national + collaborative	Australia
Stronza and Gordillo (2008)	Kapawi Reserve; Madidi National Park; Bahuaja Sonene National	II	Not stated Kapawi Reserve, Ecuador 1995: Madidi National Park 2000: Bahuaja Sonene National Park, Peru	1991-2000	Not able to ascertain Ecuador 5 years;, Bolivia; 8 years;, Peru	2003	Cooperatively managed Collaborative management (various forms of pluralist influence)	Bolivia Ecuador Peru
Torri (2011)	Sariska Tiger Reserve	IV	1955	1955	46	2001 (pilot) 2007	Government: federal/ national	India
Yasuda (2011)	Benoue National Park	II	1968	1968	40	2004, 2009	Private: for profit	Cameroon

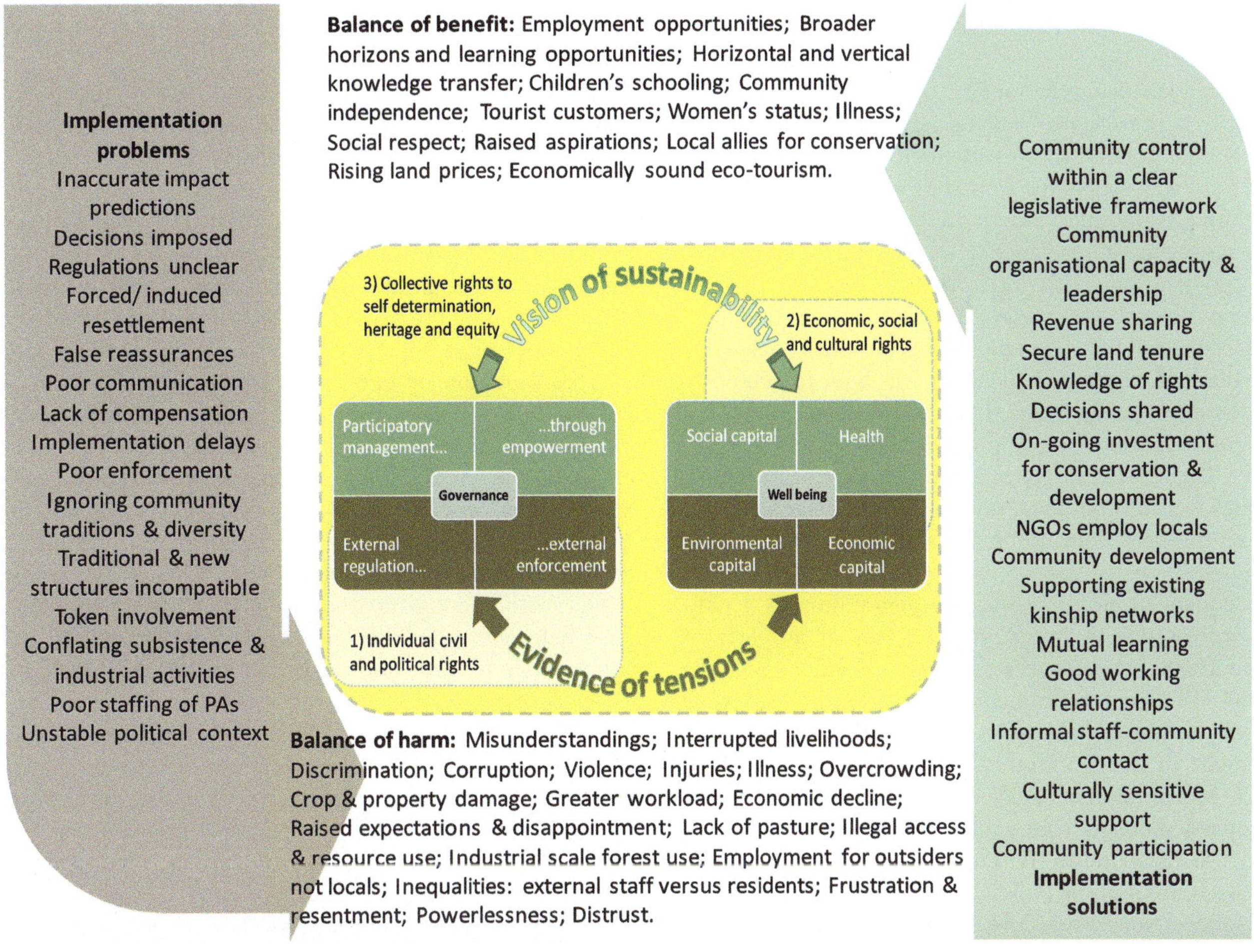

Figure 11 Summary findings from the qualitative synthesis.

managing the area prior to its official status, should play an important role in management [24].

The second source of tension is the lack of clarity in regulations and maps pertaining to protected areas. Mehring et al. [26] investigated regulatory institutions in two villages in an Indonesian park established in 1982. In one, new regulations about forest land and products drawn up by the mayor and customary organisation were neither written down, nor completely implemented. There was support for state zoning of the Park to allow traditional access to the forest for local people, but disagreement about the zone boundaries. Effective village sanctions were considered important, but confusion about when to apply them appropriately arose from discrepancies between state rules and local institutions. In Cameroon, 'traditional hunting' was still allowed in 2001 the territories outside protected areas (IUCN IV) so long as the products were for personal use, and not sold [30]. However, whether 'traditional' hunting referred to the people involved, the weapons employed, or some other characteristic was not clear. Only allowing 'traditional' weapons, depending on the definition, might outlaw common traditional practices such as the use of snares (metal wire), arrows (steel tipped) or rifles. The ban and uncertainty surrounding poorly defined traditional hunting led to tension and mistrust between locals and conservation agents. In Uganda, the legal agreement protecting the Mount Elgon National Park, established in 1951, was flawed as it failed to refer accurately to maps or related by-laws, statutes or other documents [31].

Where regulations precluded living within an area, resettlement could be forced or induced. In Cameroon, forced migration and a violent confrontation prompted villagers to accept resettlement outside familiar territories, against the recommendations of earlier research [29]. Enacting laws to drive resettlement resulted in an integrated conservation and development plan that failed, leaving villagers bitter and sceptical. Expulsion to make way for the privately managed Moremi Game Reserve in Botswana included huts being burnt down as residents got loaded into trucks for relocation outside the reserve [32]. Residents were forced to relocate (for a second time) by the suspension of the provision of all social services such as water supply, health facilities, shops, schools and communication services.

Although labelled as 'voluntary', because the term 'involuntary' is politically problematic nationally and amongst international donors, the resettlement from Limpopo National Park was widely recognised as 'induced' by planning blight and economic decline [33]. Although consulted about resettlement, residents' views were then disregarded.

> "Since the park was made we were supposed to leave. Since they said that, people don't construct houses, we don't plant trees. This house was built in 2000 but it was never really finished because the park came. There were trees but we stopped planting and the old ones died. [papaya]. No one is investing, not to do things for nothing. Even now that we have accepted to leave, the park does nothing" (p443).

Inadequate or non-existent compensation was a concern expressed in many studies; for loss of property or land in 1960s [34], for access or use restrictions in India in 2007 [35], for environmental protection by owners of Swedish forests (IUCN Ib and II) in 2001 [36], for resettlement in 2001 or for loss of crops or livestock in 2007/8 Mozambique (IUCN II) [33], for personal injury or property damage from wildlife in South Africa (IUCN II) in 2001 [37], and in Tanzania (IUCN IV) where there was no compensation policy at the time of the study in 1996 [38]; or for loss of jobs or land in China (IUCN V established 1986) [39]. For this last case, some compensation had been made in the form of new homes, crop seeds, lump sum payments, and subsidized education, electricity, and water fees, but views differed on its nature and adequacy [39]. Residents of the Sariska Tiger Reserve (India, IUCN IV) were generally discouraged by staff from claiming compensation for restrictions to access or use forest products [35]. Some villagers were never told they had a right to compensation. Another, on asking for compensation, one villager was told by a forest officer:

> "If you the villagers insist in living in the forest, then be ready to accept as well all the consequences deriving from your choice. You could live elsewhere".

Opposition to the Selous Game Reserve in Tanzania grew amongst those losing crops and livestock to wildlife without compensation [38].

Poor communication between communities and authorities was typical. Residents of the Dja wildlife reserve in Cameroon (IUCN IV, established 1951) reported being "neither informed of nor invited to participate" (p.208) in their village becoming part of a protected area but informed later (without being able to give a precise date) of the existence of a conservation initiative by the authorities [30].

Such problems are not restricted to developing countries. In Norway, several residents near a IUCN II category area found the process one-sided and undemocratic because national interests took precedence over local knowledge [24]. Petrzelka and Marquart-Pyatt [27] describe the growing anger of residents and their diminishing trust in agencies to make good decisions about the management of the land after the Grand Staircase Escalante National Monument was established in Utah, USA, in 1996 with no prior consultation or publicity. Trust diminished further over the next ten years later as residents saw roads closed and cattlemen's leases rescinded, despite prior reassurances to the contrary. Restrictions on visiting the park stoked anger amongst residents who felt 'locked out of our backyard', saying that 'law enforcement is gun-toting like we're a bunch of criminals'.

Even with management of Kenyan national reserves being delegated to a local level, and rangers and wardens claiming to initiate and maintain dialogue, residents are disappointed with the processes of communication [40]. Most of the protected area staff considered their informal word of mouth network sufficient for communicating with local communities about important management decisions.

> "Of course we cannot conserve this wildlife without the help of these communities. There must be that, a good relationship between the park and the community. So we normally go to the [homes and villages], we have meetings with them, tell them that these resources are also theirs, these are their resources." (Park ranger, p55)

In contrast, most community members thought that communication between the reserves and communities was limited or non-existent, where decisions were made without opportunities to provide input or ask questions.

In addition to the poor communication between authorities and residents, were the difficulties encountered with communication between residents. The Qwaqwa National Park (IUCN IV), South Africa had been established for the purpose of ecotourism; however, this did not suit the livelihoods of stockholding families and others would have preferred the land to be subdivided for agriculture. Amongst all this disagreement, some residents were more able than others to make their voices heard, and family conflicts escalated as housing became more crowded because erecting new homes within the park was forbidden [41]. In Lore Lindu National Park, Indonesia, the weak point for communication was between villagers and their leaders [26]. Although the village leadership was active in the negotiations about the park regulations many ordinary villagers had never heard of the agreements.

In Australia, at the Purnululu National Park (IUCN II) relationships between different indigenous groups were so acrimonious that one group withdrew from the management of the park [22]. In Slovakia, local authorities near Slovensky Raj Park (IUCN II) tried to make management of the park a focus for building relationships and developing mutual trust between different groups [42]. Communication problems could be compounded by new regulatory arrangements being incompatible with traditional ways. For instance, very few Mexican farmers applied for resource use permits because the formal biosphere (IUCN VI) rules competed with customary rules [28]. The formal rules were generic and did not take into account local variation in natural resource management. These mismatches created feelings of frustration:

> "The reserve is like a beautiful woman whom you cannot touch. It does not do you any good. The hills are rich, but a poor man stays poor" [28] p205.

Contrary to tradition, only people living on the borders of Mount Elgon National Park (IUCN II) in Uganda were given rights of access [31]. The new outsiders were required to pay the 'insiders' for access, even though half the insiders thought outsiders should have equal rights of access.

Even where access was allowed, as in Permululu National Park, Australia, costs of transportation across long distances over rough ground could be prohibitive [22].

In Indonesia Mehring et al. [26] attributed some of the difficulties of incompatibility to the government's indifference to cultural and social diversities when managing the Lore Lindu National Park (IUCN II). Indigenous people respected their own traditional informal rules that suited traditional use rights and sanctions at the village level. Elsewhere, more prosperous and ethnically diverse villagers, growing more cash crops, referred not to traditional institutions but to economic power structures, where there was a widely spread laissez-faire attitude to resource use. With forest resources and agricultural land in short supply, villagers had no alternative to using the Park to extend their land. The State's formal rules interacted with traditional informal rules, leading to confusion and conflict. Migrants struggled to implement traditional informal rules, and indigenous people failed to obey state-induced laws.

Traditional land ownership rights for indigenous populations were also contested in Australia, both between local residents and protected area management, and amongst local residents of different groups living near Purnululu National Park (IUCN II) [22].

A major challenge to developing and implementing regulations to protect areas is distinguishing subsistence activities for a sustainable environment from larger scale industrial activities. As mentioned above, this challenge was seen in Cameroon where regulations failed to distinguish clearly 'traditional' hunting methods for personal consumption from commercial hunting [30]. In Masoala National park (IUCN II), Madagascar, residents acknowledged that some members of the community benefiting from illegal lemur hunting and timber harvest [43]:

> 'people [who] wanted easy money, especially the youth, so they went into the park to cut rosewood' [p160].

However, much greater damage was done by industrial scale rosewood logging for international markets, and Park agents have limited legal powers over loggers.

> "people from all over come to this area to cut rosewood, there is no other way to get money than from valuable wood" (Park resident p.160).

> '[international] demand is driving the outside buyers of rosewood, and this is a much bigger issue than lemur hunting.' (Park Manager, p.162).

Similarly in Cameroon, the impact of residents hunting in Dja Wildlife Reserve (IUCN IV) to ensure a diet that includes animal protein is minor compared with intensive industrial logging which opened up forest tracks and thereby provided access for well organised, commercial poachers to use the tracks for transporting their game to city markets [30] p.208.

Ironically, it was a combined forestry management and community development project in Ecuador that opened the eyes of indigenous people to the potential benefits of logging; when profits were not what they had hoped for, they started making deals outside the community with industrial loggers [44].

Implementing regulations that have disadvantages for local communities is challenging enough. The relationship between residents and park officials in Masoala National park officials in Madagascar (IUCN II) was further damaged by absenteeism amongst staff who, unlike many locals, had the privilege of employment yet lacked training and clear job expectations, and had little interaction with residents [43].

Implementation solutions Blunt regulations imposed by external authorities have been widely disregarded so that protected areas have continued to be exploited on domestic and industrial scales. Studies have focused on efforts to improve communication, draw on indigenous knowledge and share decisions to combine community development with environmental conservation. They have had mixed success.

The Lore Lindu area in Indonesia was established as a UNESCO Biosphere reserve in 1977 and a national park (IUCN II) in 1993. Since then participatory approaches have been advocated for managing Biospheres [45] and protected areas more widely [46]. Initial efforts to impose external regulations failed and in the late 1990s, the park authority, NGOs and village representatives began to negotiate Community Conservation Agreements [26]. Within designated zones, village conservation councils were the bridge between the Park authority and the community for planning, implementing, evaluating and reporting the results of the Agreement. Despite the village leadership being active in the negotiations, communication between the Park authorities and the whole community was poor, so many ordinary villagers had never heard of the agreements. The Agreements covered use of forest products and land and the village conservation councils were responsible for monitoring activities. The council could employ punishments or sanctions, which were usually based on village traditional rules. Insights into this system came from NGO interviewees. A collaborative management approach aimed to minimise the gap between the park management and the people, through participation of local inhabitants and integration of local rules. Respect for the rules was greater where they were 'more practical' having been locally adapted, and allowed income-generating possibilities.

In Mount Elgon National Park, Uganda (IUCN II), as in other African countries, a similar 'fortress management' or 'fence and fine policy', based on systematic evictions, exclusions and prohibition of using natural resources, met increasing resistance [31]. Lack of success with 'fence and fine' policies prompted approaches with greater participation of local people in management and changes in regulations to legitimise sustainable use. Establishing the agreements was difficult even with guidelines and training for park staff; converting staff from law enforcers to community collaborative workers was difficult. Nevertheless, meeting locals and getting to know them improved relations. Some local people acknowledged that their initial reluctance lessened as they met staff and learnt more about the resource base; while a third of respondents did not participate at all.

Once established, the agreement provided greater clarity over rights and duties, and opportunities for long term planning about livelihood strategies. However, as a legal document, the agreement was flawed as it failed to refer accurately to maps or related by-laws, statutes or other documents [31]. Subsequently people were more positive towards the park, its resources and staff. However, as the focus was on the park rather than the community, people were sometimes organised according to what resources they collected, rather than by other socially relevant criteria such as ethnicity, kinship, location, wealth etc. Contrary to tradition, only people living on the park borders were given rights of access. The new outsiders were required to pay the 'insiders' for access, even though half the insiders thought outsiders should have equal rights of access. Conflicts arose from this situation and threatened the agreement's endurance. In such sensitive situations, staff need the socio-cultural skills to understand, interpret and interact with local people about livelihoods, conflicts and challenges in appropriate ways. Reports of misuse and corruption remained common. Nevertheless, collaborative arrangements improved relations and benefited biodiversity and livelihoods.

Sletten's findings in Mount Elgon Park (IUCN II) are supported by other studies. Elsewhere in Uganda supporting community's transition from a hunter gatherer to a settled farming community in a culturally sensitive way was more likely to result in community satisfaction and personal efficacy [44]. Training and capacity building by charities and NGOs led to an increase in skills and knowledge and new income generating activities. Two NGOs working with local people helped to organise efforts around existing kinship networks and this community reported the highest states of economic development compared to other communities. At the other end of the scale these communities were willing to sacrifice their land claims to join relatives in other areas and access charitable projects there leaving the settlements struggling to maintain a viable community.

In Masoala National Park (IUCN II), Madagascar, residents who were more familiar with Park staff viewed the staff as well as the Park more favourably than residents who were unaware of staff or who had had negative interactions with Park agents [43]. Residents were confused by the different NGOs' responsibilities and changing priorities. A park manager and a local town official both considered community development as essential for maintaining a protected area. There was local support for protecting the park by providing community benefits through alternative livelihoods. However, it is unclear from this whether the benefits essential for behaviour change were the intangible empowerment benefits of community development, or the material benefits.

In Selous Conservation Programme (IUCN category IV), Tanzania, support from communities was greatest in areas where education and mobilisation campaigns had been conducted and benefits were beginning to be derived; findings suggest that the majority of villagers supported the project. The evidence showed that they were motivated to join the conservation programme by promises of socioeconomic benefits [38].

The arrival of western donors and NGOs in Caohai Nature Reserve (IUCN V), China, in 1993 changed the focus from enforcement of resource regulations towards small-scale community development and outreach programmes [47]. These included small grants and a micro-credit programme for farmers to start up micro-enterprises in the hope that they would be less reliant on the reserve's natural resources, infrastructure development, environmental education, a community based natural resource management programme, and school fees for girls from poor families. This involved two employees who had extensive prior experience of working with farmers, and required extensive training in community development, gender issues and a variety of participatory methodologies. The result was many fewer hostile confrontations between local people and nature reserve managers, the participation of local people in conservation activities and farmers contrasting the nature reserve's concern for local people with the indifference of corruption of other government agencies. Farmers now work cooperatively with the reserve to seek resolutions to their own problems, sometimes taking the initiative to raise issues about road construction, sanitation improvements, and agro-forestry projects. The transformation from conflict to cooperation has been dependent on funds from NGOs and donors, which raises questions about the project's sustainability.

Another successful example of cooperative management was on the margins of a category II park in Mozambique where land values increased exponentially. With the support of an NGO residents thrived, benefitting materially from land titles, revenues and empowered by the process of acquiring land titles and setting boundaries [48].

Participatory approaches to governance were not always successful. Almudi and Berkes [49] investigated the relationship between a local fishing community and officials responsible for the creation and maintenance of Brazil's Peixe Lagoon National Park. They took a particular interest in the factors that could empower local fishers to 'defend their rights to remain physically within the park and politically in the conservation policy process' (p.220). The authors also found that fisher communities struggled to participate in discussions essential to securing their 'long-term access to the resources for their livelihoods or to trigger the development of a PA co-management arrangement' (p.225). The following quotes were provided as examples of the fishers experiences:

The authors summarised two of the main barriers contributing to the fishers' lack of empowerment as: weak assistance for developing community organisational capacity and leadership; and lack of basic knowledge on laws and fisher rights.

Well-being

Environmental (natural) capital

- Appreciation of protected areas other than for economic benefit was found in IUCN categories II, IV and VI; and both before and after the Durban Accord
- Communities expressed a tension between appreciating the environment and wanting to protect it, and also needing to make immediate use of land or natural products
- They could be encouraged to participate in further conservation measures where they could anticipate socio-economic benefits

  Evidence synthesised from seven studies [23,25,26,28,29,38,43]

Even where residents have recognised that conserving a park (IUCN II) and its wildlife is valuable on an individual, local, national and global scale for economic, educational, recreational, aesthetic and environmental reasons, and for future generations, they still lament the economic limitations imposed by restrictions on access, extracting resources and grazing, and the dangers of wild animals [23]. Indeed, some communities refuted the need for resettlement, having managed the land (IUCN II) for centuries; this was confirmed by the authors' historical analysis and portrayal of a pristine forest whose protection was incompatible with indigenous residents as a scientific myth [29].

A range of Mexican farming styles have developed along the spectrum of reciprocal relationships between man and nature (co-production) between wilderness and urbanisation, with farmers valuing the land for its provision of food, water, wood and other products [28]. Despite appreciating protected areas for their aesthetic, environmental products, economic opportunities and spiritual values [25,28] their positive attitudes were lost when conservation regulations (IUCN IV) competed with productive activities such as cattle ranching or growing imported varieties of fruit, or with personal safety [25].

Responses to plans for balancing conservation and economic development appear to vary depending on where the benefits might be felt. For some, it was not an interest in conserving wildlife that motivated participation in a Community-based Wildlife Management Programme (IUCN IV) but promises of socioeconomic benefits to themselves [38]. Whereas other respondents were critical of a programme for promoting economic activities alongside maintaining ecosystems; here tourism was expected to benefit entrepreneurs and rich family owners of coastal lands with tourism potential [25].

Access to land

- PAs are important to communities for grazing, agriculture, hunting, foraging and spiritual homes
- Relocation and loss of control over land and resources can result in resentment, poaching and antagonism

- Participating in the process of setting boundaries and securing land rights can be empowering
Evidence synthesised from nine studies [30,33,34,40,41,44,48,50,51] from IUCN categories II after the Durban Accord, and from categories V and IV before the Durban Accord

Access restrictions to protected areas (IUCN II) had implications for grazing cattle, hunting and collecting natural products [33,40,51]. Local people would like grazing rights in the park (IUCN II), especially during drought, and the opportunity to visit traditional areas and burial areas [51]. Where staff were mostly concerned about illegal grazing; they would commonly impound livestock, and fine owners or refer them to a local judicial ward; in contrast most community members felt the harsh environment justified grazing cattle in protected areas [40]. Communities considered that threats to their environmental capital outweighed any potential economic benefits of living near the national park [50]. Outside the Park residents felt they would not have access to resources so far essential to their livelihoods: access to agricultural land, forest resources and grazing land [33]. Wealthy cattle owners were even less disposed to moving because cattle would need to compete for food and water with host villages, and cattle theft was more common outside of the park [33].

Access restrictions posed similar problems for communities in or near PAs categorised as IUCN IV. Accessing the forest was important in order to obtain resources to support local people's livelihood and for feelings about the forest as 'theirs' [30]. Authors described relocation and loss of control over land and resources resulting in resentment, poaching and antagonism [34] and overcrowding resulted from restrictions on building new homes [41].

More positive views were expressed where land values had increased exponentially on a park's margins (IUCN II). Here residents thrived, benefitting materially from land titles and revenues and felt empowered by the process of acquiring land titles and setting boundaries [48].

Resource use

- Residents appreciated protected areas for their rich products
- Resource use was common even where illegal
Evidence synthesised from seven studies [28,30,35,39,43,44,52] mostly conducted before the Durban Accord and spanning IUCN categories II, V and VI.

Residents viewed protected areas as rich sources of food and other products [28,30,39,43,44,52]. They acknowledged that illegal resource use continued despite bans [39,43,52]. Some park residents could not conceive of a balanced diet without animal protein and protected areas were still seen as the ideal place to carry out hunting throughout the year [30]. Where dependence on access to products was high, collection continued despite a ban as compensation was not always considered adequate [39].

Once introduced to community based management to harvest, process and sell timber, Ecuadorian communities who were disappointed by high start up costs and slow, small gains made deals with external commercial loggers to raise their profits [44]. In contrast, in Mexico, where conservation is widely valued, natural products were used in a sustainable way [28].

Economic capital

- Before the Durban Accord (IUCN Ib and II), concerns focused on: reduced employment and tax revenues; reliance on foreign aid without understanding its link with conservation; and unrealistic expectations of the economic benefits of tourism.
- After the Durban Accord (IUCN II), concerns focused on: the meager benefits of tourism; what benefits there were not being shared equitably; with indigenous groups or those less amenable to falling in line with new regulations missing out.
- Concerns about lack of compensation were expressed before and after the Durban Accord about IUCN categories II and IV, and about IUCN V before the Durban Accord. Conversely, in developed countries there was a growing welfare dependency.
Evidence of economic capital found in thirteen studies [23,29,34,36-41,53-56],

Before the Durban Accord, the impact of IUCN II parks on the wealth of whole areas was seen in Scandinavia, South America and Asia. Forest workers in Sweden associated environmental protection with lower levels of employment and production in commercial forestry [36]. Politicians anticipated conservation policies leading to lower tax revenue and greater emigration [55]. In Nepal communities were developing a dependence on foreign aid, with some people considering it as an expected source of income and not necessarily associating its benefits accrued with conservation efforts [23].

Some people living in or near national parks (IUCN II) were concerned about neighbours having unrealistic expectations of the economic benefits derived from tourism and eco-lodges [55]. Others were unaware that community development was one of a park's primary objectives only initially [43].

Benefits of tourism to national parks (IUCN II) were seen as meagre, and distribution of revenues from protected areas was considered inequitable or of little benefit to indigenous communities [40]. Park and eco-lodge staff tend to be wealthier than community residents [43,55]; and local people felt overlooked for employment, with opportunities favouring neighbouring communities and those deemed amenable to new regulations [29]. Similar concerns about few or unevenly shared benefits and opportunities were expressed in other, IUCN IV, protected areas [34,41].

Communities across IUCN categories, before and after the Durban Accord, often considered as inadequate the monetary or in kind compensation available for: forestry constraint; resettlement; loss of land, crops, livestock or jobs; or personal injury or property damage [33-39].

Environmental protection is associated with economic decline in high income countries. In Utah, USA, local residents saw more tourism but no economic growth as a result of establishing a protected area [27]. In Slovakia, although residents living near Slovensky Raj National park (IUCN II) anticipated multifunctional forest management as a source of employment and income generation, in practice the socioeconomic situation worsened, particularly for minority ethnic communities, with reduced employment and changes in welfare support [42]. With their lack of experience and opportunities for involvement in small businesses and local or regional planning, minority ethnic groups saw no viable economic options.

The most positive findings about social and material benefits came from an NGO funded study with one academic author and one author employed by an ecolodge, although they too shared concerns expressed above [55].

Social capital

- Changes in livelihood strategies have influenced the shape of households and the strength of social ties, and introduced new inequalities within communities.
- PAs have been established in areas inhabited by various ethnic groups. The pressures resulting from regulatory and economic changes have introduced tensions or exacerbated historical tensions between them.
  Evidence about social capital in PAs was found in eight studies [30,35,38,39,41,44,55]

Slater [41] noted that households configured themselves in a way so as to maximize livelihood diversification; sometimes to the detriment of familial relationships. Households could be separated geographically by livelihoods, or overcrowded because sharing dwellings allowed younger adults to rely on the support of older adults claiming pensions.

Residents saw a direct link between livelihood diversification and changes in cultural traditions and traditional relationships amongst local people; both positive and negative [44,55]. Making the transition to a settled agrarian life for some meant a decrease in the traditional communal work ethic and less frequent use of their native language [44,55]. Stronza and Gordillo [55] saw changes in social ties within communities when some began working in the tourism. Locals who secured employment in eco-lodges (IUCN II) talked about their work limiting participation in gatherings traditionally employed to complete community tasks, and how communities tended to charge eco-lodge employees more because they were richer – it was buying them out of their community responsibilities. Working in eco-lodges opened villagers' eyes to new opportunities and a wider social circle but this was at the cost of leaving their family and community. More direct commitment to conservation in a national park (IUCN II) also caused family pressures where the Village Scout scheme took young men away from their family and farming responsibilities [38]. Elsewhere (IUCN V), investment in communities was mentioned by park staff and by residents; this included the aim of establishing a special university training program to prepare students to fill key park management positions [39].

Villagers emphasised the importance of social relations as part of their survival strategy and expressed apprehension about the weakening and possible disappearance of these linkages as a result of displacement. These relations were particular important in times of struggle such as periods of drought [35 p.60]. Stronza and Gordillo [55] considered that communities rich in social capital may not only be better able to manage changes associated with ecotourism, but that such changes in social capital can collectively sustain local institutions, which may subsequently be critical of conservation efforts. In contrast, the community managed forests in Ecuador created new business relationships and improved the community's social standing with other indigenous groups in the area [44].

Although ethnic identities could strengthen social capital within groups, they more often created tensions between groups. Some ethnic groups were resentful as they perceived others receiving community development preferentially, even though the authors saw these community development efforts as culturally inappropriate [44]. As noted above, migrants struggled to implement traditional informal rules, and indigenous people failed to obey state-induced laws. Some of these difficulties were attributed to the government's indifference to cultural and social diversities when managing the Park [26]. At the level of implementing regulations, residents objected to the leniency of guards towards those who were wealthier or ethnically related [23].

Health

- New diseases associated with changes in lifestyles were attributed to forest evictions or changing from a nomadic to sedentary existence.

- Accidents and injuries resulted from conflicts between guards and residents; animal conflicts resulted in increased workload and exhaustion as well as injuries
- Sexual aggression was more common when women were less protected following necessary changes in working patterns or kinship ties.

  Evidence of health and protected areas was found in five studies [35,38,41,44,54]

In one study, community members evicted from forest unanimously spoke of their exposure to new disease when integration with other groups began; authors confirmed that the community, particularly children, were seriously affected by malaria which did not exist in the forest, and that HIV/AIDS is also appearing [44]. Their forced transition to an agrarian society has cut them off from their access to and knowledge of traditional medicinal plants they previously used to stay healthy. Elsewhere, villagers reported the lack of access to basic health services and Torri [35] confirmed that child mortality was high in isolated forest villages, where common illnesses, easily treatable given basic medical facilities, could lead to death.

Residents described how crop raiding by elephants lead to food shortages and greater workloads, especially amongst women who suffered more from insect-borne diseases and heat exhaustion. When elephants had damaged water pipes, women risked drowning when collecting water from unsafe sources [54]. First Peoples [44] reported women's safety being compromised as men worked further afield, and as women were drawn out of their homes for new roles and that the erosion of kinship ties maybe reducing protection from male sexual aggression [44].

Inequalities

- PAs and residents' responses to them have exacerbated existing local ethnic tensions
- Participation in PA governance has favoured people already advantaged by their socio-economic position
- PAs impact unequally on people depending their socio-economic position, such as the size of their business, their legal land tenure or their gender
- New tourism enterprises have tended to employ outsiders rather than locals

  Evidence of inequalities arising from PAs was found in 15 studies [23,26,31-33,35,36,38,41,44,50,51,54-56]

People living in and near PAs differ in many ways. Some indigenous people living near each other come from different ethnic groups. Some PAs include indigenous communities, new immigrant communities and residents long established following historical migrations. Individuals differ in terms of their occupations (e.g. subsistence farmers, day labourers, tourism employees), tenure (landowner or not), gender, education and wealth.

Differences in wealth accrued from ownership of land or livestock have been influenced by wildlife conflict [38,54] and by PA regulations about land access or livestock numbers, including whether or not people complied [41]. Smaller businesses and entrepreneurs less able to control shocks may bear the brunt of further restrictions on the forest products [36].

Some of these individual differences have arisen at least in part from inequalities between ethnic groups or from discrimination between indigenous groups. For instance, some indigenous people have discouraged immigration and excluded immigrants from community governance procedures [26]. Within and between ethnic groups, people have taken advantage of others who earn more from the presence of the PA [55] or from people struggling to make a living who sell land legitimately owned only to claim protected land illegally [26].

Misunderstandings or prejudices about the histories or abilities of communities led authorities to pursue decisions which those communities found damaging [32,33]. Authorities also discriminated within communities by exercising policies that protected landowners but not other longstanding residents [41]; or by involving people chosen for their age, wealth, education and position rather than practical relevant knowledge [31]. Ethnic discrimination has been introduced by PA legislation that forbad resource use by indigenous people but allowed resource use by predominantly white landowners [51]. Where ethnic discrimination predates establishment of PAs, historic practices of favouritism have strengthened as land has become more valuable and rare because much of it has been put aside for conservation [50]. Prejudice and nepotism have been the source of favouritism or corruption amongst PA employees [23].

Some differences have arisen from commercial or conservation enterprises providing more earning power for some, but not all, roles [56]. Some eco-tourism enterprises have a history of being poor employers of local casual labour [32]. Financial inequalities were introduced when spent funds resulted in financial support for some but not others, and bank loans have been refused on the grounds of having an address in a protected area [44].

Women were vulnerable to attack and injury from men or wildlife [38,44,54]. Female heads of households were vulnerable financially as they tried to balance paid labour with subsistence farming [41]. More may have been learnt if some women had not been reluctant to express their opinions to researchers [35].

Narrative synthesis of quantitative evidence

In this section we provide a narrative overview of all included studies providing quantitative data on impacts.

The studies are divided into six subsections; produced iteratively based on the outcomes reported in the accepted literature (see typology in Figure 12). We do not claim that these divisions are definitive or optimal but they do provide a pragmatic breakdown of a complex body of evidence. A summary of data presented in the included studies is provided as Additional file 7. A more detailed set of data extraction tables is available as supplementary material. Of the 79 studies included in this synthesis 63 were categorized as having 'high', 11 'medium' and 3 'low' susceptibility to bias. Additional file 10: Table S1 displays the 14 studies and 33 outcome measures that were categorised as having low and medium susceptibility to bias. Additional file 11 provides detail of critical appraisal and the basis for assigning susceptibility to bias for all 79 studies. In the following text we concentrate on results reported in these 14 studies. Here where we discuss studies' 'susceptibility to bias' we refer to their categorisation resulting from our critical appraisal. In some instances we also identify specific types of bias (defined in Table 5, below) and in others we refer to shortcomings of the study design and implementation that implicitly increase the studies' susceptibility to bias.

It is important to note that since the majority of studies were identified as being highly susceptible to bias, the results of these studies are not considered further, either individually or in general. These studies are unreliable both alone and in concert. Because this group of potentially biased studies is unreliable, we cannot summarise their findings any more than we can discuss individual results. However, for completeness we also consider the diversity of outcomes reported by all studies. Furthermore, we avoid vote-counting, where the sum of all negative, positive and neutral study results are calculated. Vote-counting is unreliable because it assumes that a significant finding is evidence that an effect is present and a non-significant finding is evidence that an effect is absent. This former statement is true, but the latter is not (see Borenstein et al. [21] for further details of vote-counting).

Ecosystem goods

- 17 studies identified
- Only two studies not highly susceptible to bias ('medium' susceptibility to bias)
- Nyahongo et al. [57] found that meat and fish consumption increased with proximity to Serengeti National Park
- Sarker and Røskaft [58] found residents' perception of timber and fuelwood benefits from four PAs in Bangladesh decreased with distance from the PA boundary
- High susceptibility to bias in remaining studies related to low methodological detail, confounding variables, and weak experimental design

Of the 17 studies reporting results on ecosystem goods, only two studies had designs that were not highly susceptible to bias (i.e. medium). Nyahongo et al. [57] reported the number of meat and fish meals consumed by survey respondents to be significantly negatively

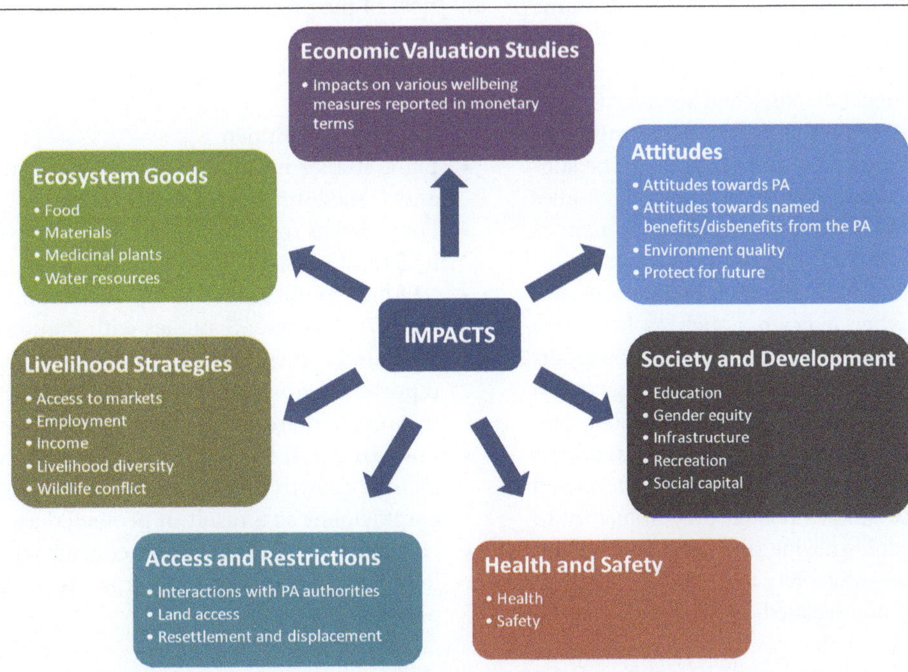

Figure 12 Typology for structuring the quantitative synthesis of impacts of protected areas.

Table 5 Definitions of bias mentioned in the narrative synthesis

Recall bias	Imperfect recollection of past events by respondents. Generally worsened by longer periods of recall
Social desirability bias	Tendency to respond to questioning in such as way as to be viewed favourably by others
Questioning bias	Questioner leading respondents to reply to questioning in a certain direction
Neyman bias	Arises from a time lag between exposure and sampling such that undetected drop-out of participants may occur before the study begins.
Attrition bias	A skew in results where participants are lost between measurements at two time points (potentially as a result of the exposure) during the study
Optimism bias	A belief by a respondent that they are less likely to experience a negative event relative to other respondents, or over-optimism on the part of analysts or interviewers, about the effects of a project.
Hypothetical bias	Failure of respondents to consider the true budget constraints in responding to financial questioning
Strategic bias	Tendency for respondents to alter their answers in an attempt to influence an event

correlated with distance from Serengeti National Park boundary. The authors' statistics account for a range of other potentially influential variables, contributing to the studies favourable susceptibility to bias rating (medium). The study's results indicate that meat consumption increased with proximity to the park at a rate of 0.218 (±0.052) meals per week per km, whilst fish consumption increased at a rate of 0.931 (±0.205) meals per week per km (assuming units in analyses are identical to units described in the methodology). Sarker and Røskaft [58] found that people inhabiting the areas surrounding four protected areas in Bangladesh identified more benefits from the protected area in the form of timber and fuelwood extraction closer to the park boundaries than further away, with an associated odds ratio of 1,000; i.e. "The odds that respondents living closer to the protected area boundary reported a higher level of benefit from timber and firewood because of the conservation programme were 1,000 times greater than those for respondents living further away".

Thirteen studies used questionnaires and semi-structured interviews for data collection, resulting in a higher susceptibility to bias since reporting by the respondents can be subject to recall or social desirability bias. Fourteen studies reported 39 different (but not all independent) outcomes related to food and materials, comprising a mixture of foods obtained by hunting and gathering, agriculture and purchased, as well as other indicators such as land area under cultivation or grazing, availability of fodder, support for agricultural development, dietary diversity, gathered fuelwood and other plant products including timber, and also change in ownership of goods. Five studies reported seven different (but not all independent) outcomes related to water resources: three studies related predominantly to water quality (e.g. households relying on least safe water resources) and two to water availability and supply. Only one study reported one outcome related to medicinal plants/animals.

Studies reporting data on common themes differed significantly in the precise outcomes measured. For example, whilst Bajracharya et al. [59], Mehta and Kellert [60], Naughton-Treves et al. [61], Rinzin et al. [62], and Okello et al. [63] all measured accessibility and quality of water resources, there was insufficient conformity to allow meaningful synthesis.

Other studies could not be used to infer reliably any effect of a protected area for a range of reasons, including; (i) they recorded change over time with no spatial comparison and no adequate basis for determining whether the changes observed were attributable to the effect of the protected area; (ii) they made a spatial comparison, but location relative to the protected area is clearly confounded with a range of other important site variables; (iii) the scale of the spatial comparison was too small to use it to infer effects of the protected area on the specific outcomes recorded in the study; (iv) time elapsed between the creation of the protected area and the study was too short to infer any effect of the protected area.

Livelihood strategies

- 43 studies identified
- Eight studies not highly susceptible to bias (1 'low' and 7 'medium' susceptibility to bias)
- Four studies report poverty-related outcomes, finding beneficial impacts of land protection in all cases
- Wildlife conflict is relatively well-studied (18 studies and 47 outcomes; 5 studies with 7 outcomes 'medium' susceptibility to bias. The majority reported significantly more problems with proximity to protected areas than further afield
- Lundgren [64] found no significant difference in income growth or forestry/tourism sector employment as a result of protected areas in Sweden
- Household income [65] and consumption [66] were found to increase with proximity to/within protected areas

Studies in this set report on access to markets, employment, income, livelihood diversity, human development measures and wildlife conflict (Additional file 7). Wildlife conflict was the most frequently reported outcome measure in this group, contributing 18 of the 43 studies and 47 of the 101 outcomes. Only five of these 18 studies were not highly susceptible to bias, all bar one [52] showing significantly greater crop and livestock loss closer to protected areas. Two of these five studies employed questionnaires to elicit perceived disbenefits, whilst three studies used observations of conflict. Lundgren [64] found no significant correlation between protected areas and income growth or employment in either tourism or forestry sectors in Sweden. Sims [66] reports higher consumption in regions with a high 'share' of PAs compared to a low 'share' in Thailand. Richardson et al. [65] found households within game management areas (GMAs) to have greater income across a number of Zambian PAs than those outside GMAs.

Four studies reported poverty-related measures (human development measure outcomes) as either poverty index or poverty headcount. All four studies found significant beneficial impacts of protected areas on poverty alleviation in Costa Rica [67], Thailand [66,67] and Bolivia [68]. The study by Andam et al. [67] involved subsequent in-depth reanalysis in two later publications [8,69]. These later analyses showed that along with protection alleviating poverty, poorer areas (measured at baseline) were found to have greatest level of poverty reduction than those that were less poor [8]; and that poverty alleviation was also associated with characteristics that reduced the efficacy of deforestation prevention (i.e. where protection had been assigned to land that was unsuitable for agriculture, near major cities and infrastructure, and where agricultural employment is low) [69].

The remaining studies were judged to have high susceptibility to bias and the following examples are only illustrative of the diversity. Foerster et al. [70] report that purchasing power is lower in villages closer to PAs in Gabon. Cardozo [71] conducted a questionnaire-based site comparison of communities inside and outside Allpahuayo-Mishara National Reserve, Peru and reported changes in income and livelihood diversity. Annual income from agriculture was lower inside the PA whereas income from domestic animals and palm products was higher. Kayser et al. [72] report greater annual transfer of money to SMEs through contracts financed by Addo Elephant National Park, South Africa. Saayman & Saayman [73] provide data on self-reported change in the business environment around South African PAs. This provides weak evidence that opportunities for and turnover of business has increased as a result of the PA.

Land access and restrictions

- Six studies identified
- All studies are highly susceptible to bias due to lack of methodological detail, non-random sample selection, spillover, questioning bias and uncontrolled confounding variables
- The majority of outcomes relate to restrictions on access or extraction
- Two outcomes related to the perception of relationships with park employees

The six studies in this set report on 11 different (but not all independent) outcomes related to land and resource access, interactions with protected area authorities, displacement and resettlement (Additional file 7). All had a high risk of bias in their study design because of a lack of detailed description of their methodologies. In cases where the description was detailed, specific risks of bias were identified; replicates were non-randomly selected [59,74] and spillover, questioning bias, and confounding variables were not accounted for [59,75,76].

Health and safety

- Nine studies identified
- Only one study identified as not highly susceptible to bias ('medium' susceptibility to bias)
- Korhonen et al. [77] found highly variable infant mortality rates in and around Ramonafana National Park in Madagascar, with slightly higher levels outside than inside the PA, although the pattern is not clear

Studies in this section look at health in the population and access to health services (Additional file 7). Studies reported a limited range of outcomes relating to health and safety, but the reliability of findings in general was compromised due to various aspects of study design which make them highly susceptible to bias, such as a lack of comparator or non-random selection of study sites or participants, or incomplete reporting, with no details about selection of study populations, validity of survey instruments or survey response rate, [73,78]. The only study in this group with 'medium' susceptibility to bias, Korhonen et al. [77], report a case-study of reproductive health from a 10-year old Integrated Conservation and Development Project (ICDP). The study was of BACI design comparing purposively selected (for representativeness) villages and municipalities within a 3 km belt around Ranomafana National Park (the 'peripheral zone') (7 villages, 6 municipalities) with those outside this zone (6 villages, 4 municipalities). Data on modern contraceptives suggest an increase in use over time, a decrease with increasing distance from the park and variability in use within 'park' villages attributed by authors to varying access to ICDP activity, but also to educational status of women, local culture and religious beliefs. Fertility levels were estimates rather than direct measures and do not indicate any change over time or differences between park and 'outside' villages. Visits to the health centre for pre-natal care increased over time in line with national trends and did not differ between park and 'outside' village residents. The data presented for infant mortality were drawn from census data pre-1999 and showed year-on-year variation and no clear differences between 'park' and 'outside' residents, although a slightly higher mean over an 8 year period outside than inside the PA. Post-1999 data were health centre data and only related to health centre births and thus may be subject to bias if a higher percentage of more problematic births occurred at the health centre as opposed to in villages.

Society and development

- 13 studies identified
- Two are not highly susceptible to bias (1 'low' and 1 'medium' susceptibility to bias)
- Sheppard et al. [79] found a greater number of infrastructural developments inside Wechiau Community Hippo Sanctuary in Ghana than outside
- Korhonen et al. [77] found no difference in the percentage of girls in primary schools either over time or inside relative to outside Ramonafana National Park in Madagascar, but a possible slight increase in the percentage of girls in the third grade over time was found inside the PA

Studies in this set include measures of education, empowerment, infrastructure, recreation and social capital. Thirteen studies reported data on 54 development-related outcome measures. Two studies were classed as having low (1) or medium (1) susceptibility to bias. Sheppard et al. [79] found a greater number of developments in various aspects of infrastructure inside Wechiau Community Hippo Sanctuary in Ghana than outside. Korhonen et al. [77] found no difference in the percentage of girls in primary schools over time and no difference inside relative to outside Ramonafana National Park in Madagascar, but they did find a slight increase in the percentage of girls in the third grade over time inside the PA (but not outside).

All remaining studies were judged to be of 'high' susceptibility to bias. There is weak evidence for improvement in education provision following PA establishment in terms of increased number of schools and perceptions of improvement [62,72,78]. A questionnaire based on reported change conducted by Saayman and Saayman [73] in multiple South African PAs produced variable responses (in most, but not all, the majority agreed) to statements such as 'participation in community activities has increased', 'the pride that the residents have in their town has improved', 'the opportunities to meet new people has increased'. Other studies report perceived improvements in infrastructure in and around PAs compared to elsewhere or before establishment [59,78] but this is not always the case [62].

Attitudes towards PA and the benefits (or otherwise) they provide

- 24 studies identified
- Only one study identified as 'medium' susceptibility to bias, all remaining studies are highly susceptible to bias due to lack of methodological detail, confounding variables unaccounted for, and spillover from protected areas into controls
- Sarker & Røskaft [58] found attitude to PAs to be negatively associated with PA proximity
- Four studies failed to identify the location of 'inside' populations, two studies failed to report the distance of controls from PAs, and in general controls were very close to PA boundaries (e.g. < 1 km and 2.5 km)
- Studies reported attitudes towards PAs, attitudes towards identified benefits/disbenefits from the PAs, and perception of environmental change

Studies reporting attitudes were separated into two distinct categories; attitudes towards the protected area or conservation, and attitudes concerning benefits or disbenefits resulting from the protected area. All but one of the 24 studies reporting attitudes fell into the category of 'high susceptibility to bias'. The main issues identified in the critical appraisal were lack of detail in the methods, not accounting for important confounding variables, and, where spatial comparators were used, potential spill-over effects due to the intervention and 'comparator' sites being close together. This latter issue was a particular problem for studies looking at the effect of 'distance from the PA boundary' on attitudes. Thirteen studies examined spatial differences in attitude with respect to distance from the protected area. However four studies failed to report the location of the intervention population inside the protected area. Two studies [58,80] regressed attitude scores against distance from the protected area, but the distances involved were not stated. Control populations in 'inside-outside' studies were generally very close to the protected area boundary, for example 2.5 km (Sekhar 1998) and < 1 km [71]; although it was 50 km in the study of Bonaiuto et al. [81].

Attitudes towards the protected area Fourteen articles reported a wide range of general attitudes towards the PA. Thirteen of these gave respondents' statements (usually a mix of positive and negative statements) and presented data on the percentage agreeing or disagreeing with each statement. One study [81] presented composite attitudinal scores made up of responses to a series of questions which were not presented in the article.

In the only study that was judged not to be highly susceptible to bias, Sarker and Røskaft [58] found that respondents from around four parks in Bangladesh had negative attitudes towards the protected areas, and that negative attitudes decreased with distance from each protected area. The remaining studies were of 'high' susceptibility to bias. Bonaiuto et al. [81] reported that regional identity and place attachment were higher inside the Tuscan Archipelago National Park in Italy, but that specific and general attitude scores towards the protected area were lower relative to a control group of respondents 50 km away. Jim and Wu [82] noted that a higher proportion of people living on the boundary of Shimentai Nature Reserve in China 'disliked' the park than those living 4 km from its boundary. Finally, Shrestha and Alavalapati [80] observed a positive correlation between positive attitude and distance from Koshi Tappu Wildlife Reserve in Nepal.

In contrast to the above studies which found a positive relationship between distance from the park and attitudes, Cardozo [71], Gubbi et al. [83], Infield and Namara [84] and Sekhar [85] reported higher positive and lower negative attitudes inside protected areas than outside. Other studies found no statistically significant or observable difference in attitudes between inside and outside protected area [86-89] or over time [90].

Attitudes towards named benefits/disbenefits resulting from the park Thirteen studies reported data concerning respondents' attitudes towards named benefits or disbenefits resulting from the protected area. Again, all studies were classed as highly susceptible to bias. Ite [89] found fewer respondents close to Cross River National Park in Nigeria to believe that they have benefited from the protected area than those 5 km away, although a third group of respondents 7.5 km from the protected area showed an intermediate perception of benefits. Jim and Wu [82] reported no significant difference in the perception of benefits from Shimentai Nature Reserve, China, between respondents inside and those 4 km from the protected area, whilst significantly more respondents inside than outside claimed to have felt losses as a result of the reserve.

Other studies found no evidence that respondents felt that either negative or positive impacts resulted from the protected area [75,86,88,91,92].

A smaller category of studies reported respondents' perceptions of environmental change as a result of the protected area; with respondents in one study predominantly not perceiving a change [90] and two studies reporting the majority to have perceived an increase in environmental 'appearance' [73,93]. However, these studies were of low quality predominantly due to a lack of detailed methodology and shortcomings of the experimental design. For example, Cihar and Stankova [90] lacked a true before-and-after comparison and generated 'before data' by asking informants to recall the past. This is clearly open to recall bias.

Economic valuation studies

- 10 studies identified
- One study judged as not highly susceptible to bias ('medium')
- Four groups of studies found: cost-benefit analyses; stated preference studies; stated preference combined with a distance comparator; and reported direct financial losses from a PA (e.g. fines/foregone income)
- These studies do not have real comparators (with the exception of the distance comparator): instead they are hypothetical, and as a result are highly susceptible to bias (e.g. optimism bias)
- Studies are too heterogeneous and open to bias to permit meaningful quantitative synthesis of valuations

Studies in this section reported welfare impacts in monetary terms. Economists usually hold that individual well-being is not directly and cardinally measurable, nor comparable between individuals or time periods e.g. [94]. However, changes in an individual's well-being as a result of a PA's existence can be expressed in terms of the amount of money needed to render that individual indifferent to the existence of the PA (the aggregation of such monetary amounts across individuals is common in applied economics, but deeply problematic).

Ten studies were included that estimated well-being impacts of protected areas in monetary terms. Nine of these were categorized as having high susceptibility to bias, whilst one was categorised as medium susceptibility to bias.

Shrestha et al. [95] used a contingent valuation survey with a stratified random sample of 160 households within c. 6 km of Koshi Tappu Wildlife Reserve, Nepal, to estimate their willingness to accept the PA in terms of foregone resources. They found substantial local one-time costs of 11,776.70 Nepali Rupees per household (1994/1995).

A major reason for excluding economic studies (see Additional file 6) was that they measured the well-being impacts of ecosystems within PAs, but did not isolate the impact of the PA itself. One study [96] carried out a contingent valuation survey of willingness to accept compensation for costs of the protected area's presence on traditional pasture land, with respondents within the protected area and further away. The remaining studies included only hypothetical or 'modelled' comparators. In all cases this is done more or less explicitly by the analysts themselves, but many studies also required respondents to mentally construct hypothetical comparators, in order to answer stated preference surveys. Such constructed comparators can be useful and indeed essential when "real" comparators (RCTs, BACI etc.) are unavailable (they may also be used in conjunction with such research designs). However, they are vulnerable to a number of potential biases, such as optimism bias, strategic bias, and hypothetical bias (see Table 5 for definitions).

Without real comparators, direct evidence is lacking on the effects of the PA on individuals. Instead these must be predicted, using whatever information and opinion is available to the analyst or respondent. Numerous assumptions must necessarily be made, and will not always be explicitly stated. This leaves these studies open to well-recognised biases. For example, cost-benefit analyses are known to suffer from optimism bias, especially when conducted by groups with an interest in the project: in the case of PAs this may be the government or conservationists more generally. For example, the results of Kremen et al. [97] are heavily dependent on optimistic assumptions made about the efficacy of development interventions planned to accompany the PA: no evidence is presented on whether these interventions indeed had the effects assumed by the authors, since the analysis was conducted ex ante.

Stated preference studies are known to suffer from both hypothetical bias and strategic bias on the part of respondents. Hypothetical bias may lead respondents to overstate their willingness to pay (WTP) for goods or services provided by a PA, because they fail to consider their true budget constraint. Respondents may also behave strategically: beneficiaries may overstate their willingness to pay for a PA, in order to increase the likelihood of its establishment if they suspect they will not be required to contribute to it, or understate their WTP if they suspect that this will result in lower user fees. Those who expect to lose from PA establishment may overstate the amount that they would require to receive in compensation for the establishment of the PA (i.e. their willingness to accept (WTA) the PA) to reduce the likelihood of its establishment, or increase compensation payments. Alternatively surveys may under-estimate opportunity costs if the activities concerned are considered sensitive or of dubious legality, and are likely to be under-reported. The results of stated preference studies are also known to be sensitive to the information provided by surveyors, and the precise formulation of the questions. This renders them vulnerable to the same optimism bias noted above.

Methodologically, the included studies fell into three groups: cost-benefit analyses, stated preference studies, and reported direct financial losses from a PA. One study, Kremen et al. [97] carried out an ex ante cost-benefit analysis of the establishment of Masoala NP, Madagascar, disaggregating costs and benefits by local, national and international groups. Both the effects of the PA and the counter-factual (no PA) were modelled, though little detail is presented and the evidence upon which the modelling is based is often rather weak. They estimated that local populations would lose from the establishment of the park, but that this would become a net gain if development projects associated with the park succeeded in raising local incomes. At the national level, there would be a net loss due to the protected area.

Seven studies used stated preference techniques to elicit estimates of welfare gains or losses. Four studies [7,98-100] used contingent valuation to estimate regional or national populations' willingness to pay for existing protected areas in India, Brazil, China and Greece respectively, all four studies indicating generally positive welfare impacts of the PAs on these broad populations. A fifth study, Ascuito et al. [101], similarly estimated local willingness to pay for a fire prevention programme in an existing protected area again finding positive welfare impacts. Two studies [95,96] used contingent valuation to estimate local populations' willingness to accept restrictions on livelihoods imposed by existing protected areas in Ethiopia and Nepal respectively, indicating negative welfare impacts of the PA. In the case of Jemal [96], surveys were carried out with respondents inside and further away from the PA, and found that people within the PA were less willing to accept compensation than those further away, suggesting negative welfare impacts increased with proximity to the PA. Finally, Abbot and Mace [102] present data on fines levied on local people for illegally harvesting fuel-wood in Lake Malawi NP. These fines were levied by the PA but no information is provided on areas outside the PA.

Inequalities

Assessing the impact of PAs on health or social inequalities would require either individual sound studies with justifiable subgroup analyses, or a set of comparable studies which describe in detail the socioeconomic position of the populations studied [103,104]. Neither was available from the extant literature.

Meta-synthesis of qualitative and quantitative evidence

In attempting to bring together the findings of the qualitative and quantitative reviews it is important to reflect on the differences in their philosophies. The qualitative synthesis is essentially formative and attempts to form a picture of how PAs are perceived to impact on human well-being. As such it can form a template for empirical investigation and hypothesis testing. The synthesis of quantitative evidence is more summative and attempts to test hypotheses of impact. In consequence we should not expect the meta-synthesis to be a simple matching of similar studies or outcomes. In this section we summarise the findings of the qualitative synthesis and ask whether the quantitative evidence of impacts can inform the questions raised by these findings or whether it suggests something different.

Governance

The qualitative synthesis reveals a number of factors that can lead to negative views and impacts of PA establishment: lack of clarity in regulations and boundaries; discrepancies between state rules and local institutions; forced migration, inadequate or non-existent compensation; and poor communication between communities and authorities and government indifference to cultural and social diversities. Negative views on impact of management can arise from poor relationships between residents and park officials. Views on how to lessen negative impacts or achieve positive impacts include: rules that are locally adapted or based on traditional rules; greater clarity over rights and duties; planning focussed on community livelihoods as well as the park; appropriate capacity building; and empowerment through the process of acquiring land titles and setting boundaries. The existence of these views enables hypotheses to be generated on how to achieve change in impact. The synthesis of quantitative measures of impact shows that these hypotheses are yet to be tested. What is absent from the evidence base is a quantitative comparison of costs and benefits to local people of different forms of PA governance.

Well-being

Environmental (natural) capital The qualitative synthesis presents a range of positive and negative attitudes among local populations towards PAs. Alongside an appreciation and desire to protect the environment were concerns about reliance on those same areas to maintain economic livelihoods. Although the qualitative literature provides evidence of difference views, quantitative evidence to estimate the scale and reach of those views was not extractable since all but one study were highly susceptible to bias,

Access to land The qualitative synthesis revealed two very different scenarios in terms of access to land. The first is resentment at loss of access and the second is benefit from acquisition and value of land on the PA margin. All quantitative studies of impact of PAs on land access and restrictions were highly susceptible to bias and so the current evidence does not allow the magnitude of these scenarios to be assessed.

Resource use A range of positive and negative views were found concerning PAs as a source of natural resources and ecosystem goods. In the synthesis of quantitative evidence of impact only two studies were not highly susceptible to bias, showing that meat and fish consumption was greater in proximity to a PA and that timber and fuelwood benefits were more frequently appreciated nearer another PA.

Economic capital Views expressed on impacts of PAs on economic capital are generally negative, with the exception of some views on the benefits of ecotourism. In contrast the quantitative evidence of impact from three studies on livelihood strategies was neutral to positive in terms of poverty reduction. In particular, there were concerns in Sweden amongst foresters about sustaining employment and amongst politicians about sustaining tax revenue in the presence of regulations. However, these concerns were not upheld by a quantitative assessment of impact in the same country. All but one of the economic valuation studies suffered from high susceptibility to bias and therefore add limited reliable quantitative evidence to this issue.

Social capital The qualitative synthesis suggests development associated with PAs can exacerbate ethnic tensions through perceived preferential treatment of some communities. There may be a relationship between existing social capital and ability to adapt to new circumstances. Quantitative evidence of impact on social capital is mixed. There is some evidence of positive impact of land protection on poverty alleviation and on housing and infrastructure but also of increasing incidence of wildlife conflict.

Health Views expressed on health of local populations are predominantly negative, including exposure to disease, wildlife conflict and women's safety. Quantitative studies of impact of PAs on health and safety are notable by their absence.

Discussion

Historical accounts of establishment of PAs provide evidence that substantial negative impacts on local populations can occur and have occurred. Forced displacement of communities is a recurring theme in the narrative concerning negative impacts of PAs e.g. [4]. This review does not seek to question this historical narrative. Community development and infrastructural improvements in close proximity to PAs has also been documented and suggests that PA establishment can be positive (i.e. win-win solutions for biodiversity and human well-being are possible). Establishment of PAs will inevitably lead to impacts on local, and possibly regional, populations but the challenge is to improve our capacity to predict which factors will influence the balance of positive and negative impacts. In this review we have attempted to provide an assessment and characterization of the range of positive and negative impacts in the period following the Rio Summit and establishment of the CBD. It was not the objective of this review to revisit this history prior to 1992. The review also attempts to collate evidence on the factors that modify impact, either in a positive or negative direction.

Comparison of qualitative and quantitative evidence

Although the quantitative evidence is insufficient to draw conclusions about the scale of either positive or negative impacts of protected areas on well-being, it was possible to synthesise understandings from qualitative studies about how positive and negative changes in well-being can arise from establishing and implementing regulations to protect the natural environment, with or without simultaneous investment in community development.

The qualitative synthesis has identified a number of themes in the ways governance of protected areas affect human populations well-being and how PAs are viewed. Some of these themes, such as the impact of land protection on forestry sector employment in Sweden, are reflected in the impacts assessed in studies considered in the quantitative review. Other themes, however, have not been rigorously assessed in the quantitative literature, for example, novel diseases resulting from changes in lifestyle, increased workload and heat exhaustion due to crop raiding, safety risks felt by women as men worked further afield and as women themselves were drawn out of their homes for new roles.

Reasons for heterogeneity

The identification of variables that influence whether positive or negative impacts will occur would be desirable for supporting decision making on the process of establishment and subsequent management of PAs. Unfortunately the nature of the evidence provides little opportunity to analyse differences in impact among different PAs (see limitations below).

Mode of governance is commonly viewed as a key variable determining impacts of PAs and this is supported by the qualitative synthesis in which many narratives are available on different aspects of governance. However, rigorous tests of governance as an effect modifier are absent. A similar lack of quantitative evidence is apparent for the following questions:

- What practices repeatedly lead to negative impacts, and which ones seem to be recurrently improving people's well-being?
- Are some of these practices becoming more/less common with time?
- Are any costs or benefits associated with particular types of PAs (e.g. size or location)?

Strengths and limitations of the review

This is the first systematic review of which we are aware that attempts to identify and synthesise, in a transparent manner, the findings from international studies of people's views about the impact of protected areas on their lives. To reduce the likelihood of missed studies, sensitive searches of bibliographic databases were supplemented by other methods to seek out less easily found literature such as unpublished reports from topic relevant websites. Studies providing thick descriptive data spanning the different categories of IUCN protected areas and before and after the Durban Accord offered an excellent source for synthesising understanding about how protected areas impact on people's lives.

During the reading and re-reading of individual studies we found that data often encapsulated the complexity of living in or near protected areas, touching on a multiplicity of interrelated themes. Within the limitation of time and resources for this study we have only been able to present these themes fairly superficially without exploring fully all their interconnections.

The disparate, fragmented literature limits our ability to test the comprehensiveness of the search. In reviewing such a broad and interdisciplinary question it has been a significant challenge to test all the possible sources of relevant material; nor is it simple to measure what proportion of the relevant articles we have been able to access with the time and resources available. Limiting our search to English-language articles may be significant.

Diversity of the literature has limited any assessment of extent of publication bias. Whilst we have attempted to minimise publication bias by employing a systematic search strategy, we have no way of testing for publication bias in the literature we obtained. The selective nature of many studies, in terms of the type of impacts investigated, is also a potential source of bias, since researchers may 'cherry-pick' (possibly inadvertently) those impacts most likely to show a particular effect.

Limitations of the evidence base

Although the quality of the studies was sufficient to draw out their findings to explain how different impacts may arise, many of the studies failed to report adequately their methods of data collection and analysis.

There is a potential in any studies looking at the quantitative impact of protected areas that respondents will bias their responses in an effort to influence protected area governance. Some questionnaire-based studies attempted to minimise this strategic bias by clearly stating that interview and questionnaire results would be used solely for research purposes, other studies did not acknowledge this potential bias or attempt to reduce it. Of the 305 outcome measures extracted from 79 included studies, 92 outcome measures (30%) involved 'reported changes'. For these data the comparator is implicit in the respondent's reply; they are reporting a change over time due to the protected area. Whilst these results are relevant (externally valid) to the review in hand, they can be susceptible to significant recall bias and questioning bias (elements of internal validity). Several studies attempted to retrospectively elicit opinions and attitudes towards protected area establishment many years after the event, which is similarly open to substantial sources of bias.

For these reasons, results in the form of reported changes which involve significant recall should be viewed with caution, and studies critiqued in depth for potential sources of bias. We attempted to account for these sources of bias during critical appraisal using our 'susceptibility to bias' scoring system.

Sixty-six of the 79 studies accepted following critical appraisal collected data in the form of self-reported measures. Fifty-six of these articles failed to provide details of the questionnaires given to respondents, and only two articles provided a copy of the survey instrument in full [76,105]. Without details of the questioning involved in these surveys it is difficult to assess questioning bias.

Variation in all the question elements (PECO; population, exposure, comparator, outcome) and the high degree of specificity in outcome measures identified in this review provides problems for synthesis. In particular, studies based on self-reported measures commonly asked very specific questions that could not then be synthesised along with other similar outcomes. Similarly, a high degree of variability in the choice and design of comparators prevented synthesis. In some studies, the inside-outside comparison was open to many confounding factors that cloud the link between protected area presence and impacts. This highlights the difficulty of balancing minimisation of spillover effects, whereby the comparator population is close enough to feel the effects of the exposure, and control of non-target variables. Whilst some studies accounted for this problem by including confounding variables in statistical models, many others did not. Furthermore, very few studies examined differences in environmental conditions between the comparator and exposure populations.

High susceptibility to bias in most studies limits ability to attribute outcomes/impacts to presence of PAs. Forty-five studies were excluded during critical appraisal due to flaws in experimental design and data analysis, or due to a lack of methodological detail. However, many studies included after the first stage of critical appraisal also failed to account for confounding variables, selected replicates in a non-random manner, and used opportunistic methodology. The most frequently occurring factor that affected the susceptibility to bias score in included studies, however, was a failure to appropriately report their methodology. Significant details such as recall period, response rate, item pool balance and order, sample selection process, sample size, and sample location were not disclosed in a large number of cases. Together, these factors limit the ability to attribute the reported impacts to protected areas.

There is a lack of primary studies estimating impact of PAs on human well-being using direct measurement techniques in a BACI format. In addition to a generally high susceptibility to bias, very few studies employed robust comparators over appropriate time frames in order to maximise evidence linking protected areas to observed human well-being impacts. Only one included study used a full BACI design to account for spatial and temporal confounding variables. Only three studies used direct data collection.

We found a surprisingly small number of studies on health of populations. Only nine studies reported data on human health impacts of protected areas. This is surprising since the majority of articles in this review purport to measure human well-being. Difficulties in ethical approval for human study may account in part for the paucity of health studies.

Review conclusions

Implication for policy/management

The evidence base provides a range of possibilities to inform but little evidence to support decision making on how to maximise positive impacts of PAs on human well-being. The diversity of studies and of outcomes measured, together with the diversity (or lack of clear signal) in the data suggests that impacts of PAs are highly context dependent. However, the evidence base is insufficient to provide any power with which to predict impacts on well-being from a knowledge of their context. It logically follows that there is an insufficient evidence base to identify circumstances/variables/effect modifiers that might lead to greater or lesser impact. At present, the available evidence base is failing to inform policy on the progress (or lack of it) being made, since 1992, toward lessening negative and promoting positive impact of PAs on human well-being.

Implication for research

The nature of the research reported to date forms a diverse and fragmented evidence base that is insufficiently developed to reliably inform future policy decisions (recognising that many included studies did not set out to address the review question). Many studies appear to have been conducted opportunistically and lack baseline measures. There is no evidence of a strategic approach or strategic investment to this field of research beyond individual research group initiatives. If a sufficient evidence base is to be formed then there is a need for concerted programme of research rather than an uncoordinated short term opportunistic approach.

The diversity of outcome measures and the consequent difficulty for synthesis suggests a need for use of standard indicators of human well-being that allow comparison among studies and meaningful synthesis of evidence.

Comparative research needs to progress from PA/no PA to PA type A/PA type B comparisons. Comparisons should be made between potential proximate causes of positive or negative impacts when the ultimate cause is PA establishment/management. This review suggests some of the candidates to be governance models, existing social capital, cultural diversity and poverty index.

It would be helpful to research efforts for funders to find consensus on minimum standards for methodologies, for both qualitative and quantitative evidence, that provide improved quality and thus reliability of data. The large proportion of included studies that suffered high susceptibility to bias is an indicator of such a need and also an indicator that scarce research resources are not being used effectively.

Recommended study design

In order to better assess the impacts of protected areas on human well-being we make the following recommendations for future research study design and reporting;

> *Methodological detail;* Studies must report sufficient details regarding the location of sample sites (in relation to the protected area boundaries in particular), the degree of replication, the data collection tool (e.g. quote questions posed to respondents in questionnaires), the method of sample selection (e.g. random or purposeful), and the times and duration of sampling. This is not an exhaustive list, and sufficient detail must be provided to allow the sampling to be repeated. Where information cannot fit within published articles these details should be provided in supplementary material.
>
> *Baseline assessment;* Where changes following establishment or change in protected area governance are being investigated, adequate baselines must be assessed. Although this is difficult and requires planning prior to the intervention, full 'before-after-control-intervention' (BACI) study design is vital to account for confounding temporal and spatial confounding factors. By assessing baselines, any differences between intervention and comparator populations can be compared relative to the starting conditions to strengthen the evidence towards causation.

Matched controls

'Control' or 'comparator' populations are vital to enable conclusions to be drawn about impacts in the absence of the intervention. A reliable comparison requires that as many other variables describing the environment are held constant or matched between comparator and intervention populations, allowing only the intervention to change in an ideal situation. In practice this is very difficult (and why baseline assessment is important), and there is often a payoff between maximising similarity and minimising spillover (the overflow of impacts from the intervention into the nearby comparator). Statistical tests can help to confirm similarity across intervention and comparator populations, and descriptive variables can be included in models that test for the significance of the intervention in order to account for differences that might occur.

Replication

Care must be taken to ensure that there is an appropriate trade-off between a study's accuracy and its precision. When combining many studies in a synthesis, more accurate results are preferable to more precise ones. For example, a study that measures daily resource extraction over a year in ten households from one intervention and one comparator village is less likely to reflect the true impact of the intervention than a study that measures daily resource extraction over a month from 12 intervention and 12 comparator villages. This spectrum is not clear-cut, however, and the allocation of resources to pseudoreplication (improving precision) and true replication (improving accuracy) must be considered carefully. Indeed, the scale at which conclusions will be drawn defines what is pseudoreplication and what is true replication, and this definition may be different for the author and the systematic reviewer.

Statistics

Statistics, both in summarising results and analysing patterns, must be used with great care. We recommend that a statistician be consulted during experimental design in order to optimise design for analysis. The use of models that account for changes in non-target variables across temporal and spatial scales are recommended, but tests for differences in confounders between intervention and comparator populations are also appropriate. Where information can be presented in summary statistics (e.g. mean/median and standard deviation/confidence intervals) this will aid future meta-analysis.

Additional files

Additional file 1: Search Strategy. Details of the development of the finalised search string, including scoping and testing against list of recommended key articles, and recording of search results from all databases, including duplicate removal and library file creation.

Additional file 2: Unobtainable and Un-translated. A list of articles that could not be obtained in full text, along with a list of foreign language articles that were not translated for consideration in the quantitative review.

Additional file 3: Web site searches. A list of organisational web sites searched for material, along with search terms, hits returned and actions taken for the quantitative review.

Additional file 4: Bibliographic Searches. Details of five relevant review article bibliographies searched for supplemental material in the quantitative review.

Additional file 5: Coding Tool. The coding tool used to assess articles at full text for the qualitative review, including the coding of outcomes used in both reviews.

Additional file 6: Excluded. Lists of articles excluded at full text assessment from both the qualitative and quantitative reviews along with reasons.

Additional file 7: Narrative Synthesis. Full narrative synthesis tables for 79 included studies in quantitative review.

Additional file 8: Supplementary Descriptive Statistics for 'Qualitative Synthesis'. Descriptive statistics for the 305 articles included in the systematic map of the first stage of analysis of the qualitative review.

Additional file 9: Qualitative Synthesis. The full synthesis of 30 studies considered in the qualitative review, from which the summary synthesis was produced.

Additional file 10: Table S1. Summary table of included studies and their measured outcomes in the quantitative review that were scored as having 'low' or 'medium' susceptibility to bias (Susc. to Bias).

Additional file 11: Details of the information considered during critical appraisal from each of the 79 studies included in the quantitative review.

Competing interest

There are no potential conflicts of interest to report.

Authors' contributions

ASP managed and planned the conduct of the SR. SO managed the conduct of the qualitative section. SD and NRH undertook searches. Quantitative Review; SD and NRH screened for inclusion. NRH carried out data extraction and critical appraisal. JPGJ, JRH, NH and TK assisted in planning, interpretation and synthesis. Qualitative Review; MB, KD, HH SO and CV undertook screening, data extraction and synthesis. All authors contributed to writing and approved the final manuscript.

Acknowledgments

We thank Sandra Diaz and Tom Hammond for their help in administering the progress of this review.

Sources of support

This review was funded by the Scientific and Technical Advisory Panel of the Global Environment Facility.

Author details

[1]Centre for Evidence-Based Conservation, School of Environment, Natural Resources and Geography, Bangor University, LL57 2UW Bangor, Gwynedd, UK. [2]Evidence-informed Policy and Practice Information and Coordinating Centre (EPPI-Centre), Institute of Education, University of London, 20 Bedford Way, WC1H 0NR London, UK. [3]School of Environment, Natural Resources and Geography, Bangor University, LL57 2UW Bangor, Gwynedd, UK.

References

1. Jenkins CN, Joppa L: **Expansion of the global terrestrial protected area system.** *Biol Conserv* 2009, **142:**2166–2174.
2. Bertzky B, Corrigan C, Kemsey J, Kenney S, Ravilious CCB, Burgess N: *Protected Planet Report 2012: Tracking progress towards global targets for protected areas.* Cambridge, UK: IUCN, Gland, Switzerland and UNEP-WCMC; 2012.
3. Brockington D, Igoe J, SCHMIDT-SOLTAU K: **Conservation, human rights, and poverty reduction.** *Conserv Biol* 2006, **20:**250–252.
4. Dowie M: *Conservation refugees: The hundred-year conflict between global conservation and native peoples.* Cambridge USA: MIT Press; 2009.
5. Turner WR, Brandon K, Brooks TM, Gascon C, Gibbs HK, Lawrence KS, Mittermeier RA, Selig ER: **Global biodiversity conservation and the alleviation of poverty.** *BioScience* 2012, **62:**85–92.
6. Kramer RA, Sharma NP, Munasinghe M: *Valuing tropical forests: methodology and case study of Madagascar.* Madagascar: World Bank Publications; 1995.
7. Adams C, Seroa da Motta R, Ortiz RA, Reid J, Ebersbach Aznar C, de Almeida Sinisgalli PA: **The use of contingent valuation for evaluating protected areas in the developing world: Economic valuation of Morro do Diabo State Park, Atlantic Rainforest, São Paulo State (Brazil).** *Ecol Econ* 2008, **66:**359–370.
8. Ferraro PJ, Hanauer MM, Sims KRE: **Conditions associated with protected area success in conservation and poverty reduction.** *Proc Natl Acad Sci* 2011, **108:**13913–13918.
9. West P, Igoe J, Brockington D: **Parks and peoples: the social impact of protected areas.** *Annual Reviews in Anthropology* 2006, **35:**251–277.
10. Sutherland WJ, Adams WM, Aronson RB, Aveling R, Blackburn TM, Broad S, Ceballos G, Cote IM, Cowling RM, Da Fonseca GAB, *et al*: **One hundred questions of importance to the conservation of global biological diversity.** *Conserv Biol* 2009, **23:**557–567.
11. Adams WM, Aveling R, Brockington D, Dickson B, Elliott J, Hutton J, Roe D, Vira B, Wolmer W: **Biodiversity conservation and the eradication of poverty.** *Science* 2004, **306:**1146–1149.
12. Dudley N: *Guidelines for Applying Protected Area Managment Categories.* Gland Switzerland: World Conservation Union; 2009.
13. Pullin AS, Bangpan M, Dalrymple S, Dickson K, Healey JR, Hockley N, Jones JPG, Knight TM, Oliver S: **Human well-being impacts of terrestrial protected areas?** *CEE protocol 11-009 Collaboration for Environmental Evidence: wwwenvironmentalevidenceorg/SR11009html*; 2012.
14. Neuman L: *Social Research Methods: Qualitative and Quantitative Approaches.* New York: Allyn and Bacon; 1997.
15. Gough D, Thomas J, Oliver S: **Clarifying differences between review designs and methods.** *Systematic Reviews* 2012, **1:**28.
16. Oliver SR, Rees RW, Clarke-Jones L, Milne R, Oakley AR, Gabbay J, Stein K, Buchanan P, Gyte G: **A multidimensional conceptual framework for analysing public involvement in health services research.** *Health Expect* 2008, **11:**72–84.
17. Ritchie J, Spencer L: **Qualitative data analysis for applied policy research.** In *Analyzing Qualitative Data.* Edited by Bryman A, Burgess RG. London: Routledge; 1994:305–329.
18. Pope C, Ziebland S, Mays N: **Qualitative research in health care: Analysing qualitative data.** *BMJ: British Medical Journal* 2000, **320:**114.
19. Thomas J, Harden A, Newman M: **Synthesis: combining results systematically and appropriately.** In *Introducing systematic reviews. Edited by Gough D, Oliver S, J. T.* London: Sage Publications; 2012.
20. Vasak K: *Human Rights: A Thirty-Year Struggle: the Sustained Efforts to give Force of law to the Universal Declaration of Human Rights. UNESCO Courier 30:11.* Paris: United Nations Educational, Scientific, and Cultural Organization; 1977.
21. Borenstein M, Hedges LV, Higgins JPT, Rothstein HR: *Introduction to meta-analysis.* Chichester, UK: Wiley; 2011.
22. Strickland-Munro J, Moore S: **Indigenous involvement and benefits from tourism in protected areas: a study of Purnululu National Park and Warmun Community, Australia.** *J Sustain Tour* 2013, **21:**26–41.
23. Allendorf TD, Smith JLD, Anderson DH: **Residents' perceptions of Royal Bardia National Park, Nepal.** *Landscape and Urban Planning* 2007, **82:**33–40.
24. Haukeland JV: **Tourism stakeholders' perceptions of national park management in Norway.** *J Sustain Tour* 2011, **19:**133–153.
25. Castillo A, Magana A, Pujadas A, Martínez L, Godínez C: **Understanding the interaction of rural people with ecosystems: A case study in a tropical dry forest of Mexico.** *Ecosystems* 2005, **8:**630–643.
26. Mehring M, Seeberg-Elverfeldt C, Koch S, Barkmann J, Schwarze S, Stoll-Kleemann S: **Local institutions: Regulation and valuation of forest use—Evidence from Central Sulawesi, Indonesia.** *Land Use Policy* 2011, **28:**736–747.
27. Petrzelka P, Marquart-Pyatt S: **"With the Stroke of a Pen": Designation of the Grand Staircase Escalante National Monument and the Impact on Trust.** *Hum Ecol* 2013, **41:**285–297.
28. Gerritsen PRW: *Diversity at stake: a farmers' perspective on biodiversity and conservation in western Mexico.* Mexico: Wageningen Universiteit Holanda; 2002.
29. Diaw MC, Tiani AM: **Fences in our heads: A discourse analysis of the Korup resettlement stalemate.** *J Sustain For* 2010, **29:**221–251.
30. Nguiffo S: **One forest and two dreams: The constraints imposed on the Baka in Miatta by the Dja Wildlife Reserve.** In *Indigenous people and protected areas in Africa.* Edited by Nelson J, Hossack L, Moreton-in-Marsh. UK: Forest Peoples Programme; 2003:195–214.
31. Sletten M, Vedeld P, Kaboggoza J: *To co-operate or not to co-operate? a study of collaborative management planning in Mount Elgon National Park.* Uganda: NORAGRIC working paper; 2008.
32. Bolaane M: **The impact of game reserve policy on the River BaSarwa/Bushmen of Botswana.** *Social Policy & Administration* 2004, **38:**399–417.

33. Milgroom J, Spierenburg M: **Induced volition: Resettlement from the Limpopo National Park, Mozambique.** *J Contemp Afr Stud* 2008, **26:**435–448.
34. Mbaiwa JE: **Wildlife resource utilisation at Moremi Game Reserve and Khwai community area in the Okavango Delta, Botswana.** *J Environ Manage* 2005, **77:**144–156.
35. Torri MC: **Conservation, relocation and the social consequences of conservation policies in protected areas: Case study of the Sariska Tiger Reserve.** *India. Conservation and Society* 2011, **9:**54.
36. Keskitalo ECH, Lundmark L: **The controversy over protected areas and forest-sector employment in Norrbotten, Sweden: forest stakeholder perceptions and statistics.** *Soc Nat Resour* 2009, **23:**146–164.
37. Spenceley A, Goodwin H: **Nature-based tourism and poverty alleviation: impacts of private sector and parastatal enterprises in and around Kruger National Park, South Africa.** *Current Issues in Tourism* 2007, **10:**255–277.
38. Songorwa AN: **Community-based wildlife management (CWM) in Tanzania: Are the communities interested?** *World Dev* 1999, **27:**2061–2079.
39. Stone M, Wall G: **Ecotourism and community development: case studies from Hainan, China.** *Environ Manage* 2004, **33:**12–24.
40. Bruyere BL, Beh AW, Lelengula G: **Differences in perceptions of communication, tourism benefits, and management issues in a protected area of rural Kenya.** *Environ Manage* 2009, **43:**49–59.
41. Slater R: **Between a rock and a hard place: contested livelihoods in Qwaqwa National Park, South Africa.** *Geogr J* 2002, **168:**116–129.
42. Bizikova L, Nijnik M, Kluvanková-Oravská T: **Sustaining multifunctional forestry through the developing of social capital and promoting participation: a case of multiethnic mountain communities.** *Small-Scale Forestry* 2012, **11:**301–319.
43. Ormsby A, Kaplin BA: **A framework for understanding community resident perceptions of Masoala National Park, Madagascar.** *Environ Conserv* 2005, **32:**156–164.
44. First Peoples Worldwide: *Okiciyab: To help each other-Promoting best practices in Indigenous community Development.* 2006. Final report, June.
45. ESCO: *Biosphere reserves: The Seville Strategy and the Statutory Framework of the World Network.* Paris: UNESCO; 1996.
46. World Parks Congress: *The Durban Action Plan.* Durban, South Africa: World Parks Congress; 2003.
47. Herrold-Menzies M: **From Adversary to Partner: The Evolving Role of Caohai Nature Reserve in the Lives of Reserve Residents.** *Canadian Journal of Development Studies/Revue canadienne d'études du développement* 2006, **27:**39–50.
48. Lunstrum E: **Mozambique, neoliberal land reform, and the Limpopo National Park.** *Geogr Rev* 2008, **98:**339–355.
49. Almudi T, Berkes F: **Barriers to empowerment: Fighting eviction for conservation in a southern Brazilian protected area.** *Local Environment* 2010, **15:**217–232.
50. Davis A: **' Ha! What is the benefit of living next to the park?' Factors limiting in-migration next to Tarangire National Park, Tanzania.** *Conserv Soc* 2011, **9:**25–34.
51. Hoole A, Berkes F: **Breaking down fences: Recoupling social–ecological systems for biodiversity conservation in Namibia.** *Geoforum* 2010, **41:**304–317.
52. Hartter J: **Attitudes of rural communities toward wetlands and forest fragments around Kibale National Park, Uganda.** *Hum Dimens Wildl* 2009, **14:**433–447.
53. Bedunah DJ, Schmidt SM: **Pastoralism and protected area management in Mongolia's Gobi Gurvansaikhan National Park.** *Dev Chang* 2004, **35:**167–191.
54. Ogra MV: **Human–wildlife conflict and gender in protected area borderlands: a case study of costs, perceptions, and vulnerabilities from Uttarakhand (Uttaranchal), India.** *Geoforum* 2008, **39:**1408–1422.
55. Stronza A, Gordillo J: **Community views of ecotourism.** *Ann Tour Res* 2008, **35:**448–468.
56. Yasuda A: **The Impacts of Sport Hunting on the Livelihoods of Local People: A Case Study of Bénoué National Park, Cameroon.** *Soc Nat Resour* 2011, **24:**860–869.
57. Nyahongo J, Holmern T, Kaltenborn BP, Roskaft E: **Spatial and temporal variation in meat and fish consumption among people in the western Serengeti, Tanzania: the importance of migratory herbivores.** *Oryx* 2009, **43:**258.
58. Sarker A, Røskaft E: **Human attitudes towards the conservation of protected areas: a case study from four protected areas in Bangladesh.** *Oryx* 2011, **45:**391–400.
59. Bajracharya SB, Furley PA, Newton AC: **Impacts of community-based conservation on local communities in the Annapurna Conservation Area, Nepal.** *Biodivers Conserv* 2006, **15:**2765–2786.
60. Mehta JN, Kellert SR: **Local attitudes toward community-based conservation policy and programmes in Nepal: a case study in the Makalu-Barun Conservation Area.** *Environ Conserv* 1998, **25:**320–333.
61. Naughton-Treves L, Alix-Garcia J, Chapman CA: **Lessons about parks and poverty from a decade of forest loss and economic growth around Kibale National Park, Uganda.** *Proc Natl Acad Sci* 2011, **108:**13919–13924.
62. Rinzin C, Vermeulen WJV, Wassen MJ, Glasbergen P: **Nature Conservation and Human Well-Being in Bhutan An Assessment of Local Community Perceptions.** *J Environ Dev* 2009, **18:**177–202.
63. Okello MM, Buthmann E, Mapinu B, Kahi HC: **Community opinions on wildlife, resource use and livelihood competition in Kimana Group Ranch near Amboseli, Kenya.** *Open Conservation Biology Journal* 2011, **5:**1–12.
64. Lundgren T: **Environmental protection and impact on adjacent economies: evidence from the Swedish mountain region.** *Growth and Change* 2009, **40:**513–532.
65. Richardson RB, Fernandez A, Tschirley D, Tembo G: **Wildlife Conservation in Zambia: impacts on rural household welfare.** *World Dev* 2012, **40:**1068–1081.
66. Sims KRE: **Conservation and development: Evidence from Thai protected areas.** *J Environ Econ Manag* 2010, **60:**94–114.
67. Andam KS, Ferraro PJ, Sims KR, Healy A, Holland MB: **Protected areas reduced poverty in Costa Rica and Thailand.** *Proc Natl Acad Sci* 2010, **107:**9996–10001.
68. Canavire-Bacarreza G, Hanauer MM: **Estimating the impacts of bolivia's protected areas on poverty.** *World Dev* 2012, **41:**265–285.
69. Ferraro PJ, Hanauer MM: **Protecting Ecosystems and Alleviating Poverty with Parks and Reserves: 'Win-Win'or Tradeoffs?** *Environ Resour Econ* 2011, **48:**269–286.
70. Foerster S, Wilkie DS, Morelli GA, Demmer J, Starkey M, Telfer P, Steil M: **Human livelihoods and protected areas in Gabon: a cross-sectional comparison of welfare and consumption patterns.** *Oryx* 2011, **45:**347–356.
71. Cardozo M: **Economic displacement and local attitude towards protected area establishment in the Peruvian Amazon.** *Geoforum* 2011, **42:**603–614.
72. Kayser D, Sobrevila C, Ledec G: *From Planning to the Implementation of a Successful Conservation and Socio-Economic Model.* South Africa: Report from The World Bank and Addo Elephant National Park; 2011.
73. Saayman M, Saayman A: **Regional development and national parks in South Africa: Lessons learned.** *Tour Econ* 2010, **16:**1037–1064.
74. Ninan KN: **Non Timber Forest Products and Biodiversity Conservation: a Study of Tribals in a Protected Area in India.** In *Conserving and Valuing Ecosystem Services and Biodiversity: Economic, Institutional and Social Challenges.* Edited by Ninan KN. London: Earthscan; 2012:99–112.
75. *ETFRN NEWS 39/40: Globalisation, localisation and tropical forest management.* http://www.etfrn.org/etfrn/newsletter/pdf/etfrnnews3940.pdf.
76. Phamtrong N, Swan S: *An investigation into the drivers of forest encroachment in Bidoup Nui Ba National Park, Vietnam.* Vietnam: Report to WWF Greater Mekong - Vietnam Country Programme; 2009.
77. Korhonen K, Rahkonen O, Hemminki E: **Implications of integrated nature conservation for human reproductive health: a case study from Ranomafana National Park, Madagascar.** *Dev South Afr* 2004, **21:**603–621.
78. Mishra C: **Socioeconomic transition and wildlife conservation in the Indian Trans-Himalaya.** *Journal of the Bombay Natural History Society* 2000, **97:**25–32.
79. Sheppard DJ, Moehrenschlager A, McPherson JM, Mason JJ, Pattenden J, Khandker S, Koolwal G, Biggs B, King L, Basu S: **Ten years of adaptive community-governed conservation: evaluating biodiversity protection and poverty alleviation in a West African hippopotamus reserve.** *Environ Conserv* 2010, **37:**270–282.
80. Shrestha RK, Alavalapati JRR: **Linking conservation and development: An analysis of local people's attitude towards Koshi Tappu wildlife reserve, Nepal.** *Environ Dev Sustain* 2006, **8:**69–84.
81. Bonaiuto M, Carrus G, Martorella H, Bonnes M: **Local identity processes and environmental attitudes in land use changes: The case of natural protected areas.** *J Econ Psychol* 2002, **23:**631–653.
82. Jim C, Xu SSW: **Stifled stakeholders and subdued participation: interpreting local responses toward Shimentai Nature Reserve in South China.** *Environ Manage* 2002, **30:**327–341.
83. Gubbi S, Linkie M, Leader-Williams N: **Evaluating the legacy of an integrated conservation and development project around a tiger reserve in India.** *Environ Conserv* 2008, **35:**331.

84. Infield M, Namara A: **Community attitudes and behaviour towards conservation: an assessment of a community conservation programme around Lake Mburo National Park, Uganda.** *Oryx* 2001, **35:**48–60.
85. Sekhar NU: **Crop and livestock depredation caused by wild animals in protected areas: the case of Sariska Tiger Reserve, Rajasthan, India.** *Environ Conserv* 1998, **25:**160–171.
86. Fiallo EA, Jacobson SK: **Local Communities and Protected Areas: Attitudes of Rural Residents Towards Conservation and Machalilla National Park, Ecuador.** *Environ Conserv* 1995, **22:**241–249.
87. Harada K: **Attitudes of local people towards conservation and Gunung Halimun National Park in West Java, Indonesia.** *J For Res* 2003, **8:**271–282.
88. Hartter J, Goldman AC: **Life on the Edge: Balancing Biodiversity, Conservation, and Sustaining Rural Livelihoods around Kibale National Park, Uganda.** *Focus Geogr* 2009, **52:**11–17.
89. Ite DUE: **Community perceptions of the Cross River national park, Nigeria.** *Environ Conserv* 1996, **23:**351–357.
90. Cihar M, Stankova J: **Attitudes of stakeholders towards the Podyji/Thaya River Basin National Park in the Czech Republic.** *J Environ Manage* 2006, **81:**273–285.
91. Alexander SE: **Resident attitudes towards conservation and black howler monkeys in Belize: the Community Baboon Sanctuary.** *Environ Conserv* 2000, **27:**341–350.
92. Rugendyke B, Son NT: **Conservation costs: Nature-based tourism as development at Cue Phuong national park, Vietnam.** *Asia Pacific Viewpoint* 2005, **46:**185–200.
93. Saayman M, Saayman A, Ferreira M: **The socio-economic impact of the Karoo National Park.** *Koedoe* 2009, **51:**1–10.
94. Just RE, Hueth DL, Schmitz A: *The Welfare Economics of Public Policy: A Practical Approach to Project and Policy Evaluation.* Cheltenham, UK: Edward Elgar; 2004.
95. Shrestha RK, Alavalapati JRR, Seidl AF, Weber KE, Suselo TB: **Estimating the local cost of protecting Koshi Tappu Wildlife Reserve, Nepal: a contingent valuation approach.** *Environ Dev Sustain* 2007, **9:**413–426.
96. Jemal H, Bogale A, Hagedorn K: **Welfare loss for pastoralists due to wildlife protection areas: the case of Awash National Park, Ethiopia.** *Quarterly Journal of International Agriculture* 2008, **47:**217–238.
97. Kremen C, Niles J-O, Dalton MG, Daily GC, Ehrlich PR, Fay JP, Grewal D, Guillery RP: **Economic incentives for rainforest conservation across scales.** *Science* 2000, **288:**1828–1832.
98. Han F, Yang Z, Wang H, Xu X: **Estimating willingness to pay for environment conservation: a contingent valuation study of Kanas Nature Reserve, Xinjiang, China.** *Environ Monit Assess* 2011, **180:**451–459.
99. Maharana I, Rai S, Sharma E: **Environmental economics of the Khangchendzonga National Park in the Sikkim Himalaya, India.** *GeoJournal* 2000, **50:**329–337.
100. Jones N, Iosifides T, Evangelinos KI, Florokapi I, Dimitrakopoulos PG: **Investigating knowledge and perceptions of citizens of the National Park of Eastern Macedonia and Thrace, Greece.** *International Journal of Sustainable Development & World Ecology* 2012, **19:**25–33.
101. Asciuto A, Schimmenti E: **Comparison between Willingness-to-Pay expressed by a panel of forestry experts and by a sample of non-expert respondents in a pilot survey conducted for a contingent valuation study.** *Mediterranean Journal of Economics, Agriculture and Environment* 2005, **4:**42–51.
102. Abbot JIO, Mace R: **Managing protected woodlands: fuelwood collection and law enforcement in Lake Malawi National Park.** *Conserv Biol* 2001, **13:**418–421.
103. Kavanagh J, Oliver S, Lorenc T, Caird J, Tucker H, Harden A, Greaves A, Thomas J, Oakley A: **School-based cognitive-behavioural interventions: A systematic review of effects and inequalities.** *Heal Sociol Rev* 2009, **18:**61–78.
104. Petticrew M, Tugwell P, Kristjansson E, Oliver S, Ueffing E, Welch V: **Damned if you do, damned if you don't: subgroup analysis and equity.** *J Epidemiol Community Health* 2012, **66:**95–98.
105. Reid R, Stone M, Whitely T: *Economic value of wilderness protection and recreation in British Columbia.* (Victoria BC): FRDA Working Paper; 1995.

8

Does the effectiveness of forest protected areas differ conditionally on their type of governance?

Biljana Macura[1,2*], Laura Secco[1] and Andrew S Pullin[2]

Abstract

Background: In context of protected areas (PAs), governance can be defined as a set of processes, procedures, resources, institutions and actors that determine how decisions are made and implemented. Current governance modes of forest PAs are multilevel and complex, with a variety of actors, different levels of power sharing, various formal and informal rules and vested interests. However, there is no systematic information on how different local governance modes and day-to-day decision-making processes within forest PAs may cause a change in PA effectiveness in terms of producing desired conservation outcomes. In this review we aim to assess relative effectiveness of forest PAs worldwide with respect to different governance modes and linking them to the multiple outcomes in order to discern the effective governance strategies for biodiversity and forest conservation. We will evaluate effectiveness of forest protected areas by multiple outcome measures: 1) attitudes of local stakeholders towards forest protected area governance, 2) conservation-related behaviour of local stakeholders, 3) ecological parameters such as forest cover, biodiversity level, density, overall forest condition and/or health as well as 4) existence of local spillover effects defined as social, institutional and ecological effects in surrounding social-ecological systems.

Methods: To gather empirical evidence on the effectiveness and outcomes of different governance arrangements of forest protected areas, we will search electronic databases, organizational websites, use web search engines and perform a bibliographic search. Articles will be filtered by title, abstract and full text. To facilitate understanding of predictors and conditions for effective PAs, both qualitative and quantitative data from the relevant studies will be extracted and integrated. Multivariate statistical analyses will be performed to give insights into association between different governance modes, their characteristics and PA effectiveness. Finally, the review will report on data gaps and potential for future empirical research.

Keywords: Evidence synthesis, Systematic review, Biodiversity conservation, Assessment, Evaluation, Conservation success, Participation, Ecological outcomes

Background

Forests contain roughly 90% of terrestrial biodiversity and they provide a wide variety of ecosystem services, contributing to the livelihoods of more than 1 billion people [1]. Yet, forest degradation and deforestation are advancing at alarming rate, especially in the tropics [2] and are putting at risk a high diversity of species and habitats sustained by forest ecosystems [3].

Establishment of *in situ* conservation strategies, such as protected areas (PAs), has been the major response to a global demand for conservation of biodiversity and ecosystem services [4] and more specifically, to the reduction of tropical deforestation [5]. Accordingly, there has been a year-on-year increase in the number of PAs and they are today covering more than 12% of the total world's land surface [6] and 13.5% of the world's forests [3].

Nevertheless, the effectiveness of biodiversity and forest conservation measures[a] is under question as the rate of biodiversity loss is not decelerating [7]. There is evidence that PAs are decreasing the deforestation rate [5,8], estimated through measures of land clearing prevention [9] and decreasing the incidence of forest fires

* Correspondence: biljana.macura@studenti.unipd.it
[1]Department of Land, Environment, Agriculture and Forestry (TESAF), University of Padua, Viale dell'Università 16, 35020, Agripolis, Legnaro (PD), Italy
[2]Centre for Evidence-Based Conservation, School of Environment, Natural Resources and Geography, Bangor University, Gwynedd, LL57 2UW, UK

[10]. However, some authors argue that many of the claimed positive conservation effects might be a function of a PA location i.e. low accessibility of protected land, but not the effect of actual protection measures [5,11]. Additionally, increasing deforestation and pressures on the resources in social-ecological systems that surround PAs, might diminish conservation efforts inside PAs, through effects of ecological isolation and landscape fragmentation [8,12].

However, there is no systematic information on how different local governance modes and day-to-day decision-making processes within forest PAs may cause a change in PA effectiveness in terms of producing desired conservation outcomes. In context of PAs, governance can be defined as "a set of processes, procedures, resources, institutions and actors that determine how decisions are made and implemented" ([13]:105). It is about power, relationships, accountability and responsibility exercised by organisations and actors [14,15]. Conservation governance arrangements are becoming multilevel and complex [16]. Governments are not the only source of environmental decision-making authority and there is a shift from administrative to collaborative state [17]. Power to make and enforce decisions is distributed among diverse social actors [18], including indigenous, mobile and local communities, local governments, NGOs and the private sector [19]. The change of the scale of governance has been occurring (mostly in the developing countries), and the authority and responsibility to make and enforce decisions are shifted from nation-state to lower-level authorities (decentralization) or to institutions outside the state (devolution) [20,21].

Following the main trends in conservation governance and based on the power distribution and scale of decision-making, type of different actors involved and level and nature of their collaboration, four broad modes of PAs governance can be identified [15,19]: 1) governance by government, 2) shared governance or co-management, 3) private governance and 4) governance by communities and indigenous people. These governance modes are briefly described in the following paragraphs as each of them may deliver different social and ecological outcomes.

1) Government PAs are governed by the centralised governmental agency (ministry or park agency reporting directly to the government) that enforces decisions, has authority, responsibility and accountability for managing PAs [15]. Government agencies are often considered as legitimate actors that can deliver public benefits and are accountable directly to the public [22]. Nevertheless, some authors argue that this 'old' [23] hierarchical type of governance is not able to handle size and complexity of PAs [19]. Moreover, state PAs with top-down and exclusionary conservation approach, frequently present in developing world, are being increasingly reported to produce unequal distribution of rights, power and benefits and create social conflicts [24].

2) Co-managed or multi-stakeholder PAs are governance modes where a governmental agency and other stakeholders, such as local/mobile/indigenous communities that depend on the area culturally or for their livelihoods, or user associations, private entrepreneurs and landowners share power and responsibility, make and enforce decisions. Formal decision-making authority might be vested in one agency (often governmental body), but that agency is required by law to collaborate with other stakeholders [25]. This collaborative partnership may be materialised through many forms: from consultation to decision-making carried out by consensus 15]. Co-management is frequently labeled as managing relationships, not resources [26,27]. However, it is argued that the partnerships in co-management arrangements can be problematic as nature of power sharing makes less powerful partners, such as indigenous people, disadvantaged [28].

3) Private PAs where private landowners, individuals, NGOs and other not-for profit and for-profit organisations make and enforce decisions, have control and/or ownership over resources. PAs can be governed by private and non-governmental actors, that might be perceived more efficient than bureaucratic structure of governmental agencies, also providing technical and financial support, bringing new ideas and capacity building [22]. However, the legitimacy and accountability of private parties is always limited and questionable, especially due to the vested interests of funding agencies and reluctance of governments to give authority or legal recognition to private parties [22,25]. Moreover, since designation of a private PA is a voluntary act, providing long-term security for conservation may pose a challenge [25].

4) Community conserved areas[b] are governed and voluntarily conserved by indigenous groups, local and mobile communities through customary laws. Authority and responsibility is vested within communities through a variety of ethnic governance or locally arranged rules and organisation that can be very complex, with diverse management and ownership rights. Community conserved areas depend on the government recognition and respect of community/indigenous rights over the territory [15]. However, community based conservation are criticised to be vulnerable to external drivers and not being able to deal with larger scale biodiversity processes (e.g. management of migratory species) [29].

Fifth, hybrid governance type may be added to this classification as in the reality borders between governance modes are blurred [30] and this is especially because of complex land and resource ownership rights, diversity in management authority and funding sources [31].

Nevertheless, the importance of the local political processes within PAs is frequently ignored in the conservation effectiveness literature. Therefore, to improve PAs governance and their conservation outcomes, there is need for more clear information on how differences in local governance modes and decision-making processes may cause variability in the outcomes and thus, in the effectiveness of forest PAs.

Four previous Systematic Reviews have addressed the various aspects of community-based conservation, synthesising and assessing primary literature on: 1) development as a conservation tool [32]; 2) the effect of local cultural context [33] and 3) broader social - political context on community based management [34], and 4) community forest management as a mechanism for supplying global environmental benefits and improving local welfare [35]. There are two more Systematic Reviews that have a wider conservation focus on terrestrial PAs and their 1) effectiveness in maintaining biodiversity and reducing habitat loss [36] and 2) securing human-well being (in preparation) [37]. This Systematic review is complementary to previous ones, looking from the governance perspective on effectiveness of forest protected areas worldwide; and determining the links between governance processes and multiple conservation outcomes.

Due to high complexity and variety of conservation practices, we will focus our analysis on conservation of forest resources only and on governance of forest PAs. To be defined as a PA, conservation governance arrangements have to: 1) have geographical limits or boundaries; 2) predominantly aim to achieve conservation benefits, but not excluding other related benefits (e.g. social benefits); 3) be designated and managed by legal gazetted means or by non-gazetted, but officially recognized NGO policies or customary laws; 4) have a body of governing rules; and 5) have a clearly identified organization or individual with a governance authority [25,38]. We define forest PAs as "a subset of all protected areas that includes a substantial amount of forest as defined for the purposes of Forest Protected Areas. This may be the whole or a part of a protected area" ([39]:19). This IUCN's definition excludes commercial plantations and forest managed for industrial purposes within the less strictly protected categories [39].

Objectives of the review

We aim to assess relative effectiveness of different governance regimes within forest PAs by contrasting different governance characteristics and processes on the basis of multiple measures of success[c].

There are recently quite a few studies that have estimated effectiveness of PAs (e.g. [40-42]), but they have focused solely on the tropics and only few of them assessed PAs effectiveness integrating multiple performance measures [43-45]. Apart from biodiversity conservation, PAs have various multifaceted and context-dependent objectives [43], in both, ecological and the social-economic domain. Examining PAs effectiveness in terms of biodiversity conservation only, might lead to restricted conclusions as it disregards local conflicts and resistance expressed through negative attitudes towards conservation policies and practices and anti-conservation behaviour of local stakeholders; it does not take into account institutional, economical or political changes in surrounding social-ecological systems influenced by PAs (e.g. employment opportunities or migration level) and it may reinforce a fortress conservation mentality [43-50].

In this review, we look at the following outcome measures:

A) Outcome measures within forest PA boundaries:
   1) Attitudinal success measured through (difference/change in) attitudes of local stakeholders[d] towards focal PA, authority and/or management practices
   2) Behavioural success measured through (difference/change in) level of conservation-oriented behaviour necessary to decrease the threats to natural resources (e.g. decrease in level of illegal activities, poaching, etc.)
   3) Ecological success measured through (difference/change in) deforestation rate, biodiversity level, maintenance of forest cover and forest density, condition, health.

B) Outcome measures outside of forest PAs boundaries:
   4) Spillover effects in surrounding social-ecological systems i.e. social, institutional and ecological changes on the local level including [5]: displacement of deforestation and agricultural pressures, preventive clearing at the nearby private land to prevent protective regulation, establishment of private reserves, better law enforcement at the neighbouring land, reforestation initiatives, new employment opportunities and similar. The spillover effects will be included into analysis only if there are reported baseline data against which these effects might be defined and measured [51]. Because of practical reasons, these changes will be recorded only at the local level that might be at a lowest administrative unit where a PA is located (e.g. municipality).

This review aims at answering following primary question:

Does the effectiveness of forest protected areas differ conditionally on their type of governance?

Elements of the primary question are shown in the Table 1.

Table 1 Elements of the systematic review question

Setting	Perspective	Interest, phenomena of	Comparison	Evaluation
Forest protected areas	1. Local Community 2. PA Authority/ Management staff	1. Governmental PAs 2. Co-managed PAs	1. With different governance regimes	1) Attitudinal success measured through (difference/change in) attitudes of local stakeholders towards focal PA, authority and/or management practices
		3. Private PAs 4. Community conserved areas		2) Behavioural success measured through (difference/change in) level of conservation-oriented behaviour necessary to decrease the threats to natural resources
		5. Hybrid forms		3) Ecological success measured through (difference/change in) deforestation rate, biodiversity level, maintenance of forest cover and forest density, condition, health, etc
				4) Spillover effects: social, institutional and ecological changes in surrounding social-ecological systems

Secondary question is:

Which characteristics of decision-making process influence the outcomes of forest protected areas?

Based on the aforementioned trends in PA governance, we selected following analytical variables that might describe governance processes:

1) Scale of decision-making:
   1.1) Level of decentralization i.e. level of implementation of "subsidiarity" principle (central decision-making, decentralization or devolution);
2) Individual versus multi-actor decision-making:
   2.1) Diversification of stakeholders' categories (one versus multi-actor);
   2.2) Nature of stakeholders' participation (pro-active, consultancy, passive, none);
3) Collaboration among stakeholders in decision-making:
   3.1) Nature of collaboration (formal, informal, none);
   3.2) Level of collaboration (horizontal/internal, vertical/external, multilevel);

We assume that governance processes i.e. how decisions are made and implemented, influence level of conservation effectiveness, its ecological and social outcomes. Using theory of change approach, we hypothesise that: (**H1**) making decisions at lowest level possible, (**H2**) collective or multi-actor decision making, (**H3**) high level of proactive participation in day-to-day decision-making and (**H4**) multilevel collaboration among stakeholders can lead to more positive attitudinal, behavioural and ecological success of PAs and decrease negative spillover effects around them.

Methods

Search strategy

We will search for all available evidence relevant to the questions, whether published or unpublished, including both peer reviewed papers and relevant grey literature.

Publication databases

The general search will be conducted using the following online databases:

ISI Web of knowledge
Scopus
PubMed
Agricola
International Development Research Center (IDRC) digital library
Scienceindex
Public library of science
Directory of Open Access Journals
COPAC
Social Sciences research network
Index to Theses Online
CAB Abstracts

Web search engines

Due to repeatability, the web search will be mainly used for reference cross-checks. Following web search engines will be used:

1) www.scholar.google.com
2) http://scientific.thomsonwebplus.com/
3) www.scirus.com (web sources only)

Only the first 50 hits of each search will be screened.

Organisational website search

Specific searches will be conducted using the following websites of organisations specialised in the field of (forest) PA management and governance. Where possible, only publication sections of the websites will be used for search. List of websites was compiled from previous Systematic Reviews on effectiveness of PAs and community-based conservation [32-37] and completed by including websites of organisations well-known in the field of natural resource governance, forestry and PAs.

http://www.agter.org//
http://www.capri.cgiar.org/
http://www.catie.ac.cr/Magazin_ENG.asp?CodIdioma=ENG
http://www.cbnrm.net/
http://www.cgiar.org/
http://www.cifor.org/
http://www.cof.orst.edu/org/istf/ftpp.htm
http://www.communityforestryinternational.org/
http://www.conservation.org
http://www.cooperationcommons.com/
http://www.culturalsurvival.org/
http://cfs.nrcan.gc.ca/publications
http://community.eldis.org/
http://www.conservationgateway.org/Pages/Advanced-Search.aspx/
http://seslibrary.asu.edu/seslibrary/search
https://dec.usaid.gov/dec/home/Default.aspx#
https://www.gov.uk/government/organisations/department-for-international-development
http://www.eci.ox.ac.uk/publications/index.php
http://www.eldis.org/
http://www.etfrn.org
http://www.fao.org/documents/en/search/init/,
http://www4.fao.org/faobib/
http://www.firstpeoples.org/
http://www.forest-trends.org/publications.php
http://forests.org/
http://www.ifad.org/
http://www.iied.org
http://iog.ca/
http://www.indiana.edu/~workshop/publications/index.php
http://www.iucn.org/knowledge/publications_doc/
http://www.iufro.org/publications/
http://www.lib.umn.edu/cgi-bin/forestry/index.cgi
http://www.nzdl.org/fast-cgi-bin/library?a=p&p=about&c=envl
http://wwf.panda.org
http://povertyandconservation.info/en/bibliographies
http://protectedareasandgovernance.groupsite.com
http://www.rainforestportal.org/
http://www.tropicalforests.ox.ac.uk
http://www.un.org/Depts/dhl/
http://www.undp.org/content/undp/en/home/librarypage.html/, http://sgp.undp.org/
http://web.undp.org/gef/gef_library.shtml
http://www.unep-wcmc.org/
http://www.unep.org,
http://ekh.unep.org/
http://www.wcs.org
http://web.worldbank.org

Bibliographic searches

Reference lists of relevant review studies will be searched for relevant primary articles.

Search terms

The following English search terms and their various combinations using Boolean operators (AND, OR), wild-cards (for any group of characters (*) or for a single character ($)) will be used to perform search in the databases and Internet search engines. Search strings will be adapted to different formats and requirements of databases and search engines to be explored. Specifically, if a website does not allow for complex search strings and Boolean operators, we will use simple search terms such as "protected area", "governance", "park", "reserve", "biodiversity", "conservation".

1) Search string for PA governance and management regimes
   "NGO*" OR (non$governmental and organi$ation) OR "private nature reserve*" OR "privat*" OR "governme*" OR "community conserved area*" OR "indigenous" or ("comanag*" or "co-manag*") OR "collaborative" OR "decentrali*" OR "devolut*" OR "joint management" OR (delegat* AND authorit*) OR ("integrated and conservation and development") or "ICDP*" or "governance" or "self-governance" or "institution*" or "rule*" or "norm*" or "polit*" or "polic*" or "paper park*" OR "participat*" or "accountab*" or "legitima*" or "compliance" or "enforcement*" or "coercion*" or "trust*" or "conflict*" or "exclusion*" or "access" or "local elite*" or "elite capture" or "revenue$sharing"
   AND
   "protected area*" OR "nature reserve*" OR park* OR "monument*" OR "wilderness area*" OR "world heritage site*" or "sanctuar*" or "refug*" or "biosphere reserve*" or "protected landscape" or "management area*" or "sacred forest*" or "sacred grove*"
   AND
   forest*
2) Search string for social outcomes
   "attitude*" OR "behavi*" OR "perception*" OR "belief*" OR "perspective*" OR "opinion*" OR "view*"
3) Search string for ecological outcomes
   "conserv*" or "deforest*" or "degrad*" or "biodiversity" or desert* or "threaten" or "leakage*" or ("spillover*" or "spill-over*") or "reforest*" or "afforest*" or ("re-growth" or "regrowth") Or "forest clearance" or "land use change" or "land cover change" or "loss*"

We will combine search strings as follows: 1 AND (2 OR 3).

Search strings developed above are the result of numerous iterations performed in ISI Web of Knowledge database. Full record of iterations has been kept and will be further developed while advancing the search. Citations will be imported into an Endnote library and online systematic review software EPPI-reviewer 4.0 [52]. Duplicates will be deleted.

Study inclusion criteria

Relevant documents will be selected by application of inclusion criteria. Inclusion criteria will be first applied to the document title, after to the abstract and in the final phase, to the whole document.

To filter studies based on abstracts, two reviewers will apply inclusion criteria. Repeatability of the application of inclusion criteria will be inspected using Kappa statistics on a sample of abstracts to assess the level of agreement between two reviewers. In case of kappa < 0.6, inclusion criteria will be discussed, re-interpreted and adjusted if necessary. After this procedure is done, only one reviewer will apply inclusion criteria to the rest of the studies.

Relevant subject populations: Biodiversity indicators within and human populations living in and/or around forest PAs.

Relevant interventions/phenomena of interest: Forest PAs under government, co-managed or joint, private and community modes of governance worldwide.

Relevant comparators: comparisons among different interventions (governance regimes). They will follow the appropriate study design explained below. Studies without relative comparators may be included into analysis as well. Comparators reported within the qualitative study can be created using perceptions or reconstructing the memories of respondents. If present in the study, constructed comparators where external data sets or models are applied to develop scenarios for comparison will be also included into our analysis.

Relevant outcomes:

1) Changes or differences in attitudes of local stakeholders towards focal PA governance, authority and/or management practices;
2) Changes or differences in level of conservation-oriented behaviour necessary to decrease the threats to natural resources;
3) Changes or difference in deforestation rate, biodiversity level within a forest ecosystem, maintenance of forest cover and forest density, condition, health (including fires);
4) Social, institutional and ecological changes on the local level that may include for example leakage (i.e. increased pressures on resources shifted outside a focal forest PA) or policy side effects (i.e. positive or negative impacts of a policy instrument on non-focal sectors and activities).

To be included into our analysis, a study has to report on at least two types of outcomes.

Relevant types of study design: Empirical studies using qualitative, quantitative or mixed methods that can be designed as control-intervention site comparisons/case control study, cohort study, case series, cross sectional study, interrupted time series, Before-After/Control-Intervention (BACI design), randomized control trials/control trials.

In case of multiple evidence sources for one PA, data will be combined but the most recent evidence will be prioritised.

Language: Studies published in English.

Following studies will be excluded:

- Studies with a focus on PAs that do not meet the previously mentioned definition of Forest PAs [39]. This definition is provided in the IUCN Guidelines on use of PA management categories and we will follow and consult it for further clarifications and detailed interpretation.
- Studies with a focus on conservation of a single or a group of species within forest PAs.

Potential reasons for heterogeneity and effect modifiers

Set of effect modifiers (predictor variables) that can cause variation in the outcomes are expected to be as follows:

- Governance and decision-making characteristics: scale of decision-making; individual versus multi-actor decision-making; nature of stakeholders' participation; level and nature of collaboration among stakeholders;
- Resource ownership;
- Level of resource access and use by the local actors;
- Presence of a local leader;
- Source of PA funding;
- National context: corruption and illegality, development level, income inequality;
- Human population size in and around PAs;
- Type of ecosystem and climatic conditions;
- Proximity of the forest PA to the urban areas roads, settlements;
- PA size;
- Time since PA establishment;

More effect modifiers may be recorded and extracted from the primary studies.

Study quality assessment

Under study quality assessment we refer to aspects of study design important both for reducing susceptibility to bias and ensuring validity with respect to the question. Depending on the methodology of a study, two quality assessment strategies may be applied:

1) Quantitative studies: Quantitative studies will assessed based on the score assigned to each of following criteria: 1) appropriateness of control cases and presence of valid counterfactual, 2) controlled for and/or minimized confounding factors, 3) study design category (from highest to lowest score): randomized control trial, non-randomized control trial, BACI (before/after/control/impact) design, interrupted time series study, case control study, cohort study, case series, cross sectional study, 4) methodology: clarity and completeness of reporting [53]. We expect that the (non)randomized control trial studies and full BACI design might be less represented in the PA literature [36], as it is difficult to meet these study design requirements in conservation policy assessment due to various reasons (non-random allocation of conservation interventions across the landscape, counterfactual thinking is not widespread in conservation assessment exercises, evaluation is usually not a built-in component of a conservation project design, etc. (see [11]:483)).
2) Qualitative studies: Qualitative studies will be assessed using Harden's methodology [54] applied in Rees et al. [55] and Pullin et al. [37]. This assessment tool uses eight study validity criteria focusing on 1) study design and methods (rigour of sampling, data collection and analysis); 2) findings (how well presented data support findings, quality of findings); 3) use of methods to assess the respondents' perspectives and experiences. A score range will be assigned to each of these criteria. A Qualitative Appraisal Tool [56] may be combined for additional assessment details and to provide guidance for a more structured quality appraisal exercise. This tool is a checklist composed of the ten questions connected to study rigour, credibility and relevance of findings.

Depending on the variability of study quality, decision for the study inclusion may be based on the overall summary score assigned to each study.

Data extraction

Data will be extracted from included studies and recorded in a spreadsheet with pre-determined coding. Extracted information across all included studies will be as follows.

Study characteristics

Objectives and focus of the study;
Study design and methodology for data collection;
Reported study biases;
Study conclusions including underlying factors of social / ecological change reported.

Governance characteristics

1) Scale of decision-making (variable with 3 levels: decision-making out of state (devolution), decision-making vested in lower level/local authorities (decentralization), centralized decision-making);
2) Individual versus multi-actor decision-making described through i) Diversification of stakeholders' categories (2 levels: one versus multi-actor); ii) Nature of stakeholders' participation (4 levels: pro-active, consultancy, passive, none);
3) Collaboration among stakeholders in decision-making described through i) Level of collaboration (3 levels: formal, informal, none); ii) Nature of collaboration (3 levels: horizontal (internal), vertical (external), multilevel);

Institutional, social, economical and political context in which PA governance is embedded

Resource ownership (state, local, private, mixed);
Level of resource access and use by the local actors measured through 1) IUCN PA management category (1 to 6); 2) Local community dependency on the forest resources (3 levels: high, moderate, low);
Presence of a local leader (yes/no);
Source of PA funding (4 levels: international, national/governmental, local/communal, private);
National context: corruption and illegality (Governance index score), country development level (Human Development Index score), income inequality (GINI score)
Human population size around PAs (high, medium, low);
Proximity to the urban areas, roads, settlements (high, medium, low);

Comparator type (if any)
Outcome (independent variables)

1) Attitudinal success measured through (level of changes/difference in) attitudes of local stakeholders towards focal PA governance, authority and/or management practices (3 levels: high, moderate, low);
2) Behavioural success measured through (level of changes/difference of) level of conservation-oriented behaviour necessary to decrease the threats to natural resources (3 levels: high, moderate, low);

3) Ecological success measured through (level of changes/difference in) deforestation rate, biodiversity level, maintenance of forest cover and forest density, condition, health (3 levels: high, moderate, low);
4) Spillover effects in surrounding social-ecological systems i.e. social, institutional and ecological changes/differences on the local level that may include leakage or policy side effects (3 levels: high, moderate, low);

Additional variables/controls:

Time since PA establishment (in years);
PA size (in km^2);
Type of ecosystem and climatic conditions (temperate, tropical, boreal);

Data synthesis

Synthesis will encompass narrative and summary findings of each study and it will be presented in a table and visualised graphically. Attitudinal, behavioural, ecological success and spillover effects will be estimated based on the aforementioned criteria of performance and inferred from the (valid) evidence reported in included studies (using descriptive levels: low, moderate, high).

In order to discern the underlying conditions and determinants of PA success, qualitative and quantitative information to be extracted from the empirical studies will be integrated by pre-determined coding (as shown above under section "Study Quality Assessment") and creation of ordinal/categorical variables that will be used in multivariate statistical analyses. Independent variables in the analyses will be 4 measures of success: attitudinal, behavioural, ecological and spillover effects. Dependent variables will be governance characteristics, institutional, economical, political and social setting (effect modifiers). The analyses will be done separately for each governance mode. Finally, comparisons will be done at the final phase and based on the regression results.

We will not infer conclusions about the comparisons between governance regimes if original studies had different counterfactual outcomes i.e. we will not contrast studies that estimated counterfactual of no protection versus counterfactual of a different governance mode. Data extraction and synthesis will be additionally refined during the review process.

In case of missing data in the included studies, we will contact authors and request relevant information.

This review will report methodologies for assessment of forest PAs governance effectiveness, data gaps and potential for future empirical research.

Endnotes

[a] Under effective conservation we mean positive and measurable effects of conservation policies and practices on biodiversity and target ecosystems, populations, species or habitats.

[b] Community conserved areas have been relatively recently internationally recognized as a PA at IUCN World Parks Congress in 2003 (Durban) and at the COP VII of CBD in 2004 (Kuala Lumpur) ([57]:1);

[c] The outcome measures are adopted from Systematic Reviews by Brooks et al. [32,34,53] that employed set of ecological, attitudinal, behavioral and economic measures of success to estimate effectiveness of conservation intervention. However, as Pullin and colleagues [37] in their review focused mainly on social-economical analysis of terrestrial PAs, this review does not assess the economical effectiveness to avoid the potential overlap

[d] A stakeholder in this study refers to those who affect and those affected i.e. both actively and passively involved individuals, groups or organizations in a PA governance (after [58]).

Competing interests

The authors declare that they have no competing interests.

Authors' contributions

BM designed the protocol, wrote and revised the manuscript. LS and AP contributed to the design and development of the protocol, discussed and commented drafts, provided theoretical and methodological guidance for the protocol. All authors read and approved the final manuscript.

Acknowledgments

This research is funded by the FONASO Erasmus Mundus fellowship. Authors would like to thank to reviewers for their valuable comments.

References

1. World Bank: *Forests Sourcebook: Practical Guidance for Sustaining Forests in Development Cooperation.* Washington, DC; 2008. http://siteresources.worldbank.org/EXTFORSOUBOOK/Resources/completeforestsourcebookapril2008.pdf.
2. FAO: *Global Forest Resources Assessment 2010. Main report. FAO Forestry Paper 163.* Rome: FAO; 2010. http://www.fao.org/docrep/013/i1757e/i1757e.pdf.
3. Schmitt CB, Belokurov A, Besançon C, Boisrobert L, Burgess ND, Campbell A, Coad L, Fish L, Gliddon D, Humphries K, Kapos V, Loucks C, Lysenko I, Miles L, Mills C, Minnemeyer S, Pistorius T, Ravilious C, Steininger M, Winkel G: *Global Ecological Forest Classification and Forest Protected Area Gap Analysis. Analyses and recommendations in view of the 10% target for forest protection under the Convention on Biological Diversity (CBD). 2nd revised edition.* Freiburg: Freiburg University Press; 2009.
4. Millennium Ecosystem Assessment (MEA) Ecosystems and Human Well-being: *General Synthesis.* Washington: Island Press; 2005.
5. Andam KS, Ferraro PJ, Pfaff A, Sanchez-Azofeifa GA, Robalino JA: **Measuring the effectiveness of protected area networks in reducing deforestation.** *PNAS* 2008, **105:**16089–16094.
6. Jenkins CN, Joppa L: **Expansion of the global terrestrial protected area system.** *Biol Conserv* 2009, **142:**2166–2174.
7. Butchart SHM, Walpole M, Collen B, van Strien A, Scharlemann JPW, *et al*: **Global biodiversity: indicators of recent declines.** *Sci* 2010, **328:**1164–1168.
8. Naughton-Treves L, Holland M, Brandon K: **The role of protected areas in conserving biodiversity and sustaining local livelihoods.** *Annu Rev Environ Resour* 2005, **30:**219–252.
9. Bruner AG, Gullison RE, Price RE, da Fonseca GAB: **Effectiveness of parks in protecting tropical biodiversity.** *Sci* 2001, **291:**125–128.

10. Nelson A, Chomitz KM: *Protected Area Effectiveness in Reducing Tropical Deforestation: A Global Analysis of the Impact of Protection Status.* Washington: World Bank Independent Evaluation Group; 2009.
11. Ferraro PJ, Pattanayak SK: **Money for Nothing? A call for empirical evaluation of biodiversity conservation investments.** *PLoS Biol* 2006, **4**:482–488. [http://www.plosbiology.org/article/info%3Adoi%2F10.1371%2Fjournal.pbio.0040105]
12. Sanchez-Azofeifa GA, Daily G, Pfaff A, Busch C: **Integrity and isolation of Costa Rica's national parks and biological reserves: examining the dynamics of land-cover change.** *Biol Conserv* 2003, **109**:123–135.
13. Secco L, Pettenella D, Gatto P: **Forestry governance and collective learning process in Italy: Likelihood or Utopia?** *Forest Policy and Economics* 2011, **13**:104–112.
14. Institute on Governance: *Governance principles for protected areas in the 21st century: a discussion paper.* Ottawa; 2002. http://iog.ca/wp-content/uploads/2012/12/2003_June_pa_governance2.pdf.
15. Borrini-Feyerabend G, Johnston J, Pansky D: **Governance of protected areas.** In *Managing protected areas: a global guide.* Edited by Lockwood M, Worboys G, Kothari A. London: Earthscan; 2006:116–145.
16. Berkes F: **Community-based conservation in a globalized world.** *PNAS* 2007, **104**:15188–15193.
17. Koontz TM, Thomas CW: **What do we know and need to know about the environmental outcomes of collaborative management?** *Public Adm Rev* 2006, **66**(suppl):111–121.
18. Lemos MC, Agrawal A: **Environmental governance.** *Ann Rev Environ Resour* 2006, **31**:297–325.
19. Borrini-Feyerabend G: **Governance of protected areas: innovations in the air….** *Policy Matters* 2003, **12**:92–101.
20. Sikor T, Barlosius E, Scheumann W: **Introduction: public-private relations and key policy issues in natural resource governance.** In *Public and private in natural resource governance: A false dichotomy? Edited by: Sikor T.* London: Earthscan; 2008:1–21.
21. Agrawal A, Gupta K: **Decentralization and participation: the governance of common pool resources in Nepal's Terai.** *Wold Dev* 2005, **33**:1101–1114.
22. Baral N, Stern MJ: **Looking back and looking ahead: local empowerment and governance in Annapurna Conservation Area, Nepal.** *Environ Conserv* 2009, **37**:54–63.
23. Peters GB: **Governance and comparative politics.** In *Debating Governance. Authority, Steering, and Democracy.* Edited by Pierre J. Oxford: Oxford University Press; 2000:36–53.
24. Kothari A: **Protected area and people: the future of the past.** *Parks* 2008, **17**:23–34.
25. Dudley N: *Guidelines for Applying Protected Area Management Categories.* Gland, Switzerland: IUCN; 2008.
26. Natcher DC, Davis S, Hickey CG: **Co-management: Managing relationships, not resources.** *Hum Organ* 2005, **64**:240–250.
27. Berkes F: **Evolution of co-management: Role of knowledge generation, bridging organizations and social learning.** *J Environ Manage* 2009, **90**:1692–1702.
28. Nadasdy P: **Reevaluating the co-management success story.** *Arct* 2003, **56**:367–380.
29. Berkes F: **From community-based resource management to complex systems.** *Ecol Soc* 2006, **11**:45 [http://www.ecologyandsociety.org/vol11/iss1/art45/]
30. Rhodes M: **The Scientific Objectives of the NEWGOV Project: A Revised Framework.** In *NEWGOV Consortium Conference: 30–31 May 2005.* Florence: European University Institute; 2005.
31. Eagles PFJ: **Governance models for parks, recreation and tourism.** In *Transforming Parks and Protected Areas: Policy and Governance in a Changing World.* Edited by Hanna KS, Clark DA, Slocombe DS. London: Routledge; 2008:39–61.
32. Brooks JS, Franzen MA, Holmes CM, Grote MN, Borgerhoff Mulder M: *Development as a conservation tool: Evaluating ecological, economic, attitudinal, and behavioural outcomes. CEE review 05–014 (SR20). Collaboration for Environmental Evidence*; 2006. www.environmentalevidence.org/SR20.html.
33. Waylen KA, Fischer A, McGowan PJK, Thirgood SJ, MilnerGulland EJ: *The effect of local cultural context on communitybased conservation interventions: evaluating ecological, economic, attitudinal and behavioural outcomes. CEE review 09–019 (SR80). Collaboration for Environmental Evidence*; 2010. www.environmentalevidence.org/SR80.html.
34. Brooks JS, Waylen KA, Borgerhoff-Mulder M, Brosius P: *The effect of non-local socio- political context on community-based conservation interventions: evaluating ecological, economic, attitudinal and behavioural outcomes. CEE protocol 09–021 (SR82). Collaboration for Environmental Evidence*; 2010. www.environmentalevidence.org/SR82.html.
35. Bowler D, Buyung-Ali L, Healey JR, Jones JPG, Knight T, Pullin AS: *The evidence base for community forest management as a mechanism for supplying global environmental benefits and improving local welfare. CEE review 08–011 (SR48). Collaboration for Environmental Evidence*; 2010. www.environmentalevidence.org/SR48.html.
36. Geldmann J, Barnes M, Coad L, Craigie I, Hockings M, Burgess B: *Effectiveness of terrestrial protected areas in reducing biodiversity and habitat loss. CEE review 10–007. Collaboration for Environmental Evidence*; 2013. http://www.environmentalevidence.org/SR10007.html.
37. Pullin AS, Bangpan M, Dalrymple S, Dickson K, Healey J, Hockley N, Jones J, Knight T, Oliver S: *Human well-being impacts of terrestrial protected areas. Systematic review protocol 11–009. Collaboration for environmental evidence*; 2012. http://www.environmentalevidence.org/SR11009.html.
38. Kothari A: **Community conserved areas: towards ecological and livelihood security.** *Parks* 2006, **16**:3–13.
39. Dudley N, Phillips A: *Forests and Protected Areas: Guidance on the use of the IUCN protected area management categories. Best Practice Protected Area Guidelines Series No. 12.* Gland and Cambridge: IUCN; 2006. http://data.iucn.org/dbtw-wpd/edocs/PAG-012.pdf.
40. Porter-Bolland L, Ellis EA, Guariguata MR, Ruiz-Mallén I, Negrete-Yankelevich S, Reyes-García V: **Community managed forests and forest protected areas: An assessment of their conservation effectiveness across the tropics.** *For Ecol Manage* 2011, **268**:6–17.
41. Gaston KJ, Jackson SF, Cantu-Salazar L, Cruz-Pinon G: **The ecological performance of protected areas.** *Annu Rev Ecol Evol Syst* 2008, **39**:93–113.
42. Rodrigues ASL, Andelman SJ, Bakarr MI, Boitani L, Brooks TM, Cowling RM, Fishpool LDC, da Fonseca GAB, Gaston KJ, Hoffmann M, Long JS, Marquet PA, Pilgrim JD, Pressey RL, Schipper J, Sechrest W, Stuart SN, Underhill LG, Waller RW, Watts MEJ, Yan X: **Effectiveness of the global protected area network in representing species diversity.** *Nat* 2004, **428**:640–643.
43. Murray GD: **Multifaceted Measures of Success in Two Mexican Marine Protected Areas.** *Soc Nat Res Int J* 2005, **18**:889–905.
44. Granderson AA: **Enabling multi-faceted measures of success for protected area management in Trinidad and Tobago.** *Eval Program Plann* 2011, **34**:185–195.
45. Ferraro PJ, Hanauer MM, Sims KRE: **Conditions associated with protected area success in conservation and poverty reduction.** *PNAS* 2011, **108**:13913–13918.
46. Wilshusen P, Brechin S, Forwangler C, West P: **Reinventing a square wheel: Critique of a resurgent "protection paradigm" in international biodiversity conservation.** *Soc Nat Resour* 2002, **15**:17–40.
47. Hawken IF, Granoff IME: **Reimagining Park Ideals: Toward Effective Human-Inhabited Protected Areas.** *J Sustain For* 2010, **29**:122–134.
48. Brechin SR, Wilshusen PR, Fortwangler CL, West PC: **Beyond the square wheel: toward a more comprehensive understanding of biodiversity conservation as social and political process.** *Soc Nat Resour* 2002, **15**:41–64.
49. Brechin SR, Murray G, Mogelgaard K: **Conceptual and Practical Issues in Defining Protected Area Success: The Political, Social, and Ecological in an Organized World.** *J Sustain For* 2010, **29**:362–389.
50. Hutton J, Adams WM, Murombedzi JC: **Back to the Barriers? Changing Narratives in Biodiversity Conservation.** *Dev Stud* 2005, **32**:341–370.
51. Ewers RM, Rodrigues ASL: **Estimates of reserve effectiveness are confounded by leakage.** *Trends Ecol Evol* 2008, **23**:113–116.
52. Thomas J, Brunton J, Graziosi S: *EPPI-Reviewer 4.0: software for research synthesis. EPPI-Centre Software.* London: Social Science Research Unit, Institute of Education, University of London; 2010.
53. Brooks JS, Waylen KA, Borgerhoff Mulder M: **Assessing community-based conservation projects: a systematic review and multilevel analysis of attitudinal, behavioural, ecological an economical outcomes.** *Environ Evid* 2013, **2**:1–34. http://www.environmentalevidencejournal.org/content/pdf/2047-2382-2-2.pdf.
54. Harden A: **The quality of qualitative evidence: a review of assessment tools.** In *Seventh Annual International Campbell Colloquium: 14–16 May 2007.* London: EPPI-Centre; 2007.
55. Rees R, Oliver K, Woodman J, Thomas J: *Children's views about obesity, body size, shape and weight: a systematic review.* London: EPPI-Centre, Social Science Research Unit, Institute of Education, University of London; 2009.

56. CASP appraisal tools: **Qualitative Appraisal Tool.** *Solut Pub Health* 2006 [http://www.civilservice.gov.uk/wp-content/uploads/2011/09/Qualitative-Appraisal-Tool_tcm6-7385.pdf]
57. Kothari A: **Community Conserved Areas: editorial.** *Parks* 2006, **16**:1–2.
58. Grimble R, Wellard K: **Stakeholder methodologies in natural resource management: a review of concepts, contexts, experiences and opportunities.** *Agric Syst* 1997, **55**:173–193.

9

What is the impact of active management on biodiversity in forests set aside for conservation or restoration? A systematic review protocol

Claes Bernes[1*], Bengt Gunnar Jonsson[2], Kaisa Junninen[3], Asko Lõhmus[4], Ellen Macdonald[5], Jörg Müller[6] and Jennie Sandström[2]

Abstract

Background: The traditional approach to limiting impacts of forestry on biodiversity is to set aside forest areas of particular conservation interest, either as formally protected reserves or on a voluntary basis. Many set-asides are left more or less untouched, but some of them have a history of disturbances such as wildfires, forest grazing, coppicing or small-scale felling. Such areas may gradually lose the qualities that were to be safeguarded unless the disturbances are re-introduced (e.g. by burning) or replaced with alternatives (e.g. gap-felling). Active management of forest set-asides may be particularly relevant in areas where the biota has been impoverished by intensive and large-scale harvesting. Here, biodiversity may not be able to recover adequately without restoration measures such as gap-felling or creation of dead wood.

In recent years, interest in active management of forest set-asides has increased, but opinions differ among conservationists on how such management should be balanced against non-intervention. The topic of the proposed systematic review has therefore met approval among stakeholders in Sweden, where it is currently an issue of high concern.

Methods: The review will examine primary field studies of how various forms of active management have affected biodiversity in boreal or temperate forests set aside for conservation or restoration. The primary focus will be on forest types represented in Sweden. In some cases, useful insights about management options may also be provided by studies of interventions in commercially managed forests. Non-intervention or alternative forms of active management will be used as comparators. Relevant outcomes include assemblage diversity (species richness, diversity indices), abundance of different functional or taxonomic groups of organisms, population viability of target species, and indicators of forest biodiversity such as forest structure and amounts of dead wood.

The relevant scientific literature may turn out to be very heterogeneous, however. Numerous combinations of management forms and biodiversity outcomes can be conceived, and it remains to be seen whether any such combination is covered by sufficiently many studies to allow a meaningful meta-analysis. Nonetheless, it should be feasible to achieve a useful narrative synthesis of the available evidence.

Keywords: Biodiversity, Early-successional habitat, Forest conservation, Forest restoration, Forest set-aside, Forest reserve, Habitat management, Prescribed burning, Thinning, Dead wood

* Correspondence: claes.bernes@eviem.se
[1]Mistra Council for Evidence-Based Environmental Management
Royal Swedish Academy of Sciences, SE-104 05 Stockholm, Sweden
Full list of author information is available at the end of the article

Background

Conservation and restoration of forest biodiversity

Globally, forest cover has declined over many millennia, but extensive regions in northern Eurasia and North America still remain mostly forested. Here, the area covered by forest is currently stable or even expanding. However, this stability masks impacts by large-scale forestry and other industrial activities/developments that have transformed forest ecosystems from being governed mostly by natural processes to being under strong human influence [1]. The transition has had far-reaching consequences for forest structures and dynamics, and it has been accompanied by a significant loss of forest biodiversity at all levels [2]. Genetic diversity, species richness and ecosystem variability have all decreased.

The traditional way of compensating for such consequences is to identify forest areas that have (or may develop) high natural values and set them aside as reserves, with restrictions on how they can subsequently be used. Protection of this kind is still seen as an indispensable tool for nature conservation – current global targets state that at least 17% of the total land area should be protected for the benefit of biodiversity [3]. Such a target constitutes a major challenge, however, since the proportion of forests that still have high natural values falls far short of the 17% level in many regions.

Of the 'western taiga' forest in Sweden, for instance, it has been estimated that only 2 million hectares, about 10% of its original extent, fulfils the habitat criteria of the European Species and Habitat Directive [4]. Old-growth characteristics in particular (such as large old trees and dead trees) have been drastically reduced in forests throughout the world, and significant parts of the biodiversity that depends on such characteristics could go extinct even in countries that remain largely forested [5].

In existing reserves too, past silvicultural use has in some cases impaired forest habitats to such a degree that they are unable to support a diverse flora and fauna. For example, the extensive forest-reserve networks that have been established in Estonia and Germany are still dominated by mid-aged second-growth forests, and 10% of the area covered by the Estonian network has also been degraded by artificial drainage [6,7].

In some areas still untouched by large-scale forestry, e.g. the montane natural forests that predominate among Sweden's protected forest areas, vegetation is shaped primarily by small scale internal dynamics, i.e. the ageing and death of fully grown trees and the establishment and growth of new seedlings. In most forest reserves in such areas, biodiversity can be preserved under a *non-intervention management* approach (also called *passive management* or *free development*), intended to allow internal dynamics to continue undisturbed.

In many protected forests, however, the remaining biodiversity values are legacies of past disturbance regimes that nowadays are suppressed. In these cases, one may have to weigh the consequences of doing nothing against the uncertainties of active intervention. This situation is common e.g. in the boreal pine forest, which in its natural state is shaped by recurring fires that create an abundance of dead wood and keep the stands relatively sparse and mixed with a significant broadleaved component. In northern Europe, forest fires are now very rare. Pine forests in this region have therefore become denser, with an increasing preponderance of spruce ([8] and Hedwall & Mikusinski, in prep.).

In other forest set-asides, the existing conservation interest is partly a result of earlier forest grazing (using the forest as pasture for livestock), coppicing with standards, small-scale felling or similar human influences. Since these activities were usually discontinued several decades ago, the forest has become denser and more shaded in such areas too, to the detriment of a large number of species that are dependent on a semi-open forest landscape. In Estonia, for example, the area of wooded grasslands has plummeted from 700,000 ha in 1940 to less than 10,000 ha today. As a consequence, many species are now threatened there [9].

Reserves in areas of these kinds may need some form of *active management* to conserve the characteristics that were the reason for setting them aside. Such management could, for instance, involve burning, thinning, girdling of trees, resumption of forest grazing, removal of undesired species or blocking of drainage ditches.

Active management designed to achieve a favourable conservation status can be particularly relevant in regions where forests have already been degraded. In such areas, the creation of a network of forest reserves with high-quality habitats may call for a combination of landscape-scale planning and stand-scale *restoration.* While, in this review, 'conservation' mainly refers to management intended to preserve existing natural values, 'restoration' may be defined as management intended to recreate former values that have been lost.

In some places, it could be enough to allow exploited forest areas to gradually return to a near-natural state through non-intervention, but elsewhere active measures may be required to sustain or accelerate the natural recovery [10]. Nowadays, a multitude of restoration activities are performed to counteract losses of forest biodiversity (e.g. [11]). An ambitious goal set in the EU Biodiversity Strategy is that 15% of degraded ecosystems are to be restored by 2020.

In Sweden, the Forest Agency has estimated that, in the long term, some 500,000 ha of forest land needs to be restored if there are to be, once again, sufficient areas of those biologically valuable forest types that are currently in short supply in the country. Since 1999, Sweden's Environmental Code allows for reserves to be designated with the intention of recreating lost natural values. It is thus possible to establish reserves also in forest areas that recently have been clear-cut, and which therefore can be acquired at low cost.

Solutions of this nature could be particularly useful in historically deforested regions, such as Britain. There, the best option for protecting old-forest biota may be to manage mature planted forests in a way that facilitates the development of natural forest characteristics [12].

Finally, it should be noted that both conservation and restoration efforts may have to adjust to broader changes in the environment (e.g. climate change and variations in land use intensity).

Swedish perspectives on active management of forest set-asides

Of the financial resources available for conservation of Swedish forests, the largest share by far is still devoted to setting aside additional areas of valuable forest as reserves, rather than managing reserves that already have been established. In recent years, though, interest in active management of forest reserves has increased.

Several Government bills dealing with environmental or silvicultural issues have underscored the importance of maintaining the influence of fire in Swedish forests and of 'managing care-demanding forests with high natural and cultural values so that these assets are preserved and enhanced' (e.g. [13]). Various forms of active management of forest reserves are recommended in many of the action plans for threatened species that have been developed by the Swedish Environmental Protection Agency (SEPA) and county administrative boards. Several counties have also adopted regional strategies for prescribed burning of protected forests (e.g. [14]). During the years 2006–2010, the county administrative boards spent a total of €13 million on the management of forest reserves [15].

Prescribed burning and similar activities also take place at a substantial scale in forest areas that have been voluntarily set aside by forest companies and other forest owners. Interest in such measures among Swedish forest owners grew when, in 1997, it became possible for them to obtain forest certification. One criterion for certification is that at least 5% of a forest holding is voluntarily excluded from production; another is that 5% of harvested areas on dry or mesic soils are burned.

As a consequence of this development, prescribed burning of forest is now carried out over at least 1,000 ha annually in Sweden. This is of the same order of magnitude as the total present-day extent of wildfires in the country. EU-funded LIFE projects have contributed significantly to the reintroduction of forest fires and other restoration measures in Natura 2000 areas in both Sweden and Finland.

However, opinions currently differ among conservationists on how active management of forests set aside for conservation or restoration should be balanced against non-intervention. The Swedish Society for Nature Conservation [16] recently criticised SEPA for largely continuing with a traditional non-intervention policy. Questioning whether such a policy is in line with the Swedish Environmental Quality Objectives, where measures to preserve biodiversity are given high priority, the Society 'felt that parts of SEPA instead have acted as if impressions of "virginity", "primordialism" or "naturalness" were more important as goals for the protection and management of reserves than the conservation of biodiversity'.

SEPA has now published a revised strategy for the management of forest reserves [15]. According to the agency, non-intervention should be 'the main management strategy in objects in which internal dynamics should prevail'. Reserves in montane forests, wet forests and a number of other specific forest types are mentioned as examples of such objects, but the agency also declares that 'non-intervention contributes to perceptions of wilderness and virginity, which makes it an important management option in other forest types as well'. However, the strategy also states that various forms of active management should be given high priority in parts of the country. These management options include burning, removal of spruce, grazing, use of leaf fodder and haymaking, and maintenance or restoration of flood regimes and drained soils.

While regarding SEPA's new management strategy as acceptable, the Swedish Society for Nature Conservation maintains that day-to-day decisions taken by the agency still reflect more traditional views on how forest reserves should be managed (Kristoffer Stighäll, pers. comm.). During our consultations with stakeholders (see Additional file 1), similar criticism has also been expressed by representatives of county administrative boards.

Scientific basis

The SEPA strategy for the management of protected forests is one of many policy statements, guidelines and recommendations on forest conservation that have been published over the past decades, in Sweden and elsewhere. One common feature of much of this literature – regardless of whether it mainly recommends non-intervention, active management or both – is a lack of references to scientific evidence.

Until quite recently, few well-designed experiments had been made to investigate how forest set-asides are affected by various kinds of management. Earlier, management policies were mainly based on historical accounts or palaeoecological studies of land use, natural disturbance regimes and environmental qualities in the past (to the extent that they were supported by any evidence at all).

Sutherland [17] summarised the situation regarding habitat management as follows: 'Land managers should regularly carry out randomized, replicated, controlled and monitored experiments but in practice very rarely do. [...] Instead, the norm is to alter the management (often making a number of changes simultaneously), not have a control area, subjectively decide if the changes have been successful and pass on the opinions verbally.'

Even today, the evidence on which to base conservation or restoration of forests remains sparse, but there has been some improvement over the last few years. A review of management options for protected areas has been provided by Götmark [18], and a recent scoping exercise by the Mistra Council for Evidence-Based Environmental Management (Mistra EviEM) identified a limited but significant set of studies of how various ways of managing forest set-asides affect biodiversity.

Most of the studies found during the scoping exercise had been conducted in Fennoscandia or in the United States. The treatment most commonly investigated was prescribed burning, often in combination with partial cutting. Only a minority of the studies dealt with old-growth forest with no history of large-scale harvesting. Several of them described the restoration of certain old-growth characteristics (e.g. presence of dead wood or uneven-aged structure) to stands that earlier had been under even-aged management and still to some extent were used for commercial timber production. The investigated outcomes differed from case to case. Most studies focused on one or several groups of plants, fungi or animals, but some instead examined forest structure, amounts of dead wood or other environmental qualities related to biodiversity.

The only completed systematic review related to this subject has been conducted by Kalies *et al.* [19], who studied how thinning and burning (including wildfires) affects vertebrates (mainly birds) in *Pinus ponderosa* and mixed conifer forests in the southwestern United States.

Scarcity of scientific studies of how biodiversity is affected by active management in forests set aside for conservation or restoration is clearly still an obstacle for systematic reviews within this field.

Objective of the review

The primary aim of this systematic review is to clarify how the biodiversity of forest set-asides is affected by various forms of active management.

Primary question

What is the impact of active management on biodiversity in forests set aside for conservation or restoration?

Components of the primary question

Population/subject: Forests set aside for conservation or restoration of biodiversity

Intervention: Active management (e.g. prescribed burning, thinning, grazing, creation of dead wood, introduction or removal of species)

Comparator: Non-intervention or alternative types of intervention

Outcomes: Measures or indicators of biodiversity

It may be noted that in this context, the primary comparator (non-intervention) can be regarded as a management option as relevant to study in detail as any kind of active management.

Since the review is based on Swedish initiatives and funding, it will focus on types of forest that are represented in Sweden. These types include boreal and temperate forests, i.e. vegetation zones that extend over many parts of the world. Any study of how biodiversity has been affected by active management of a forest set-aside within these zones may be useful as input to the review. This also means that the results of the review should be relevant not only for Sweden but for other regions as well.

The scientific literature on the topic of the review may turn out to be very heterogeneous. Numerous combinations of management forms and biodiversity outcomes can be conceived, and it remains to be seen whether any such combination is covered by sufficiently many studies to allow a meaningful meta-analysis. Nonetheless, it should be feasible to produce a useful narrative synthesis of the currently available evidence on how different forms of management affect different aspects of biodiversity. Such a synthesis would also be able to reveal the most important knowledge gaps and specify the most urgent research needs within this field.

As an alternative, it would also be possible to produce a systematic map of the subject area [20]. Such a map would give an overview of the evidence base, and it might identify subtopics that could then be addressed by full systematic reviewing.

Methods

Search terms

The review team conducted a scoping exercise to assess alternative search terms, testing them against a set of about 20 articles known to be relevant. A preliminary search string was later modified and amended based on suggestions received during the public review. This resulted in the following final selection of search terms:

Subject: forest*, woodland*, "wood* pasture*", "wood* meadow*"

Forest type: boreal, boreonemoral, hemiboreal, nemoral, temperate, conifer*, deciduous, broadlea*, "mixed forest", spruce, "Scots pine", birch, aspen, beech, "*Quercus robur*", Swed*

Intervention: conserv*, restor*, rehabilitat*, "active management", (prescribed OR control* OR experiment*) AND (burn* OR fire*), thinn*, (partial OR selecti* OR gap OR retention) AND (felling OR cutting OR harvest*), "green-tree retention", *introduc*, remov*, graz*, girdl*, ditch*, flood*, fenc*, exclos*, pollard*, coppic*

Outcomes: *diversity, species AND (richness OR focal OR target OR keystone OR umbrella OR red-list* OR threatened OR endangered OR rare), "species density", "number of species", indicator*, abundance, "dead wood", "woody debris", "woody material", "forest structure", habitat*

The terms within each of the categories 'subject', 'forest type', 'intervention' and 'outcomes' will be combined using the Boolean operator 'OR'. The four categories will then be combined using the Boolean operator 'AND'. An asterisk (*) is a 'wildcard' that represents any group of characters, including no character.

The search terms within the 'forest type' category have been chosen to optimise the likelihood of finding relevant studies in Sweden or in forests elsewhere that are dominated by tree species commonly occurring in Sweden. However, the terms should also be capable of identifying a satisfactory share of relevant studies carried out in other boreal and temperate forest types throughout the world.

The selection of search terms above translates to the following search string:

English:

forest* OR woodland* OR "wood* pasture*" OR "wood* meadow*"

AND (boreal OR boreonemoral OR hemiboreal OR nemoral OR temperate OR conifer* OR deciduous OR broadlea* OR "mixed forest" OR spruce OR "Scots pine" OR birch OR aspen OR beech OR "*Quercus robur*" OR Swed*)

AND (conserv* OR restor* OR rehabilitat* OR "active management" OR ((prescribed OR control* OR experiment*) AND (burn* OR fire*)) OR thinn* OR ((partial OR selecti* OR gap OR retention) AND (felling OR cutting OR harvest*)) OR "green-tree retention" OR *introduc* OR remov* OR graz* OR girdl* OR ditch* OR flood* OR fenc* OR exclos* OR pollard* OR coppic*)

AND (*diversity OR (species AND (richness OR focal OR target OR keystone OR umbrella OR red-list* OR threatened OR endangered OR rare)) OR "species density" OR "number of species" OR indicator* OR abundance OR "dead wood" OR "woody debris" OR "woody material" OR "forest structure" OR habitat*)

Searches will also be made using the following French and German counterparts to the English search string (although they in some cases will have to be simplified):

French:

forêt* OR forestièr* OR "pâturag* bois*" OR "bois prés" OR "prairie boisée"

AND (boréale OR boréonémorale OR hémiboréale OR némorale OR tempéré* OR conifère OR résineus* OR feuillu* OR "forêts mixtes" OR épinette OR "pin sylvestre" OR bouleau OR peuplier OR hêtre OR "*Quercus robur*" OR Suèd*)

AND (conserv* OR restaur* OR réhabilit* OR ((dirigé* OR expériment*) AND (brûlag* OR incend* OR feu)) OR éclairci* OR CPRS OR ((partiell OR jardinag* OR retention OR troué*) AND (coupe OR récolte* OR exploit*)) OR *introduit* OR éliminat* OR enlève* OR pâtur* OR annel* OR drain* OR canalis* OR inond* OR clôtur* OR exclos* OR têtard OR taillis)

AND (*diversité OR ((espèce OR taxon) AND (richesse OR vises OR cibles OR clé OR disparition OR menacé* OR rare)) OR "densité des espèce" OR "nombre des espèce" OR indicatrice* OR abondance OR "bois mort" OR "debris ligneux" OR "matériel ligneux" OR "structure de la forêt" OR habitat*)

German:

(Wald OR Wälder)

AND (boreal* OR boreonemoral* OR hemiboreal* OR nemoral* OR temperat* OR gemäßigt* OR Schwed*) OR Nadelwald OR Laubwald OR Mischwald OR Fichte* OR Waldkiefer OR Birke* OR Aspe* OR Buche* OR "*Quercus robur*")

AND (Naturschutz* OR Renaturierung* OR ((kontrolliert* OR Experiment*) AND Feuer) OR Durchforstung* OR Femelschlag OR Femelhieb OR Femelnutzung OR Einzelstammnutzung OR Einzelbaumnutzung OR Einzelbaumernte OR

Lochhieb OR Einführung* OR Entnahme* OR Beweidung OR Ringeln* OR Entwässerung* OR Zaun* OR *zäunung OR Ausschluss*)
AND (Diversität OR Biodiversität OR Artenvielfalt OR (Art* AND (gefährdet OR bedrohte OR seltene OR "Rote* Liste")) OR Artenzahl OR Zielart* OR Schlüsselart* OR Schirmart* OR Artendichte OR Indikatorart* OR Abundanz OR Totholz OR Waldstruktur* OR Habitat)

No time, language or document type restrictions will be applied.

At some of the websites mentioned below, searches will also be made for relevant literature in Finnish, Russian and Swedish. The search terms will have to be customised to each of these websites, since few of them accept long and complex search strings. Final search strings used for each search will be recorded in an Appendix, together with search dates.

Publication databases

The search aims to include the following online databases:

1) Academic Search Premier
2) Agricola
3) Biological Abstracts
4) GeoBase + GeoRef
5) JSTOR
6) Libris
7) Nauchnaya elektronnaya biblioteka
8) Scopus
9) SwePub
10) Web of Science
11) Wiley Online Library

Search engines

Internet searches will be performed using the following search engines:

Google (www.google.com)
Google Scholar (scholar.google.com)

In each case, the first 200 hits (based on relevance) will be examined for appropriate data.

Specialist websites

Websites of the specialist organisations listed below will be searched for links or references to relevant publications and data, including grey literature.

Ancient Tree Forum (www.ancient-tree-forum.org.uk)
Bureau of Land Management, US Dept. of the Interior (www.blm.gov)
Environment Canada (www.ec.gc.ca)
European Commission Joint Research Centre (ec.europa.eu/dgs/jrc)
European Environment Agency (www.eea.europa.eu)
Food and Agriculture Organization of the United Nations (www.fao.org)
Finland's environmental administration (www.ymparisto.fi)
International Union for Conservation of Nature (www.iucn.org)
Metsähallitus (www.metsa.fi)
Natural Resources Canada (www.nrcan.gc.ca)
Nordic Council of Ministers (www.norden.org)
Norwegian Environment Agency (www.miljødirektoratet.no)
Norwegian Forest and Landscape Institute (www.skogoglandskap.no)
Norwegian Institute for Nature Research (www.nina.no)
Parks Canada (www.pc.gc.ca)
Society for Ecological Restoration (www.ser.org)
Swedish County Administrative Boards (www.lansstyrelsen.se)
Swedish Environmental Protection Agency (www.naturvardsverket.se)
Swedish Forest Agency (www.skogsstyrelsen.se)
Swedish University of Agricultural Sciences (www.slu.se)
UK Environment Agency (www.environment-agency.gov.uk)
United Nations Environment Programme (www.unep.org)
United States Environmental Protection Agency (www.epa.gov)
US Forest Service (www.fs.fed.us)

Other literature searches

As a check of the comprehensiveness of our searches, relevant articles and reports will also be searched for in bibliographies of literature reviews. Moreover, each member of the review team will use national and international contacts to get information on current research related to the topic of the review, and also to find non-peer-reviewed literature, including reports published in e.g. Swedish, Finnish, Estonian or Russian. In addition, unpublished data (e.g. records of prescribed fires and other interventions) may be available from forest owners, consultants or local authorities involved in the management of forest set-asides. Stakeholders will be asked to suggest suitable contacts.

Study inclusion/exclusion criteria

Articles found by searches in databases will be evaluated for inclusion at three successive levels. First they will be assessed by title by a single reviewer. In cases of uncertainty, the reviewer will tend towards inclusion. As a check of consistency, a subset of 100 titles will be assessed by all members of the review team. If the main reviewers' assessments tend to differ from those of the other team members, the team will discuss and agree on how the inclusion criteria should be interpreted.

Next, each article found to be potentially relevant on the basis of title will be judged for inclusion on the basis of abstract. The reviewer will tend towards inclusion in cases of uncertainty. A subset consisting of at least 100 of the articles will be assessed by at least two reviewers. A kappa statistic [21] relating to the assessments will be calculated. If this statistic indicates that the reviewers are inconsistent in their assessment ($\kappa < 0.6$), discrepancies will be discussed and the inclusion criteria will be clarified or modified.

Finally, each article found to be potentially relevant on the basis of abstract will be judged for inclusion by a reviewer studying the full text. Again, reviewers will tend towards inclusion in cases of uncertainty. Final decisions on whether to include doubtful cases will be taken by the review team as a whole.

Studies or datasets found by other means than database searches will be entered at the last stage of this screening process.

A list of studies rejected on the basis of full-text assessment will be provided in an appendix together with the reasons for exclusion.

Each study must pass each of the following criteria in order to be included.

- *Relevant subjects:* Forests in the boreal or temperate vegetation zones. Any ecosystem with a tree layer will be regarded as forest, which means that studies of e.g. wooded meadows and urban woodlands may be included.
  Studies of forests subject to large-scale commercial forestry may also be included, although this review deals with the management of forests set aside from forestry of that kind. The reason is that certain practices commonly applied in commercial forestry (e.g. thinning) may be useful as management options for set-asides too.
- *Relevant types of intervention:* Active management which is used or can be used to conserve or restore forest biodiversity. Relevant management types include but are not restricted to prescribed burning, thinning, livestock grazing, blocking of drainage ditches, introduction or removal of species, and creation of dead wood e.g. through girdling of trees. Although stakeholders suggested that studies of wildfires should be included, the review team decided not to do so. Wildfire is in most cases not a management option, although it may be possible to choose whether to suppress a fire or not, and while there is an extensive literature on the effects of unplanned and uncontrolled fires (e.g. [22,23]), their consequences for biodiversity cannot be assumed to be identical to those of prescribed burning. Studies have been made of prescribed burning too, and we judge that they will be sufficient for the needs of this review.
  We will also exclude studies of restrictions of public access and creation of artificial habitats such as provisioning of artificial nest sites for vertebrates (the latter management option has already been extensively reviewed, e.g. by Newton [24]).
- *Relevant type of comparator:* Non-intervention or alternative types of intervention.
- *Relevant types of outcome:* Measures or indicators of biodiversity in the terrestrial environment. Relevant measures of biodiversity include e.g. species richness, diversity indices and abundance of different taxonomic or functional groups of organisms, and also the population viability of target species. Forest structure and amounts of dead wood are two examples of relevant indirect biodiversity indicators.
- *Relevant type of study:* Primary field studies. This criterion excludes e.g. modeling studies and review papers.
- *Language:* Full text written in English, French, German, Danish, Norwegian, Swedish, Finnish, Estonian or Russian.

Potential effect modifiers and reasons for heterogeneity

To the extent that data are available, the following potential effect modifiers will be considered and recorded:

Geographical coordinates
Altitude
Climate (and climate change)
Mean age of forest stand
Dominant tree species
Stand productivity
Type and areal extent of forest set-aside
Type, areal extent, duration and intensity of management activities
Time elapsed from management activities to final data sampling
Landscape aspects (such as degree of isolation)
History of land use and protection

When assessing the outcomes of conservation efforts in set-aside forest areas, the land-use legacy must never be overlooked. It will also be necessary to consider conditions in the surroundings of such areas. Many reserves lie like scattered islands in a 'sea' of efficiently managed but species-poor timber production forest. This fragmentation of less exploited, species-rich natural forest can make it difficult to maintain biodiversity within the 'islands' set aside from forestry.

A final list of modifiers and causes of heterogeneity to be recorded will be established as the review proceeds.

Study quality assessment

Studies that have passed the relevance criteria described above will be subject to critical appraisal. Based on assessments of their validity, they will be categorised as having high, medium or low susceptibility to bias (e.g. due to presence or absence of replication or data on variance and potential effect modifiers). Studies with high susceptibility to bias will be excluded from the review.

The exact criteria to be applied when assessing study quality will be developed as the review proceeds, but factors such as methodological description, duration between intervention and data sampling, accounting for potential effect modifiers, replication and statistical treatment are likely to feature prominently among them.

Comparison of how different kinds of management affect biodiversity can in principle be made both temporally and spatially. Studies with a 'BA' (Before/After) design compare data collected at the same site prior to and following an intervention. Other studies may be based on comparison of different parts of a forest, some that have been subject to a certain kind of active management and some that have not. These may be termed as 'CI' (Comparator/Intervention) studies, or 'BACI' (Before/After/Comparator/Intervention) if they present data collected both before and after the intervention.

In practice, studies with a CI or BACI design are likely to be more useful than BA studies in the context of this review. This is because of the long time-scale of forest ecosystem changes, which means that a forest set-aside that has been subject to some kind of active management may also be affected by other influences (e.g. climate change, nitrogen deposition, or ecological succession following earlier land-use changes) before management effects have had time to develop fully. Such influences can be controlled for in CI and BACI studies, but not in BA studies.

Detailed reasoning concerning critical appraisal will be recorded in a transparent manner. In general, the quality of a study will be assessed by one reviewer, but final decisions on how to judge doubtful cases will be taken by the review team as a whole.

A list of studies rejected on the basis of quality assessment will be provided in an appendix together with the reasons for exclusion.

Data extraction strategy

Outcome means and measures of variation (standard deviation, standard error, confidence intervals) will be extracted from tables and graphs, using image analysis software when necessary. As a rule, we will extract data on major functional or taxonomic groups, but not on individual species or genera. Where time-series of data are available, we will only extract the most recent results (plus pre-intervention data from BA and BACI studies). Data on interventions and other potential effect modifiers will also be extracted from the included articles, but climatic data will be downloaded from the WorldClim database [25].

It may in some cases be useful to ask authors of relevant articles to supply data in digital format. This will primarily be done where useful data have been published in graphs from which they are difficult to extract accurately enough, or when it is known or assumed that considerable amounts of relevant but unpublished data may be available in addition to the published results. If raw data are provided, summary statistics will be calculated by us.

Data synthesis and presentation

A narrative synthesis of data from all studies included in the review will describe the quality of the results along with the study findings. Tables will be produced to summarise these results. Where studies report similar outcomes, meta-analysis may be possible. In these cases effect sizes will be standardised and weighted appropriately. Details of the quantitative analysis will only be known when full texts have been assessed for their contents and quality.

If meta-analysis of effect sizes is possible, it will take the form of random-effects models, and meta-regression will be performed where effect modifiers cause significant heterogeneity between studies. Subgroup analysis of categories of studies will also be performed where sufficient studies report common sources of heterogeneity. Publication-bias and sensitivity analysis will be carried out where possible. Overall management effects will be presented visually in plots of mean effect sizes and variance.

As an alternative to a full systematic review (or in addition to such a review), we may compile a systematic map of the evidence base. Such a map would be produced in an easily searchable format and provide basic data on the studies that we have found to be relevant, including study locations, forest types, types of intervention and types of reported outcomes.

Additional file

Additional file 1: Identification of topic and stakeholders.

Competing interests
The authors declare that they have no competing interests.

Authors' contributions
All authors participated in the drafting, revision and approval of the manuscript. All authors read and approved the final manuscript.

Acknowledgements
We are grateful for valuable comments and advice received from a considerable number of stakeholders, both at meetings in the initial phase of this project and during the public review of the protocol.
The review will be financed by the Mistra Council for Evidence-Based Environmental Management (EviEM).

Author details
[1]Mistra Council for Evidence-Based Environmental Management Royal Swedish Academy of Sciences, SE-104 05 Stockholm, Sweden. [2]Department of Natural Sciences, Mid-Sweden University, SE-851 70 Sundsvall, Sweden. [3]Metsähallitus Parks & Wildlife Finland, c/o UEF, P.O. Box 111FI-80101 Joensuu, Finland. [4]Institution of Ecology and Earth Sciences, Tartu University, Vanemuise 46, 51014 Tartu, Estonia. [5]Department of Renewable Resources, University of Alberta, 751 General Services Building, Edmonton, AB T6G 2H1, Canada. [6]Department of Conservation and Research, Bavarian Forest National Park, Freyunger Str. 2, 94481 Grafenau, Germany.

References
1. Schelhaas MJ, Nabuurs GJ, Schuck A: **Natural disturbances in the European forests in the 19th and 20th centuries.** *Glob Chang Biol* 2003, **9:**1620–1633.
2. Secretariat of the Convention on Biological Diversity: *Global Biodiversity Outlook 3.* Montréal: Secretariat of the Convention on Biological Diversity; 2010.
3. Convention on Biological Diversity: *Strategic Plan for Biodiversity 2011–2020,* COP Decision X/2. Aichi; 2010.
4. Eide W: *Arter och naturtyper i habitatdirektivet – bevarandestatus i Sverige 2013.* Uppsala: ArtDatabanken, SLU; 2014.
5. Hanski I: **Extinction debt and species credit in boreal forests: Modelling the consequences of different approaches to biodiversity conservation.** *Ann Zool Fennici* 2000, **37:**271–280.
6. Lõhmus A, Kohv K, Palo A, Viilma K: **Loss of old-growth, and the minimum need for strictly protected forests in Estonia.** *Ecol Bull* 2004, **51:**401–411.
7. Meyer P, Schmidt M, Spellmann H, Bedarff U, Bauhus J, Reif A, Späth V: **Aufbau eines Systems nutzungsfreier Wälder in Deutschland. Natur und Landschaft.** *Zeitschrift für Naturschutz und Landschaftspflege* 2011, **86:**243–249.
8. Linder P: **Structural changes in two virgin boreal forest stands in central Sweden over 72 years.** *Scand J For Res* 1998, **13:**451–461.
9. Sammul M, Kattai K, Lanno K, Meltsov V, Otsus M, Nõuakas L, Kukk D, Mesipuu M, Kana S, Kukk T: **Wooded meadows of Estonia: conservation efforts for a traditional habitat.** *Agric Food Sci* 2008, **17:**413–429.
10. Bernes C: *Biodiversity in Sweden.* Stockholm: Swedish Environmental Protection Agency, Monitor 22; 2011.
11. Halme P, Allen KA, Auniņš A, Bradshaw RHW, Brūmelis G, Čada V, Clear JL, Erikssom AM, Hannon G, Hyvärinen E, Ikauniece S, Iršėnaitė R, Jonsson BG, Junninen K, Kareksela S, Komonen A, Kotiaho JS, Kouki J, Kuuluvainen T, Mazziotta A, Mönkkönen M, Nyholm K, Oldén A, Shorohova E, Strange N, Toivanen T, Vanha-Majamaa I, Wallenius T, Ylisirniö AL, Zin E: **Challenges of ecological restoration: Lessons from forests in northern Europe.** *Biol Conserv* 2013, **167:**248–256.
12. Humphrey JW: **Benefits to biodiversity from developing old-growth conditions in British upland spruce plantations: a review and recommendations.** *Forestry* 2005, **78:**33–53.
13. Swedish Ministry of the Environment: *Svenska miljömål – för ett effektivare miljöarbete,* Government Bill 2009/10:155. Stockholm; 2010.
14. Östergötland County Administrative Board: *Strategi för naturvårdsbränning i sydöstra Sveriges skyddade skogsområden år 2012–2022. Bakgrund, analys och genomförande.* 2012. Meddelande 2012:13 http://www.lansstyrelsen.se/ostergotland/SiteCollectionDocuments/Sv/djur-och-natur/skyddadnatur/Naturvårdsbränning/rapp2012_13Ekologiskt_brandstrategi_orginal_webb.pdf.
15. Swedish Environmental Protection Agency: *Förvaltning av skogar och andra trädbärande marker i skyddade områden,* Rapport 6561. Stockholm; 2013.
16. Swedish Society for Nature Conservation: *Möjliggör skötsel av skyddad skog för att bevara hotade arter,* Letter to SEPA. 2012. [http://www2.naturskyddsforeningen.se/upload/Foreningsdokument/Ovrigt/Skrivelse_sk%C3%B6tsel_av%20skogsreservat.pdf]
17. Sutherland WJ: **Managing Habitats and Species.** In *Conservation Science and Action.* Edited by Sutherland WJ. Oxford: Blackwell; 1998:202–219.
18. Götmark F: **Habitat management alternatives for conservation forests in the temperate zone: Review, synthesis, and implications.** *For Ecol Managem* 2013, **306:**292–307.
19. Kalies E, Covington W, Chambers C, Rosenstock S: *How do Thinning and Burning Treatments in Southwestern Conifer Forests in the United States Affect Wildlife Density and Population Performance?* Collaboration for Environmental Evidence: CEE review 09–005 (SR66). 2010. http://www.environmentalevidence.org/Documents/Completed_Reviews/SR66.pdf.
20. Collaboration for Environmental Evidence: *Guidelines for Systematic Review and Evidence Synthesis in Environmental Management,* Version 4.2, p. 37 Environmental Evidence. 2013. www.environmentalevidence.org/Documents/Guidelines/Guidelines4.2.pdf.
21. Landis JR, Koch GG: **The measurement of observer agreement for categorical data.** *Biometrics* 1977, **33**(1):159–174.
22. Kuuluvainen T, Aakala T: **Natural forest dynamics in boreal Fennoscandia: a review and classification.** *Silva Fennica* 2011, **45:**823–841.
23. Gromtsev A: **Natural disturbance dynamics in the boreal forests of European Russia: A review.** *Silva Fennica* 2002, **36:**41–55.
24. Newton I: *Population Limitation in Birds.* London: Academic Press; 1998.
25. Hijmans RJ, Cameron SE, Parra JL, Jones PG, Jarvis A: **Very high resolution interpolated climate surfaces for global land areas.** *Int J Climatol* 2005, **25:**1965–1978.

10

Which components or attributes of biodiversity influence which dimensions of poverty?

Dilys Roe[1*], Max Fancourt[2], Chris Sandbrook[2], Mxolisi Sibanda[1], Alessandra Giuliani[1] and Andrew Gordon-Maclean[1]

Abstract

Background: There is an explicit assumption in international policy statements that biodiversity can help in efforts to tackle global poverty. This systematic map was stimulated by an interest in better understanding the evidence behind this assumption by disaggregating the terms and asking - as our review question - which components or attributes of biodiversity influence which dimensions of poverty?

Methods: We employed a search strategy that covered peer-reviewed and grey literature. Relevant studies included in the map were those that described an interaction by poor people with biodiversity in non-OECD countries and documented some kind of contribution (positive or negative) to different aspects of their well-being.

Results: A total of 387 studies were included in the final systematic map. Of these 248 met our additional criteria that studies should include a measure of the contribution to poverty alleviation. The studies were widely distributed geographically. Ecological distribution was less well spread, however, with the largest number of studies focussed on forests. We found studies addressing 12 different dimensions of poverty/well-being – although the most commonly studied was income. Similarly we found studies addressing all levels of biodiversity from genes to ecosystems. The largest number of studies was focussed on groups of resources – particularly non-timber forest products. In most cases, abundance was the attribute that made biodiversity important for poverty alleviation/well-being, while diversity was the least frequently noted attribute.

Conclusions: The map highlights a number of apparent gaps in the evidence base. Very few studies documented any causal link between use of biodiversity and an impact on poverty. In the majority of the studies biodiversity was framed in terms of its value as a *resource* – in the form of specific goods that can be used to generate tangible benefits such as cash, food fuel. Very few studies explored the underpinning role of biodiversity in ecosystem service delivery for poverty alleviation, and fewer investigated the benefits of diversity as a form of insurance or adaptive capacity. This is where we suggest research should be prioritised.

Keywords: Biodiversity, Nature conservation, Wildlife conservation, Poverty, Livelihoods

Background

Biodiversity conservation and poverty alleviation are both important societal goals attracting increasing international attention. At first glance they may appear to be separate policy realms with little connection. The Convention on Biological Diversity (CBD), agreed in 1992, was drafted in response to escalating biodiversity loss and provides an international policy framework for biodiversity conservation activities worldwide. Similarly, the OECD International Development Targets of 1996 - reiterated as the Millennium Development Goals (MDGs) in 2000 – focus international development efforts on global poverty alleviation.

However, there is an explicit assumption that conserving biodiversity (or reducing the rate of biodiversity loss) can help in efforts to tackle global poverty and enhance human well-being. Evidence of this assumption lies in the target that parties to the CBD agreed in 2002: "to achieve by 2010 a significant reduction of the current rate of biodiversity loss at the global, regional and national level *as a contribution to poverty alleviation* [emphasis added] and to the benefit of all life on earth" [1]. The development community also bought into this assumption: when the

* Correspondence: dilys.roe@iied.org
[1]International Institute for Environment and Development (IIED), 80-86 Grays Inn Road, London WC1X 8NH, UK
Full list of author information is available at the end of the article

Millennium Development Goals (MDGs) were formulated in 2000, for example, Goal 7 included a target to "reverse the loss of environmental resources", one indicator of which was the area of land under protection for biodiversity. Subsequently, the CBD "2010 Target" was included as a new target within MDG7 following the 2006 UN General Assembly, with additional biodiversity indicators [2].

The reduction in the rate of biodiversity loss anticipated in the 2010 target was not achieved [3,4]. This continued loss of biodiversity is lamented not just for its own sake but for its potential implications for continued human well-being and poverty reduction. The 2010 progress report on the MDGs, for example, noted "The irreparable loss of biodiversity will also hamper efforts to meet other MDGs, especially those related to poverty, hunger and health, by increasing the vulnerability of the poor and reducing their options for development" [5]. A high level meeting at the September 2010 UN General Assembly further stressed the linkage, claiming: "preserving biodiversity is inseparable from the fight against poverty" [6]. The CBD's new Strategic Plan (2011–2020), agreed at the 10th Conference of Parties in Nagoya, Japan in October 2010 continues to emphasise the link between achieving conservation goals and reducing poverty: its mission being to "take effective and urgent action to halt the loss of biodiversity in order to ensure that by 2020 ecosystems are resilient and continue to provide essential services, thereby securing the planet's variety of life, and contributing to human well-being, and poverty eradication." [7].

At some levels the relationship between biodiversity and poverty is absolute – biodiversity underpins the delivery of essential ecological services on which the whole of humanity is dependent. But there is certainly no linear relationship. For example, the Millennium Ecosystem Assessment (MA) suggested that while many millions of people have benefited from the transformation of ecosystems and exploitation of natural resources, the benefits have not been evenly or equitably distributed, with the poor being the biggest losers [8]. Other commentators have noted the dynamic and context-specific nature of the biodiversity conservation-poverty alleviation relationship [9] and have suggested that factors such as individual access to, and control over, resources; policies on poverty and biodiversity protection; population growth and density are critical in determining whether or not the existence, or use, of biodiversity leads to poverty reduction [10].

Biodiversity is defined by the CBD as "the variability among living organisms from all sources including, *inter alia*, terrestrial, marine and other aquatic ecosystems and the ecological complexes of which they are part; this includes diversity within species, between species and of ecosystems". This focus on variability is often missing, however, when assertions such as those above are made. The term biodiversity is often used to refer to the *amount* (in terms of abundance or biomass) of species or resources, or the *extent* of habitat rather than diversity *per se* [11,12]. Poverty is another term with many different definitions. The simplest usually relate to some level of material wealth – for example the Millennium Development Goal to "eradicate extreme poverty" refers to the billion-plus people whose income is less that US $1 a day. However, poor people often do not define themselves in cash income terms – indeed the concept of cash is completely meaningless for some indigenous communities who live outside of the cash economy. It has therefore become increasingly recognised that poverty is multi-dimensional. The World Bank, for example, describes poverty as 'to be hungry, to lack shelter and clothing, to be sick and not cared for, to be illiterate and not schooled' [13] while Sen discusses poverty in terms of 'capabilities' to fulfil a productive life including good health, access to education and political freedom [14]. This conceptualisation of poverty is similar to that of human well-being, and indeed, poverty can be thought of as the opposite - or absence - of well-being [8].

Our research is stimulated by an interest in better understanding the evidence behind the broad claims made by the United Nations and others about the role of biodiversity in contributing to poverty alleviation. Disaggregating the terms and exploring which particular components/attributes of biodiversity are important to poor people and in what ways will help achieve this understanding.

Objective of the review

Systematic mapping is intended to provide an overview of evidence on broad topics - to describe the nature, volume and characteristics of research in a chosen field [15,16]. Systematic maps can be used as a tool for identifying where studies would lend themselves to robust synthesis - e.g. through a systematic review; and to identify knowledge gaps to inform new research initiatives. In this case, the overall objective of our systematic map was to describe the current state of the evidence base on biodiversity – poverty links. In particular we set out to explore the degree to which it disaggregates "biodiversity" and "poverty", its representativeness, the types of linkages that have been found, and the methods that have been used to analyse them. Figure 1 provides a conceptual framework diagram for the review.

Defining the primary question components

We had originally framed our primary research question as follows: Which components or attributes of biodiversity affect (positively or negatively) which dimensions of poverty? Following comments from a reviewer of a draft

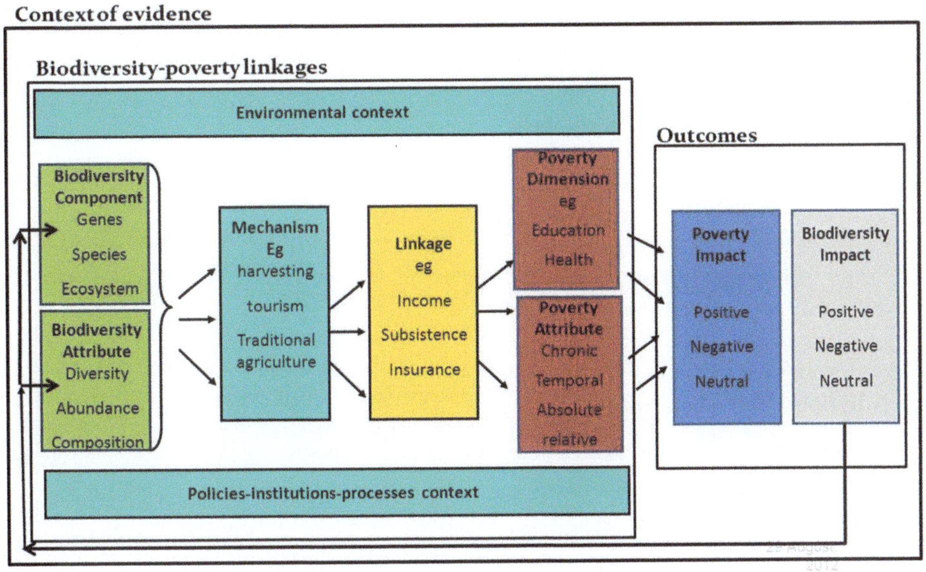

Figure 1 Conceptual framework for reviewing biodiversity-poverty linkages.

of this paper, and acknowledgment that we were not seeking solely to examine studies that had measured a quantifiable effect or impact of biodiversity on poverty we reframed the question to better reflect the broader nature of our enquiry, specifically: Which components or attributes of biodiversity *influence* (positively or negatively) which dimensions of poverty? Table 1 summarises the different components of our primary research question.

Our interpretation of the term "biodiversity" in this context merits some discussion. As we note in the Background section above, the CBD definition of biodiversity encompasses "living organisms from all sources". For the purposes of the map, however, we were predominantly interested in natural habitats and wild species – what Balmford et al. describe as "wild nature" [17] – rather than all living organisms. The line between what is wild and what is not is, however, very fuzzy. For example, we did not include mainstream agricultural crops in our interpretation of biodiversity but we did include indigenous varieties of crops and crop wild relatives, or locally domesticated wild species. Similarly, we did not include modern livestock as a component of biodiversity but we did include traditional breeds or landraces.

We also did not want to conduct a study on the health impacts of living pathogens so our interpretation of biodiversity also omitted micro-organisms, parasites and disease vectors such as mosquitoes. We recognise, however, that the impact of such living organisms on poverty is probably far more significant than that of any other component of biodiversity.

Finally, much has been written about the varied and complex inter-linkages between biodiversity and ecosystem services. In this review we have not covered the broad ecosystem service literature but only where a study has specifically linked the provision of a specific ecosystem service with particular components of biodiversity. Mace et al. [18] identify three different roles for biodiversity in ecosystem services: as a regulator of ecosystem processes; as a final ecosystem service; and as a good. Our review has encompassed studies on biodiversity in each of these roles but within that has predominantly adopted what Mace et al. would refer to as a "conservation perspective" where the focus is on "a subset of biodiversity that includes charismatic species and those on threatened species lists" [p21]. Specifically, we have focussed on a subset of biodiversity that is predominantly wild (or at least not domesticated to an industrial scale) and of a predominantly larger than micro-organism scale (but not ignoring wild genetic resources).

Table 1 Elements of the systematic map question

Subject	Intervention	Comparators	Outcomes
Human individuals, households or communities, or nation states in developing countries	Direct or indirect use or conservation of, or interaction with, one or more component of biodiversity	No use or conservation of, or interaction with, biodiversity	Positive or negative effects on the multi-dimensional poverty status of people interacting with biodiversity in developing countries

Methods

Searches

Search terms and languages

We conducted all searches in English due to the linguistic competencies of the review team and also as one mechanism for restricting the scale of the review. Because we were interested in disaggregating broad claims about biodiversity and poverty we developed an extensive set of search terms that described different components of biodiversity and different dimensions of poverty. Our research protocol [19] describes in detail the precise steps we followed to develop and test the search terms, including consultation at an expert workshop in August 2012. Additional file 1 summarises the evolution of the final search string.

Searches

Our key sources of data were two online databases of peer-reviewed publications - SciVerse's Scopus [20], and ISI's Web of Science [21] both of which cover natural and social sciences. The search string described in Additional file 1 was used firstly in Scopus to search titles, abstracts and keywords and then in Web of Science to search the field code "Topic" which includes title, abstract and keywords. The search string was slightly modified when applied to Web of Science because of the different way it structures subject areas, and hence the subjects to which the search can be limited or that can be excluded. The results from the Scopus search were combined with the Web of Science results and screened for duplicates using bibliographic software Endnote and Zotero.

To cross-check the publications database searches, Google Scholar was searched just using the terms: "biodiversity" OR "wildlife" AND "poverty" OR "livelihoods" OR "poor". The first 50 'hits' were compared with the Web of Science and Scopus search returns to test their comprehensiveness. References returned by the Google Scholar search, but not found in the Web of Science and Scopus searches, were added to the reference list.

Finally, a sample of the grey literature to include was identified through a number of steps. First, a call was issued via the mailing list of the Poverty and Conservation Learning Group (PCLG) [22]; second, a selection was made of a manageable but representative (in terms of conservation or development focus) number of international organisations that are members of the PCLG and their websites searched for relevant documents (Table 2); and thirdly the PCLG bibliographic database was searched using the search terms- "poverty" OR "livelihoods" AND "biodiversity", OR "wildlife" OR "nature" OR "species".

Table 2 List of websites searched for grey literature

Organisation	Website
A Rocha International	www.arocha.org
BirdLife International	www.birdlife.org
CARE International	www.care-international.org/
Caribbean Natural Resources Institute	www.canari.org
Catholic Agency for Overseas Development (CAFOD	www.cafod.org/uk
Centre for International Forestry Research (CIFOR)	www.cifor.org
Convention on Biological Diversity	www.cbd.int
Department for International Development (DFID), UK	www.gov.uk/government/organisations/department-for-international-development
Food and Agriculture Organisation (FAO)	www.fao.org
International Institute for Environment and Development (IIED)	www.iied.org
International Union for Conservation of Nature (IUCN)	www.iucn.org
United Nations Environment Programme	www.unep.org
United Nations Environment Programme – World Conservation Monitoring Centre	www.unep-wcmc.org
United Nations Development Programme	www.undp.org
United States Agency for International Development (USAID)	www.usaid.gov
Wildlife Conservation Society (WCS)	www.wcs.org
World Bank	www.worldbank.org
Worldwide Fund for Nature International (WWF)	http://wwf.panda.org/

Estimating the comprehensiveness of the search

Our use of disaggregated poverty and biodiversity terms in the search string was intended to ensure a comprehensive search, including articles from non-traditional sources. Overall the search yielded 10,623 peer-reviewed and grey literature documents that we then screened for relevant titles (see below). However, despite the iterative development of the search terms and our broad search, the coding of papers that occurred during the data-extraction process (also described below) revealed some gaps in the search results that must be a consequence of the search terms and strings used (for example there were very few studies documenting the livelihoods impacts of the live animal trade). A further constraint was that we ran out of time to ask our expert group to review the final list of publications which would have helped in identifying key gaps. Finally, an obvious limitation is that we only included material written in English potentially excluding a wealth of studies from non-anglophone countries and organisations.

Study inclusion criteria

The inclusion/exclusion process took place through a number of stages, following our protocol [19]. Firstly, for the peer-reviewed literature, all article titles were reviewed and articles excluded if their titles did not meet the following criteria:

- Relevant population (s): the title makes mention of biodiversity (as interpreted above) in some form (broad terms, specific components).
- Relevant outcomes: the title makes mention of some aspect of poverty (or related terms such as livelihoods or wellbeing).

Secondly, all the abstracts of the articles passing the first stage of title review were read and any which did not meet the following criteria excluded:

- Relevant intervention (s): the study makes mention of an association between biodiversity use or non-use and one or more dimensions of poverty or wellbeing.

In addition, the following exclusion criteria were applied at abstract level:

- Irrelevant countries: studies that were focussed only on OECD countries were excluded.
- Irrelevant interventions: studies that assessed the contribution of poverty to biodiversity loss/ environmental degradation.
- Irrelevant study design: theoretical studies such as models and scenarios.

At this stage we also excluded any articles that had passed the abstract screening but for which we were unable to obtain downloadable copies of the full text within the time and resources available.

The title review and abstract review stages were undertaken by two researchers and a kappa test was performed at each stage in order to check for consistency in the interpretation of the inclusion/exclusion criteria. The kappa test result was 0.804 at the title review stage (strength of agreement between the two researchers considered to be "very good") and 0.732 at the abstract review stage (strength of agreement considered to be "good"). For the grey literature the same two researchers – with proven consistency of judgement – reviewed all the titles. However it was often difficult to judge from the title whether the content of an article was relevant and so consequently many more articles were retained where there was uncertainty. The majority of grey literature articles did not have abstracts and so the abstract screening stage was omitted for these articles.

The final set of articles was exported from a Zotero library to a Microsoft Excel workbook. The workbook was structured so that each article could be analysed against a series of questions – as articulated in our data extraction strategy – and coding applied. Additional file 2 provides details of the list of questions and the coding strategy.

The final stage of inclusion/exclusion took place during data extraction. If, during the full text review and accompanying data extraction process, it became clear that studies did not meet the inclusion criteria above, then they were excluded and the reason for exclusion noted. Additionally, our data extraction framework was structured to explore in more detail any studies that actually included some measure of the contribution of biodiversity use/non-use to one or more dimensions of poverty or wellbeing.

Once all the studies had been read and data extracted by an expanded team of five researchers, the final dataset was checked for anomalies, re-coded where necessary and analysed using descriptive statistical methods. The Excel workbook showing the final set of coded studies is included as Additional file 3.

Study quality assessment

As specified in our protocol [19] we did not seek to prejudge the 'quality' of different types of evidence in our mapping process, recognising that the type of evidence required depends on the specific question being asked and on the motivation and discipline of the questioner [23]. However, by describing the methods used by each study that we map we hope that we leave future researchers, practitioners and policy analysts better able to identify what evidence is and is not relevant for the specific questions they wish to answer.

Results

Review statistics

Number and types of studies

The search of publication databases was conducted on 23 October 2012 while the search of websites and other sources for grey literature was conducted between 1st November 2012 and 31st March 2013. A total of 387 studies were judged to have met the inclusion criteria based on title, abstract, and full text review and were included in the final systematic map. Of these 248 met our additional criteria that studies should include some kind of measure of the contribution to one or more dimensions of poverty/wellbeing. Figure 2 summarises the numbers of articles that were included at each stage of the process. A bibliography of the final set of articles included in the map is provided in Additional file 4 while the list of articles rejected at full text review is included as Additional file 5.

The majority of articles were journal papers (n = 294, 76% of the full set and n = 180, 73% of the subset with a poverty measure) followed by peer-reviewed grey literature (n = 47, 12% of full set and n = 42, 17% of subset) and non-peer reviewed published grey literature (n = 30, 8% of full set and n = 16, 2% of subset). We reviewed very

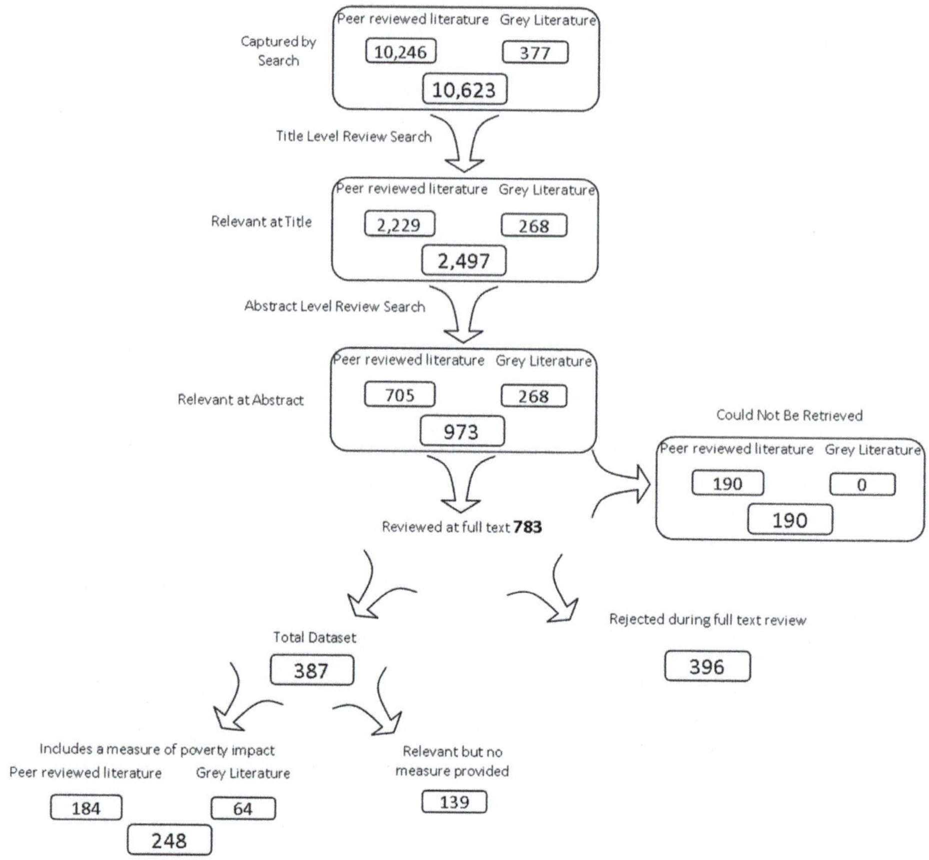

Figure 2 The number of articles retrieved in the initial search, and the numbers passing each subsequent stage of screening.

few books or conference proceedings. Some articles included multiple case studies and where these were substantive (n = 12) they were included as separate entries in the database. We set no date restrictions on our search and the earliest article retrieved was dated 1985. Figure 3 describes the number of articles published per year.

Representativeness of studies

Geographical coverage

The studies were widely distributed, covering 27 countries from Africa, 16 from Asia, 13 from Latin America and 3 from Oceania (with some studies covering more than one country). The most commonly studied region

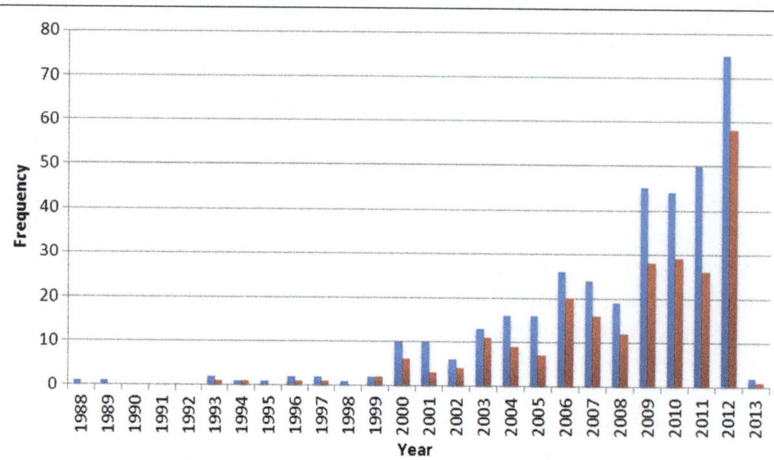

Figure 3 Trends in number of published studies per year (blue columns represent total dataset meeting primary inclusion criteria (n = 387); red columns represent subset with a measure of contribution of biodiversity to poverty (n = 248).

was Africa (n = 169, 43%) within which the most commonly studied country was South Africa (n = 31 , 18%). The next most studied region was Asia (n = 142, 36%) within which the most studied country was India (n =45, 32%). There was a smaller proportion of studies in Latin America - potentially reflecting our language bias (n = 54, 14%) and Oceania (n = 3, >1%). A further 24 studies (6%) were global in coverage. Figure 4 shows the regional distribution of studies.

Ecological coverage

Our assessment of the ecological distribution of studies employed a version of the IUCN habitats classification scheme [23] simplified by Birdlife [24]. By far the most studied habitat was forests (n = 202, 52% of the full set of papers, n = 125, 52% of the subset with a poverty measure). The second most studied (n = 88, 23% of the full set; n = 49, 17% of the subset) was terrestrial artificial habitats (this category includes, for example, agricultural land, pasture land, gardens). Wetlands (n = 40), marine (n = 31), coastlines (n = 29), savannah (n = 29) and grasslands (n = 24) were all relatively similarly studied, while the least studied habitats were artificial aquatic landscapes, mountains, deserts and introduced/exotic habitats. Figure 5 summarises the results.

Description of studies

The results presented below seek to describe the evidence base relevant to our conceptual framework in Figure 1. Moving from right to left across the framework, we describe the components and attributes of biodiversity that have been studied; the dimensions of poverty that the studies have highlighted these components of biodiversity contributing to; the mechanisms by which the link between biodiversity and poverty is made; and the outcomes - in terms of sustainability of biodiversity use and nature and scale of contribution to poverty alleviation or wellbeing. The presentation of results distinguishes between the full set of articles and the subset that includes some kind of quantification of the influence of biodiversity on poverty or wellbeing.

Components and attributes of biodiversity studied

The studies addressed different components of biodiversity – from genetic resources to ecosystems (Figure 6). We added the category "guilds" to capture studies on, for example, the role of pollinators. We made a distinction between studies that focussed on individual species and those that focussed on groups of species (for example African plains game, or "the big five"). We also distinguished studies that were focussed on particular types of resources – rather than specific species – for example non-timber forest products (NTFPs). This resources category was the most commonly studied component of biodiversity (n = 146, 38% of all papers) within which the most commonly studied type of resource was NTFPs (n = 134, 92%). The least commonly studied were guilds (n = 3, <1%) and genetic resources (n = 19, 5%). The remaining studies were relatively evenly distributed between ecosystems (n = 91, 23%), species (n = 90, 23%) and groups of species (n = 62, 16%).

In most cases, the abundance or extent of biodiversity was the attribute that made it important for poverty alleviation (n = 268, 70% of all papers) while diversity was the least frequently noted attribute (n = 83, 21%).

Dimensions and aspects of poverty studied

The Ecosystem Services for Poverty Alleviation (ESPA) programme has produced a conceptual framework for understanding poverty [25]. This identifies – from a

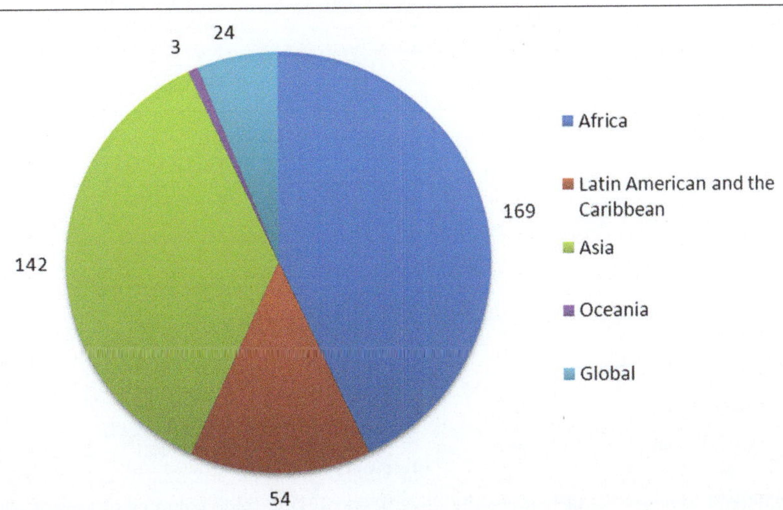

Figure 4 Geographical distribution of studies (NB some studies covered more than one location hence total number exceed n = 387 studies).

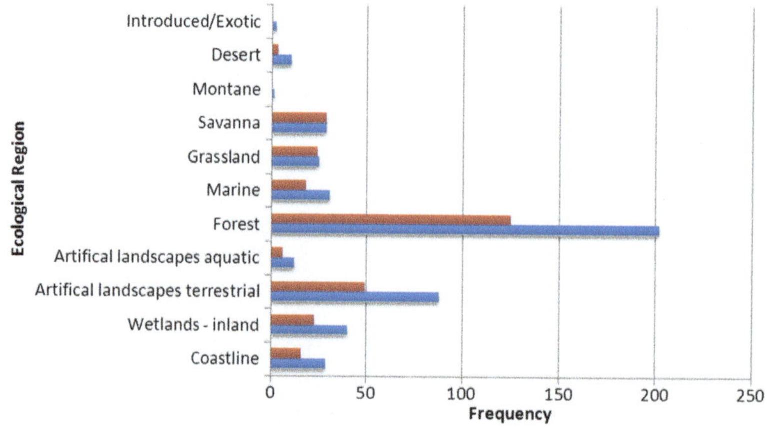

Figure 5 Distribution of studies in different ecological habitats (blue columns represent total dataset meeting primary inclusion criteria (n = 387); red columns represent subset with a measure of contribution of biodiversity to poverty (n = 248).

review of the literature – 19 commonly used dimensions of poverty. Our analysis only identified studies that addressed 11 of these, plus an additional dimension – energy security. Although most studies examined more than one dimension of poverty, the most commonly studied was income (n = 270, 70% of all papers, and n = 205, 83% of papers with a poverty measure). Other commonly studied dimensions were food security (n = 124, 32% of all papers) and asset accumulation (n = 91, 23% of all papers). The least commonly studied were energy, shelter and safe water. Figure 7 describes the frequency of study of each dimension.

The ESPA poverty framework highlights the dynamic nature of poverty and recommends that attention should be paid to both short term (temporal) poverty as well as long term, chronic poverty. It is also important to distinguish between absolute poverty – as measured against a defined poverty line – and relative poverty. Just over half (n = 196, 51%) of the papers we reviewed specified the aspect of poverty being studied. Of these 61% (n = 120) were concerned with relative poverty and 39% (n = 76) were concerned with absolute poverty. Fewer studies (n = 94, 24%) considered poverty dynamics – that is, whether it is lasting or transitory - of which the majority (n = 73, 78%) focussed on chronic (persistent) poverty and only a small minority (n = 21, 22%) focussed on temporal (transient) poverty.

Components of biodiversity associated with different dimensions of poverty

We further analysed the dataset to explore which components of biodiversity were most commonly linked with which dimensions of poverty. We have already highlighted above that income was the most frequently cited dimension of poverty that biodiversity contributes to. Table 3 describes how the types of study with which

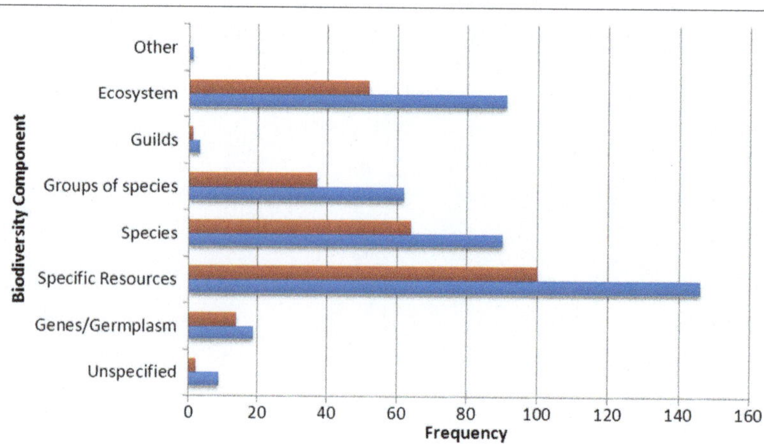

Figure 6 Frequency with which different components of biodiversity have been studied (blue columns represent total dataset meeting primary inclusion criteria; red columns represent subset with a measure of contribution of biodiversity to poverty).

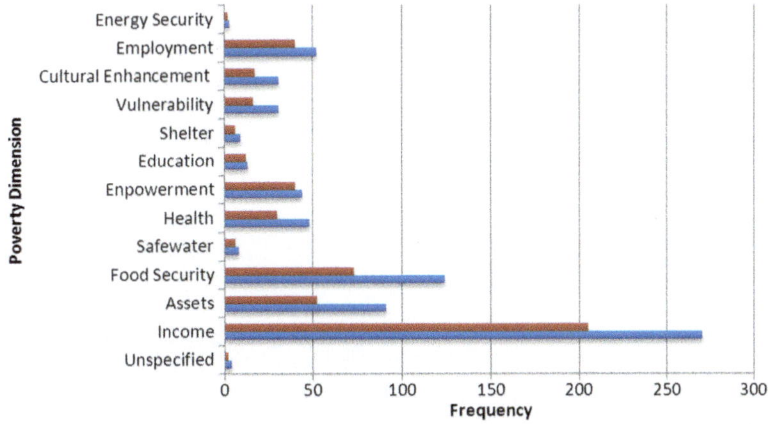

Figure 7 Dimensions of poverty studied (blue columns represent total dataset meeting primary inclusion criteria; red columns represent subset with a measure of contribution of biodiversity to poverty).

income is most commonly associated are those concerned with specific types of resources, individual species and groups of species, and ecosystems. Similarly we have already highlighted that specific resource types were the most commonly studied component of biodiversity. Further analysis shows that the most commonly cited dimensions of poverty that resources contribute to were income, assets and food security.

Type of mechanisms linking bidiversity and poverty

In our analytical framework we identified a number of "mechanisms" by which the link between biodiversity and poverty or wellbeing is achieved. We categorised these as direct use, indirect use, non-use, and "biodiversity dis-services". For the purposes of this study we used "use" and "non-use" terminology in a slightly different way to that commonly used in the concept of total economic value (TEV) [26]. Our use of the term "direct use" is consistent with TEV and refers to the consumptive use of certain species or resources such as NTFPs for subsistence or tradable purposes. Indirect use refers to the benefits derived from biodiversity being left *in situ* rather than being removed or consumed and includes, for example use of biodiversity for tourism or for soil nutrient cycling. Non-use refers to the co-benefits that can arise from biodiversity management efforts. These first three categories all refer to the benefits that people get from biodiversity but we recognise that biodiversity can also impose costs. Our final category – "biodiversity dis-services" therefore refers to the negative impacts that components of biodiversity can have on people. As noted in our introduction, we deliberately did not set out to retrieve studies on the health impacts of parasites and diseases – if we had we would expect this category to have a much higher frequency of papers. The majority of studies falling into this category thus largely referred to cases of human-wildlife conflict.

We found that the most common mechanism for linking biodiversity and poverty was through direct use of different components of biodiversity (mentioned in n = 338, 87% of all papers) (Figure 8). We further investigated the mechanism by which biodiversity and poverty are linked by identifying the precise mechanism, or form of use, described in the studies. Figure 9 illustrates the wide variety of mechanisms employed, with the most commonly identified being NTFP harvesting (noted in 203, 52% of studies). We also found that for each of the different mechanisms studied, biodiversity contributed to multiple dimensions of poverty. This was particularly pronounced in studies of fishing, NTFPs, tourism and wild plant cultivation where almost all dimensions of poverty were considered in one or more study. Additional file 6 illustrates the numbers of studies that associate different mechanisms for using biodiversity with different dimensions of poverty.

Having identified the type of mechanisms by which biodiversity was linked to poverty we also looked at *how* the mechanism worked. We differentiated between use of biodiversity for subsistence purposes only or to generate a tradeable surplus or income. We also differentiated between immediate use and longer term contributions to livelihoods – for example through maintaining productive land and ecosystems. Finally we noted any studies that described the use of biodiversity as providing an emergency lifeline or safety net and those where it had a negative effect and actually undermined livelihood security. We found that the most commonly identified processes by which biodiversity affects poverty are through generating income (n = 316, 82% of all papers) and supporting subsistence needs (n = 271, 70% of all papers). Less commonly identified processes were contributing to

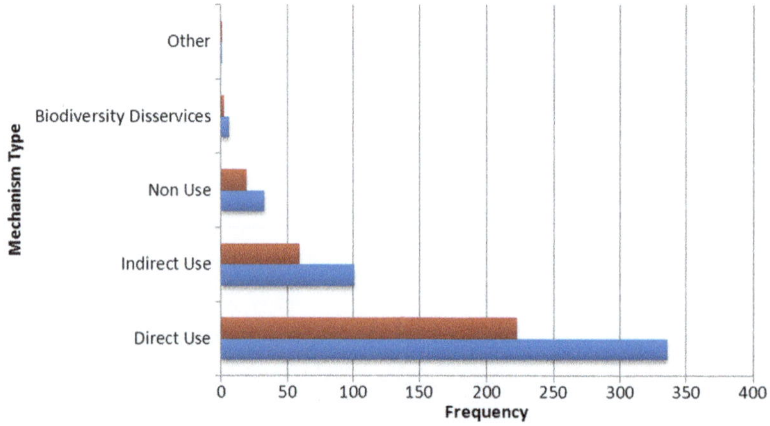

Figure 8 Mechanisms for linking biodiversity and poverty (blue columns represent total dataset meeting primary inclusion criteria; red columns represent subset with a measure of contribution of biodiversity to poverty).

longer term livelihood support (n = 78, 20% of all papers) and acting as a safety net (n = 35, 9% of all papers).

Outcomes for poverty – nature and scale

The overwhelming majority of papers (n = 326, 84%) described a positive contribution of biodiversity to poor peoples' wellbeing. However, only two-thirds of those reporting the positive contribution (n = 223, 66%) actually included any *measure* of that contribution. Measures related to almost all the dimensions of poverty were used although the most common type of measure was income-related (used in n = 223 of the 248 (90%) papers that used a measure). The specific measures used varied quite considerably and included: absolute amount of income generated per capita or household from use of biodiversity; household or per capita income generated from biodiversity as a proportion of total household or per capita income; and, income equivalent of biodiversity-related household consumption). Non-income measures included increases in food availability and intake; numbers of jobs created; improvements in health; improvements in asset productivity.

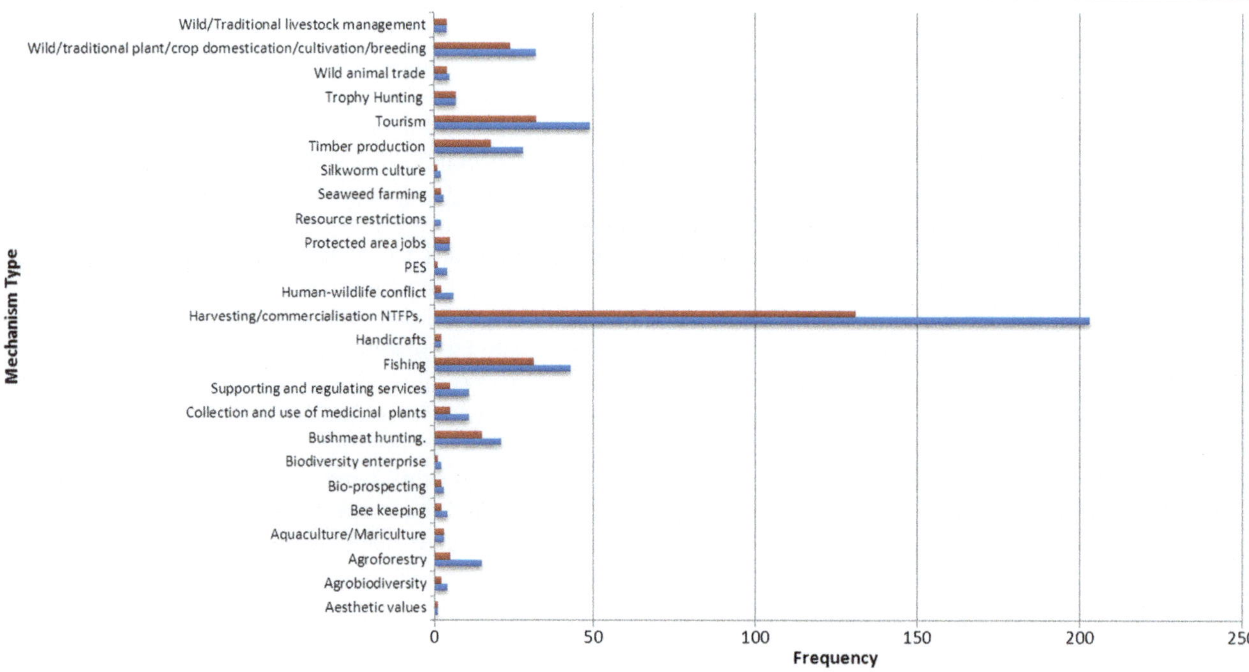

Figure 9 Different mechanisms by which people experience costs and benefits of biodiversity (blue columns represent total dataset meeting primary inclusion criteria; red columns represent subset with a measure of contribution of biodiversity to poverty).

Table 3 Numbers of studies that associate different components of biodiversity with different dimensions of poverty

	Unspecified	Genes/germplasm	Specific resources	Species	Groups of species	Guilds species	Ecosystems	Other
Unspecified	0	0	0	1	0	0	1	0
Income	2	12	83	55	31	0	42	0
Assets	0	1	21	12	7	1	18	0
Food security	2	0	33	21	12	0	10	0
Safewater	0	0	2	2	1	0	4	0
Health	1	1	14	10	1	0	8	0
Enpowerment	0	2	10	11	7	0	17	0
Education	0	1	3	4	1	0	4	0
Shelter	0	0	4	2	0	0	0	0
Vulnerability	1	1	8	2	3	0	5	0
Culture Enhancement	0	0	4	5	4	0	5	1
Employment	0	1	12	9	9	0	12	0
Energy security	0	0	1	1	0	0	0	0
Other	0	0	1	0	0	0	0	0

Over half (n = 157, 63%) of the studies that did measure the effect of biodiversity on poverty included some indication of the number of beneficiaries reached and the scale of benefits. However this was hugely variable (from less than ten to several thousand dollars per person per year) and generally not projected beyond the immediate sample size of the study. Less than one-third (n = 78, 31%) of studies included any indication of the likely duration of the effect. Of these, the majority (n = 60, 77%) considered it to be long term, 14% considered it to be short term and 9% seasonal. Just over one-third (n = 85, 34%) considered the outcomes to be achievable elsewhere – beyond the study context.

Outcomes for biodiversity - sustainability of use

Amongst the studies that employed a measure of contribution of biodiversity to poverty alleviation, only just over half (n = 142, 57%) included any consideration of whether the use of biodiversity was sustainable. Of these use was considered sustainable in 65% of cases. Most doubts about the sustainability of use were associated with direct (consumptive) use mechanisms. Biodiversity use was considered to be unsustainable in 21% of direct use studies compared to 13% for indirect use studies.

Mapping the quality of papers relevant to the question

Without making any judgements as to the superiority of one research design or method over another we categorized our subset of studies with a poverty measure according to whether they were based on primary or secondary data, and whether they had adopted an experimental, quasi-experimental or non-experimental research design. By experimental design we refer to studies that have a randomised, controlled research design whereas quasi-experimental designs have a control but are not randomised. Non-experimental research designs have no control and encompass a huge range of research methods. It was not possible to determine the research design in all the studies – some reported data without providing any details on how they were obtained. However, the majority of studies (n = 162, 65%) were based on primary data. Amongst these, the most common research design was non-experimental (n = 151, 93%). We only identified 11 (<5%) studies that were based on experimental or quasi-experimental research.

We also looked at the extent to which studies had addressed issues which we expected would moderate the biodiversity-poverty relationship. These included the governance regime, resource rights regime, land tenure, power relations, distribution of costs and benefits, and trade-offs. We found that each issue was addressed by roughly half of the studies in each case with the exception of trade-offs which was discussed by just over one-third (36%) of studies and power relations which was addressed by less than one-quarter (21%), (Table 4).

Discussion

Components of biodiversity and dimensions of poverty studied

It has been noted that studies relevant to links between poverty and biodiversity tend to treat poverty as a unidimensional issue related to income [12]. Our analysis certainly confirmed the dominance of income as a measure of poverty. However, over half the sources of evidence did measure more than one dimension of poverty, challenging the earlier characterization of the literature.

Table 4 Coverage of key issues which mediate the biodiversity-poverty relationship

Does the paper consider	Yes	No	Not applicable/not mentioned
Distributional impacts	126	121	1
The governance regime	115	107	26
The resource rights regime	128	92	28
The land tenure regime	116	105	27
Power relations	53	188	7
Possible trade-offs/costs as well as benefits	90	157	1

A large proportion of all studies either did not mention or did not specify the aspect or dynamic of poverty they were measuring, suggesting that these issues may not have been considered important by researchers. Given the well-established importance of relative versus absolute and temporal versus chronic poverty [25,27] this is an area requiring more research, or at the very least more clarity in future publications.

The dominance of studies in forest systems is interesting, given that poor people make use of biodiversity in a wide range of different ecosystems. It seems unlikely that studies based in forests would be disproportionately likely to be identified by our search, so we suggest that the relationship between poverty and biodiversity in non-forest habitats is an area requiring further study. Drylands, for example are home to a disproportionate proportion of the world's poor whose livelihoods depend on land and livestock [28]. The importance of biodiversity – for fodder, fibre and medicines – seems obvious but is poorly studied and documented [29].

Evidence was found on a wide range of components of biodiversity, with only guilds and genes/germplasm mentioned rarely. The former may be due to 'guild' being a rather specialist term, but the latter indicates a promising avenue for further research on links to poverty. It is interesting, however, to note the lack of studies that consider the importance of diversity *per se*. Given that by some interpretations biodiversity only refers to diversity, and not other attributes of living organisms, this is perhaps a cause for concern. In most cases the evidence is focused on particular species or the extent/abundance of a set of species with a particular link to poverty, and we do not know enough about issues such as whether such species could be replaced by others performing the same role for the poor, or the role of diversity in providing resilience.

It is perhaps not surprising that the most commonly cited component of biodiversity under study was NTFPs given the preponderance of studies based in forest habitats. This does seem to confirm once again that the evidence base is strongest in regard to the use made of forest products by the poor. An alternative interpretation is that the term NTFP is potentially all-encompassing, prompting Belcher to ask 10 years ago, "what *isn't* an NTFP?" [30]. We encountered a wide range of species that were classified as NTFPs. Further disaggregation and analysis would provide greater understanding of which particular species – or NTFP properties – appear to be particularly valuable for different groups of poor people in different ecological and governance contexts.

The nature of biodiversity-poverty links

The evidence base on biodiversity – poverty linkages is dominated by studies of the direct, consumptive, use of biodiversity by people. Very few studies reported negative impacts on poverty, which is partly due to our decision to exclude health impacts of pathogens. Nonetheless, it does seem that there is very little research into negative impacts of biodiversity on poverty beyond human-wildlife conflict studies. Our exercise returned remarkably few studies on issues such as live animal trade, aquaculture and agrobiodiversity. More focussed systematic maps would be useful to determine if this lack of studies is a consequence of the limitations of our search string, or indeed whether further research on these issues is needed.

Because the majority of studies that we found were focussed on the consumptive use of biodiversity by people, it is not surprising that the most common relationships between biodiversity and poverty reported were related to meeting subsistence needs and generating a source of income. Again, more focussed systematic maps would be useful to determine if the limited number of studies we found documenting the effect that biodiversity's role in contributing to long term resilience or acting as a safety net has on poverty reflects a lack of evidence or a limitation in our search.

We were surprised by the number of studies that included no information about the sustainability of biodiversity use given that this is of critical importance to any discussion of the relationship between biodiversity and poverty. Measuring sustainability is clearly challenging, in that it requires long term studies and a sophisticated understanding of ecological processes that produce biodiversity of value to the poor. This is an area clearly in need of further research.

Quality of evidence

The limited number of studies we found that had adopted experimental or quasi-experimental research designs is consistent with the findings of other researchers who have called for more controlled studies and counterfactual analysis [31]. The majority of our studies, however, describe people's everyday use of and interaction with biodiversity. They are not experiments but real world situations where ideal controls in which all variables other than the biodiversity or poverty 'treatment' are the same do not exist. Thus we did not find research design to be a good indicator of quality – although it was equally frustrating to identify case studies – particularly in the grey literature – that presented data without providing any insights into how those data were collected or validated.

We attempted to balance a focus on research design with an assessment of the extent to which key issues that have an impact of the relationship between biodiversity and poverty had been addressed in studies but again the degree to which different issues are relevant in different contexts is highly variable and thus their treatment not a comparable indicator of quality. Further debate is required as to what constitutes high or low quality evidence when attempting to evaluate complex, real-world situations rather than tightly defined interventions with suitable counter-factual sites for analysis [32-34] – particularly if full systematic reviews are conducted in the future to explore more specific questions within the topic of the link between biodiversity and poverty.

Conclusions

Both "biodiversity" and "poverty" are complex, multi-dimensional concepts. Searching for relevant literature that addresses the question "Which components of biodiversity affect which dimensions of poverty" meant that we had to cast our net wide in order to capture this complexity - as our search string detailed in Additional file 1 illustrates – but as a result also had to filter out a lot of irrelevant material. While we eliminated the irrelevant material we also recognised that our search has missed areas of relevant literature. This suggests that we may have chosen our initial keywords poorly, and also reinforces the point that no matter how objective and systematic the review, it is only ever as good as the keywords and reviewers.

Shortcomings in our coverage of the specific types of use people make of biodiversity (such as through wildlife trade) could be rectified in subsequent maps by further refinement and testing of our search terms. But in part the limitations of our search are influenced by the complexity of biodiversity and the difficulties in constructing a search string that accounts for that complexity. Thus we have identified no studies, for example, on the role of below-ground biodiversity in maintaining or improving soil productivity which in turn results in improved crop productivity which in turn contributes to increased income and improved food security. While we have sought to map the evidence base it is not clear the extent to which evidence exists and was not captured in our search, or simply does not exist.

Nevertheless, we generated a database of nearly 400 studies that document the influence (positive and negative) of one or more components of biodiversity on one or more dimensions of poverty and within that, a subset of 248 studies that have actually sought to measure that effect in some quantifiable way. While the studies are not directly comparable owing to the wide variety of metrics used as well as the different scale of analysis and study designs, collectively the map can shed light on the validity of claims that conserving biodiversity can reduce poverty. The implications for policy and research are discussed below.

Implications for policy and management

The map includes evidence on a wide range of different components of biodiversity - but particularly species and ecosystems - affecting different dimensions of poverty – particularly income, assets and food security. The overwhelming majority of studies indicates a positive influence of biodiversity on poverty. Caveats aside as to our coverage of biodiversity "dis-services" this implies that development planners should take far more seriously the importance of biodiversity in the lives of poor people. While much lip-service is paid to this relationship, "mainstream" development pathways continue to degrade the natural environment and deplete biodiversity as has been highlighted in numerous analyses – most recently the Millennium Ecosystem Assessment (MA) [35] and the study on The Economics of Ecosystems and Biodiversity [36].

Implications for research

The systematic map highlighted the difficulties in comprehensively reviewing the evidence on biodiversity-poverty in one study. We have identified a number of apparent gaps in the evidence base but given the difficulties we encountered in ensuring a comprehensive search we would recommend, in the first instance, further analysis to determine which of these are real knowledge gaps that require primary research (as opposed to gaps resulting from limitations of our search strategy). It was noticeable from our map that in the majority of the literature biodiversity is framed in terms of its value as a *resource* – in the form of specific goods that can be used to generate tangible benefits such as cash, food, fuel. Very few studies explored the underpinning role of biodiversity in ecosystem service delivery – as it is framed in the MA – and fewer really investigated the benefits of genetic diversity in terms of increasing resilience and adaptive capacity. Key areas that we identified for follow-up research are:

- Investigation into the value role of diversity over abundance of resources. The majority of the studies we identified implied that the abundance or availability of particular species or resources was more critical than their diversity. Quantifying the value of diversity and where it is particularly important in delivering ecosystem services would make a significant contribution to the biodiversity-poverty debate.

- More research on less tangible components of biodiversity. We found few studies that dealt with genetic diversity, microbes or even invertebrates. The studies that have been undertaken to date barely scratch the surface in terms of the full complement of biodiversity.
- Biodiversity-poverty trade-offs: We were surprised that more studies did not consider the sustainability of biodiversity use. More research into key factors underlying sustainability in different contexts and for different types of use, as well as consideration of thresholds and tipping points would help decision-makers balance the drive for poverty reduction with the need for biodiversity conservation.
- Long vs short term biodiversity-poverty links: The majority of the evidence we found documented the contribution of biodiversity to short term needs. More analysis is required to uncover the evidence/ generate new evidence on the role of biodiversity in poverty prevention and enhancing longer term resilience.
- Investigation into "policies and institutions that work" in enabling an effective contribution of biodiversity to poverty reduction.
- More detailed, sector-by-sector reviews on key mechanisms for generating value from biodiversity including wildlife trade, crop improvements, fishing etc. - together with an analysis of underlying conditions influencing success or failure.
- Analyses of biodiversity-poverty interactions in non-forest ecosystems particularly those that are home to significant numbers of poor people such as drylands.

One overarching issue, however, that links the policy and research implications is that we currently have no way of knowing how much of "what works" (and is therefore considered of policy relevance) is documented and therefore available for inclusion in systematic maps such as this and able to influence policy. Attention is needed to how better to integrate the documented and undocumented, the "scientific" and traditional in order to generate a much richer evidence base. This is an issue to which the newly established Intergovernmental Panel on Biodiversity and Ecosystem Services (IPBES) is expecting to pay close attention. Its draft work programme for 2014–2018 [37] includes guidance on how to address and include indigenous and local knowledge within its scientific assessments as a key early deliverable.

Additional files

Additional file 1: Evolution of search string.

Additional file 2: Data extraction questions and codes.

Additional file 3: Database of coded studies.

Additional file 4: List of references included in the map.

Additional file 5: List of references rejected at full text review.

Additional file 6: Numbers of studies linking different mechanisms for using biodiversity with different dimensions of poverty (intensity of shading depicts higher numbers of studies).

Competing interests

The authors declare that they have no competing interests.

Authors' contibutions

DR and CS jointly designed the research. DR and MF screened titles and abstracts. DR, MF, MS, AG and AGM screened full text articles. MF and CS conducted the statistical analysis of results. DR and CS drafted the article narrative and MF generated all the figures and tables. All authors read and approved the final manuscript.

Acknowledgments

The authors would like to thank the Directorate of the ESPA programme for providing the initial funding for this study from its Evidence and Impact Research Grants scheme. Additional funding for products produced under this study has been provided by UKaid from the UK Government however the views expressed do not necessarily reflect the views of the UK Government. Thanks are also due to the three anonymous reviewers of the first draft of this paper who raised a number of very important issues. We hope we have done justice to the comments provided – we feel the paper has benefitted significantly.

Author details

[1]International Institute for Environment and Development (IIED), 80-86 Grays Inn Road, London WC1X 8NH, UK. [2]United Nations Environment Programme - World Conservation Monitoring Centre (UNEP-WCMC), 219 Huntingdon Road, Cambridge CB3 0DL, UK.

References

1. SCBD: *Decision VI/26: Strategic Plan for the Convention on Biological Diversity.* Montreal: Secretariat of the Convention on Biological Diversity; 2002.
2. United Nations: *Official List of MDG Indicators.* 2008. http://mdgs.un.org/unsd/mdg/Host.aspx?Content=Indicators/OfficialList.htm.
3. Butchart SH, Walpole M, Collen B, van Strien A, Scharlemann JP, Almond RE, Baillie JE, Bomhard B, Brown C, Bruno J, Carpenter KE, Carr GM, Chanson J, Chenery AM, Csirke J, Davidson NC, Dentener F, Foster M, Galli A, Galloway JN, Genovesi P, Gregory RD, Hockings M, Kapos V, Lamarque JF, Leverington F, Loh J, McGeoch MA, McRae L, Minasyan A, *et al*: **Global biodiversity: indicators of recent declines.** *Science* 2010, **328**(5982):1164–8.
4. Mace GM, Cramer W, Diaz S, Faith DP, Larigauderie A, Le Prestre P, Palmer M, Perrings C, Scholes RJ, Walpole M, Walther BA, Watson JEM, Mooney HA: **Biodiversity targets after 2010.** *Current Opinion in Environmental Sustainability (COSUST)* 2010, **2**(1):3–8.
5. United Nations: *The Millennium Development Goals Report.* New York: United Nations; 2010.
6. United Nations: *Secretary-General, at High-Level Meeting, Stresses Urgent Need to Reverse Alarming Rate of Biodiversity Loss, Rescue 'Natural Economy.* 2010. Press Release 22 Sept 2010, http://www.un.org/News/Press/docs/2010/ga10992.doc.htm.
7. SCBD: *Decision X/2: Strategic Plan for Biodiversity 2011–2020.* Montreal: Secretariat of the Convention on Biological Diversity; 2010.
8. Millennium Ecosystem Assessment: *Ecosystems and Human Wellbeing – Biodiversity Synthesis.* Washington DC: World Resources Institute; 2005.
9. Kepe T, Saruchera M, Webster WJ: **Poverty Alleviation and Biodiversity Conservation: A South African Perspective.** *Oryx* 2004, **38**(2):143–145.

10. Tekelenburg A, ten Brink BJE, Witmer MCH: *How do Biodiversity and Poverty Relate? An Explorative Study.* Bilthoven: Netherlands Environmental Assessment Agency; 2009.
11. Leisher C, Sanjayan M, Blockhus J, Larsen SN, Kontoleon A: **Does conserving biodiversity work to reduce poverty? A state of knowledge review**. In *Biodiversity Conservation and Poverty Alleviation: Exploring the Evidence for a Link.* Edited by Roe D, *et al.* Oxford: Wiley-Blackwell; 2013.
12. Vira B, Kontoleon A: **Dependence of the poor on biodiversity: which poor, what biodiversity? A State of Knowledge Review.** In *Biodiversity Conservation and Poverty Alleviation: Exploring the Evidence for a Link.* Edited by Roe D, *et al.* Oxford: Wiley-Blackwell; 2013.
13. World Bank: *World Development Report 2000/2001: Attacking Poverty.* Washington DC: World Bank; 2001.
14. Sen A: *Development as Freedom.* London: Zed Books; 1999.
15. Clapton J, Rutter D, Sharif N: *SCIE Systematic mapping guidance.* UK: Social Care Institute for Excellence; 2009.
16. Randall NP, James K: **The effectiveness of integrated farm management, organic farming and agri-environment schemes for conserving biodiversity in temperate Europe - A systematic map.** *J Env Evid* 2012, **1:**4. doi:10.1186/2047-2382-1-4.
17. Balmford A, Bruner A, Cooper P, Costanza R, Farber S, Green RE, Jenkins M, Jefferiss P, Jessamy V, Madden J, Munro K, Myers N, Naeem S, Paavola J, Rayment M, Trumper S, Turner RK: **Economic reasons for conserving wild nature.** *Science* 2002, **297:**950–953.
18. Mace GM, Norris K, Fitter AH: **Biodiversity and ecosystem services: a multi-layered relationship.** *TREE* 2012, **27:**19–26.
19. Roe R, Sandbrook C, Fancourt M, Schulte B, Munroe R, Sibanda M: **A systematic map protocol: which components or attributes of biodiversity affect which dimensions of poverty.** *Env Evid,* **2**(1).
20. *SciVerse Scopus.* http://www.scopus.com/.
21. *Web of knowledge.* http://wok.mimas.ac.uk/.
22. *Poverty and Conservation Learning Group.* www.povertyandconservation.info.
23. *IUCN Habitats Classification Scheme (Version 3.1).* http://www.iucnredlist.org/technical-documents/classification-schemes/habitats-classification-scheme-ver3.
24. Lincoln Fishpool: *Pers Comm.* 2011.
25. Suich H. *Conceptual Framework: Poverty.* 2012. http://www.espa.ac.uk/files/espa/Povertyframework.pdf.
26. OECD: *"Total Economic Value", in OECD, Cost-Benefit Analysis and the Environment: Recent Developments.* Paris: OECD Publishing; 2006. doi:10.1787/9789264010055-7-en.
27. Hulme D, Moore K, Shepherd A: *Chronic poverty: meanings and analytical frameworks. Chronic Poverty Research Centre Working Paper no 2.* Manchester: University of Manchester; 2001.
28. Millennium Ecosystem Assessment: *Ecosystems and Human Well-being: Desertification Synthesis.* Washington DC: World Resources Institute; 2005.
29. Davies J, Poulsen L, Schulte-Herbrüggen B, Mackinnon K, Crawhall N, Henwood WD, Dudley N, Smith J, Gudka M: *Conserving Dryland Biodiversity.* Nairobi, Cambridge and Bonn: International Union for Conservation of Nature and Natural Resources, United Nations Environment Programme- World Conservation Monitoring Programme (UNEP-WCMC), and United Nations Convention to Combat Desertification (UNCCD); 2012.
30. Belcher B: **What Isn't an NTFP?** *Int For Rev* 2003, **5**(2):161–168.
31. Barrett CB, Travis AJ, Dasgupta P: **On biodiversity conservation and poverty traps.** *PNAS* 2011, **108**(34):13907–13912.
32. Sandbrook C: *What Constitutes Good Evidence in the Context of Biodiversity and Ecosystem Services?* Poverty and Conservation Learning Group Discussion Paper no 10; 2012. http://pubs.iied.org/G03715.html.
33. Adams WM, Sandbrook C: **Conservation, evidence and policy.** *Oryx* 2013, **47**(3):329–335.
34. Haddaway N, Pullin A: **Evidence-based conservation and evidence-informed policy: a response to Adams and Sandbrook.** *Oryx* 2013, **47**(3):336–338.
35. *Millennium Ecosystem Assessment.* www.maweb.org.
36. *The Economics of Ecosystems and Biodiversity.* www.teebweb.org.
37. *Intergovernmental Panel on Biodiversity and Ecosystem Services (IPBES) Draft Work Programme 2014–2018.* http://www.ipbes.net/images/IPBES%20Work%20Programme%20Review%20Draft%20-%20for%20online%20review.pdf.

11

What is the evidence that invasive species are a significant contributor to the decline or loss of threatened species? A systematic review map

Philip D Roberts[1*], Hilda Diaz-Soltero[2], David J Hemming[1], Martin J Parr[1], Nicola H Wakefield[1] and Holly J Wright[1]

Abstract

Background: Invasive species have been reported as one of the leading causes of species extinction. However, the evidence to support or contest their effects on the decline and/or extinction of threatened species has been largely unavailable and few efforts have been made to collate such information. In order to develop national and international policies addressing the loss of threatened species, and to prioritise research and mitigation efforts, the evidence needs to be systematically reviewed and mapped to determine the importance of such effects.

Methods/design: The searching of online publication databases, grey literature and other resources, such as recovery plans of endangered species, aims to gather existing evidence on whether invasive species are a significant contributor to the decline and/or extinction of threatened species. This study focuses on US species under threat. The methods used to carry out the systematic review will address the following two fundamental questions: (a) what proportion of threatened species have an invasive species as a significant contributor to their decline and (b) through what mechanisms do invasive species contribute towards the decline of native species? The pool of resources gathered will be analysed for relevance and quality using a pre-defined scoring system. A systematic map will be produced, summarising information from individual studies. Where possible, data will be extracted from included studies and synthesised using meta-analysis.

Keywords: Invasive species, Alien species, Threatened species, Impact, Decline, Extinction, Mechanism

Background

There is a broad scientific consensus that invasive species are likely to play a crucial and devastating role in species extinction. At the 2010 Convention on Biological Diversity Conference of Parties (COP10) it was stated that:

> *"Invasive species are the second biggest driving force of species extinction, after the effects of human activity (habitat loss, overexploitation, and pollution)."*

This extinction process seems likely to accelerate, with climate change acting as one of its driving factors, as it may expand the ranges of many invasive species.

However, while there are good ecological reasons for expecting that invasive species are playing a central role in the loss of native species, the evidence to support or contest that idea has not been collected or examined systematically at either a national or international level, and even the evidence source behind the COP10 statement has not been cited. For example, within the United States (USA), an internal audit of the Species Recovery Plans of candidate species for IUCN Threatened Species List, undertaken by USDA National Invasive Species Council and US Fisheries and Wildlife Service (FWS) in 2010-11, found that the information regarding invasive species that may cause the decline or extinction of threatened species is held in thousands of separate, unobtainable statements about individual species that is held in the filing cabinets of site/species experts or in disparate datasets that they manage on local machines, even though the main resource for invasive species policy and evidence is publically available (http://www.invasivespeciesinfo.gov/). Therefore, the evidence that underpins the Species Recovery Plans is either underreported observational or

* Correspondence: phildroberts@gmail.com
[1]CABI, Nosworthy Way, Wallingford, Oxfordshire, United Kingdom
Full list of author information is available at the end of the article

expert opinions that are not readily accessible to the scientific community. Wider than this, it is the experience of the review team that within the introductions of published literature the impact to a native species by an invasive, at a particular location, is often inferred, when this is not the focus of that particular paper. These unsubstantiated statements may be used as evidence in state/country/global recovery plans that it is the invasive that is the major causative agent for the decline of the native, when other factors, (i.e. change in grazing regimes/pollution etc.) may in fact be responsible for the native species' decline. Taking these elements together means that it is impossible to determine the true extent to which invasive species are impacting biodiversity from the current unreviewed evidence base. It is vital to have an answer to this central question in order to develop national and international polices to address the loss of threatened species, and to prioritise research and mitigation efforts.

Previous efforts to review the evidence have focused on a limited subset of species, such as a single taxonomic Kingdom or Order and have led to disagreements within the peer-reviewed literature. For example, Clavero and García-Berthou's [1] editorial on invasive species causing animal extinctions argued that the use of different methodologies for reporting and sourcing information had led to drastically differing conclusions to whether invasive species caused extinctions to either IUCN Red Listed Threatened Species [2] or those threatened within countries, e.g. North American fish species [3], or mammals on Australian islands [4]. Therefore, this study aims to systematically map the evidence for all US listed endangered and threatened species on the US FWS list and candidate threatened species awaiting listing, that are being impacted by an invasive species, and by what mechanism this impact is achieved. Under the US Endangered Species Act, "endangered" means "any species which is in danger of extinction throughout all or a significant portion of its range", while "threatened" is defined as "any species which is likely to become an endangered species within the foreseeable future throughout all or a significant portion of its range". Candidate species are under consideration for threatened or endangered status. The study is focused on species listed for the USA because scoping work indicated that the full IUCN list of 3,947 critically endangered species would require more resources than were available. The US FWS list includes 65% of the species on the IUCN CR list for the USA. The evidence in the literature from all taxa will be included, but there will be no attempt to balance numbers from different taxa.

The output of this study will allow the USDA and US FWS to base both Species Recovery Plans and candidate species listing nominations on the best available evidence. In addition, the evidence will be used to inform on invasive species that should be on alert lists to land managers.

Objective of review

The primary review question is: "What is the evidence that invasive species are a significant contributor to the decline or loss of US threatened species?"

To answer this main question, a group of stakeholders, at the request of the USDA, took part in a two-day meeting in Georgetown, Washington D.C., USA, to discuss invasive species and their impact in North America. The broadly based stakeholder group consisted of USDA National Invasive Species Council, The Conservation Breeding Specialist Group (CBSG) of IUCN, US FWS, the National Marine Fisheries Service (NMFS) and the review team from CABI. The group was selected to represent US stakeholders, international conservation and an independent science organisation with a specialist focus on control of invasive species. Despite their different perspectives, all stakeholders were in agreement that in order to answer such a broad main question, two sub questions (see Table 1) that would capture an impact (or any measure of change) to the threatened species and also the mechanisms by which the invasive species causes that impact would need to be investigated.

Secondary question 1: "What proportion of threatened species have an invasive species as a significant contributor to their decline?"

This question will include an analysis of the USA threatened species from the US FWS list and candidate species lists (under consideration for US FWS threatened status level). This question will map all evidence relating to both the invasive species and the threatened species, effectively coming up with species pairs (for

Table 1 Components of the systematic review subquestions (SQ)

	Population	Intervention/ exposure	Comparator/ study design/ evidence type	Outcome
SQ1	Threatened species: US FWS listed Endangered, Threatened and Candidate species	Invasive species	No invasive species: *randomised block design time series* (before and after); *Site comparison* (comparing exposure to non-exposure on adjacent sites); *Observations*	Change in status of the Threatened species population size (size/range/ density/ fecundity etc.)
SQ2	Invasive species	Threatened species: US FWS listed Endangered, Threatened and Candidate species	Biological papers, laboratory studies, genetic studies	Mechanism by which the invasive species has competitive advantage over other species

each specific invasive species and its relevant threatened species), and the impact that the invasive species has being analysed. From these individual pairs, an overall proportion of impact will be calculated.

Secondary question 2: "Through what mechanisms do invasive species contribute towards the decline of native species?"

This question will identify the mechanisms by which each of the invasive species identified in the previous question impact the native species. All the data relating to mechanisms presented in the evidence selected will be captured through the process and summarise under relevant mechanistic categories.

In addition, the review team will also highlight all the different impact scales that are available to managers and policy makers.

The essence of this approach uses the conventional population, exposure, comparator, outcome (PECO) structure, where the **population** is the native threatened species, the **exposure** is the introduction or presence of the population of the invasive species, the **comparator** is a measure within the study design, such as the population before the arrival of the invasive species, or a comparable population not exposed to the invasive species, and the **outcome** is the change in the population of the native threatened species (this may include change in distribution, abundance, density or other dynamics).

Methods

Search strategy

The following general electronic databases will be searched:

1. CAB Abstracts (through CAB Direct)
2. Web of Knowledge
3. British Library Direct
4. Science Direct
5. Directory of Open-Access Journals
6. COPAC
7. Scirus (All journal sources)
8. Scopus
9. Agricola
10. Centre for Ecology and Hydrology online database
11. JSTOR
12. ConservationEvidence.com
13. WorldWideScience.org
14. Wildlife and Ecology Studies Worldwide

The US Threatened Species Recovery Plans will also be consulted in detail. These are USA-specific plans that are known to contain references to invasive species in relation to the threatened species in question. Further references will be obtained from Smithsonian's Biological Conservation Newsletter and from four reports (March, June, Sept and Dec 2011) by Dr Jan Eldridge on invasive species threatening endangered, threatened and candidate species for the CBSG on the IUCN website.

Other specific/specialised databases will be searched as appropriate where identified or recommended by experts within the field, e.g.:

1. CABI's Invasive Species Compendium (open-access)
2. IUCN Red List of Threatened Species data
3. US Fish and Wildlife Service
4. USDA National Invasive Species Information Centre

The following resources will be examined for relevant information on invasive species and an assessment made as to whether they will contribute significant extra information pertinent to the systematic review questions that is not available in the primary resources consulted:

1. Nonindigenous Aquatic Species
2. Global Invasive Species Database
3. Pacific Island Ecosystems at Risk (PIER)

These sources will be the basis of the lists of invasive species and threatened species considered and actively sought in the systematic review. Use of specialised databases that require name-specific searches will be determined according to time capabilities. The emphasis will be on a representative, repeatable and unbiased approach that allows the clearest possible answer to the sub question.

Hand searching of key titles will also be undertaken, e.g.:

1. Aliens, a publication by IUCN

Recognised experts, practitioners and recent authors will be contacted to see if any relevant unpublished material or datasets are available for inclusion within the review.

In addition, web searches will be performed using the search engines:

Google Scholar, Microsoft Academic Search and Google (organic). The first 50 hits (.doc(x)/.txt/.xls/.pdf files, using advanced search) from each data source will be examined for appropriate evidence. No further links from the captured website will be followed unless to a .doc/.pdf file.

Email lists will be joined (CABI's Invasive Alien Species List, Science Daily and The IUCN Aliens List) to ensure relevant references published following the initial literature will be incorporated into the study. All emerging new literature will be logged in a separate EndNote library to allow for transparency of when and how all references entered the review process.

The list of search terms to be used is listed in the Appendix. The most effective search strategy will be

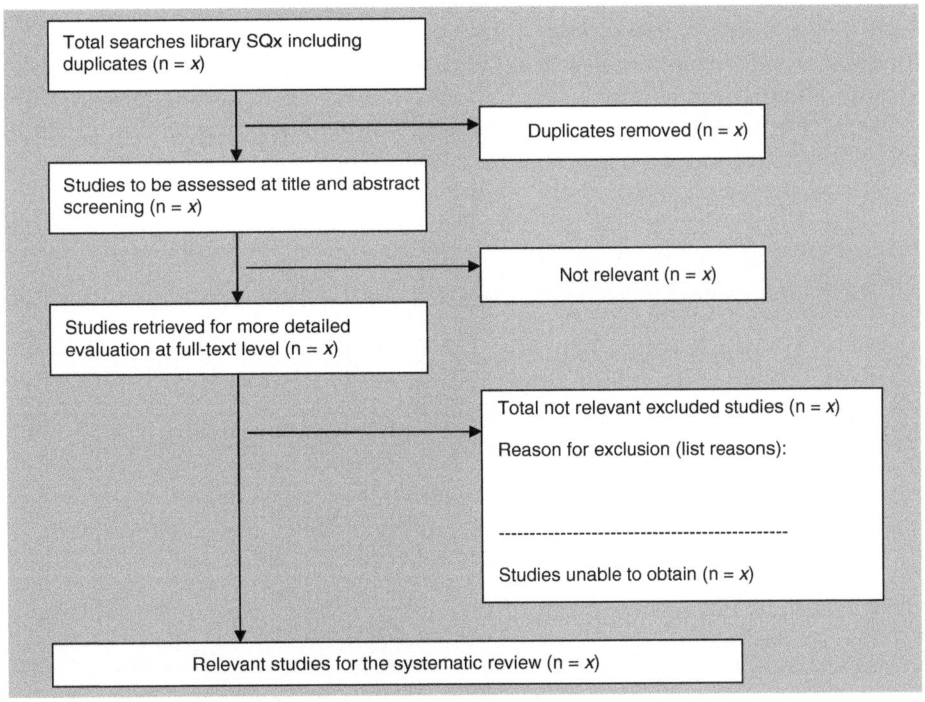

Figure 1 Literature assessment flow through the search and screening process. Values (n = x) are the number of studies at each stage, x to be replaced by the actual number in the systematic review report. (Based on QUOROM statement flow diagram) [5].

developed through small-scale scoping trials, using some of the species about which greatest concern has been raised. It is envisaged that extensive lists of full species names of potential invasive species and threatened species will need to be combined in Boolean searches to find the relevant material.

Each search will be stored in a separate EndNote Library, for record keeping and then combined into a "Total searches library SQ*x* (including duplications)". Two de-duplication processes will be undertaken, firstly using EndNote, then a second manual scan to remove duplicates with differing syntax (e.g. '2' or 'II'). This de-duplication process will create the first EndNote Library, which would be considered for applying the study inclusion criteria to at the title and abstract level (see Figure 1). As each of the secondary questions has a different search strategy to capture the potential body of evidence this process will be undertaken separately for each of the questions.

Study inclusion

The inclusion and exclusion criteria will be applied by one reviewer to all potential studies at the title and abstract level. Where there is insufficient information to make an informed decision regarding a studies inclusion, then relevance to the next stage of the review process (full-text assessment) will be assumed. A second reviewer will examine a random subset of at least 25% of the reference list (up to a maximum of 2,500 references) to assess repeatability of the selection criteria. Kappa analysis will be performed, with a rating of substantial being required to pass the assessment. Disagreement regarding inclusion/exclusion of studies will be resolved by consensus, or following assessment by a third reviewer. If the Kappa value is low, the reference list will be reassessed against adjusted inclusion and exclusion criteria. The same subset of references will be reassessed by a second reviewer with Kappa analysis. Reviewers will then consider articles viewed at the full-text level for relevance, either excluding them from, or admitting them to, the review.

For **secondary question 1,** studies will be assessed against the following inclusion/exclusion criteria:

Population

All studies that investigate one or more species that are included on the US FWS list as Endangered, Threatened or Candidate species list. These are listed in Appendix.

Exposure

Any invasive species having a documented impact (positive, negative or lack of) on the threatened species.

Comparator(s)

n/a (although for inclusion in analysis a comparator of no invasive species or a time series will be required).

Study Design

All study designs will to be included in the review. The quality of the original methodology for each study will be assessed and summarised in the final report.

For **secondary question 2** studies will be assessed against the following inclusion/exclusion criteria:

Population

Any invasive species that has been identified under secondary question 1.

Exposure

The invasive species has had an impact on a threatened species under secondary question 1.

Comparator(s)

n/a

Study Design

All study designs will be accepted into the review.

Focus of Study

The study must mention one or more mechanisms by which the invasive species impacts on the threatened species.

Exclusion for both sub questions:

a) Owing to the time constraints of the review, only English language documents will be included within the final synthesis. Any non-English documents will be identified in a separate EndNote library, which will be made available to future review teams.
b) Human pathogens are excluded from this review as an invasive species type.

For both the sub questions the definition of 'Invasive Species' as defined by the USA Executive Order 13112 of February 3, 1999 is being used. It states that an "Alien species" means, with respect to a particular ecosystem, any species, including its seeds, eggs, spores, or other biological material capable of propagating that species, that is not native to that ecosystem. It also states that "Invasive species" means an alien species whose introduction does or is likely to cause economic or environmental harm or harm to human health. Harm to animal health has been added to that definition, using the World Organisation for Animal Health (OIE) definitions as to what is meant by "harm". Threatened species includes those species that are both US state or nationally registered and those US candidate species under consideration that do not appear in the IUCN Red list.

For this Systematic Review Map, work will focus on invasive species whose establishment and spread threaten ecosystems and habitat or species [6]. For inclusion into the review, there needs to be a documented impact to a species that the USA listed on the US FWS Endangered, Threatened or Candidate species lists.

Potential effect modifiers and reasons for heterogeneity

Where information regarding the reasons for heterogeneity is presented in the studies, it will be recorded, and when possible assessed. Sources of heterogeneity within the studies may be one (or more) of the following: study design, habitat type/degradation/alteration, geographical region of USA, species taxonomic position (e.g. mammal or plant), life-history strategy of either species and trophic level.

Study quality assessment

For secondary question 1, the review team will assess the study methodologies reported in all articles accepted at the full-text level. The study quality will be scored according to a hierarchy of evidence adapted from systematic review guidelines used in medicine and public health [7] and conservation [8]; e.g. a randomised control trial would be weighed higher than a site comparison study (see Appendix for the draft coding tool).

The results of the study quality assessments will be presented in summary tables within the full review document, for full transparency.

For secondary question 2, the study type (laboratory, field, etc.) will be recorded for summary in the full review, but no formal assessment of quality will be undertaken.

Data extraction strategy

Data will be extracted by one reviewer, and a random subset of at least 25% of the selected studies will be checked by another reviewer to verify repeatability and accuracy. The electronic data extraction form will be trialled and checked with the question stakeholders. All data extracted will be presented in summary tables in the full review document.

For secondary question 1, data regarding the study characteristics (e.g. location), study methodology (e.g. timeline and sample size), the threatened species, invasive species (and whether the invasive population increases over the time of the study), and the impact will be recorded. For secondary question 2, data relating to the mechanisms used by the invasive species to impact on the threatened species will be extracted.

Data mapping and synthesis

As data synthesis may not be possible for accepted studies, summary tables and a systematic map will be produced. The map will follow the format of a searchable database, summarising study characteristics, study quality and results. This will be organised using transparent, replicable criteria for ease of updating.

Quantitative analysis will be undertaken on any data suitable for formal statistical analysis. Where possible, meta-analysis for each of the interventions will be carried out with reasons for heterogeneity assessed by meta-regression (univariate or multivariate). Subset analysis will be undertaken if there is sufficient data on taxonomic groups or specific habitats (highlighting across-species heterogeneity). If at all possible data will be captured across multiple isolated populations (e.g. mountain tops) that might allow for within species heterogeneity to be investigated.

It is envisaged that species pairs will be created between a threatened species and an invasive species. These species pairs will allow both impact and competition mechanism to be categorised against them, to produce an overall systematic visualisation of the status of invasive/threatened species relationships.

Publication bias will be assessed, although it is envisaged that this will be reduced by accepting all publication types (formal peer-reviewed journal articles through reports and species plans to observational reports), and evidence of both negative and positive relationships between threatened species and invasive species. In addition the review team's level of agreement at all stages will be assessed.

Appendix one: search terms used on databases

Secondary question 1

Search for all databases:

1. ("Invasive species" OR "Invasive alien species" OR "IAS" OR feral) AND ("threatened species" OR "endangered species" OR "candidate species")
2. ("Invasive species" OR "Invasive alien species" OR "IAS") AND (reduction OR decline OR extinction)
3. Species list (the scientific names of all US FWS Endangered, Threatened and Candidate species as listed September 2011.)

Appendix two

Table 2 shows draft quality coding tool.

Table 2 Draft quality coding tool

Generic data quality features	Specific data quality features	Quality element	Quality score
Internal validity of study	N/A	Clear aims	2
		Treatment and control groups similar at start of trial (low heterogeneity between locations, size of exp. areas, soil type, shading, plant types etc.)	2
		Outcomes measured accurately and reliably	2
		Control of confounding variables	2
		Replications of study	2
Study Design	N/A	Randomised block design time series (pre-intervention data/observations)	80
		Non-randomised block design time series	60
		Historical dataset e.g. >10 years (not from current experimental observation)	50
		Data without comparator	30
		Descriptive, field observations	20
		Expert opinion	10
Measurement of outcomes	Quantitative measurements of intervention	Population size, range, density, fecundity–pop. counts/mark-recapture/nest counts/dropping counts	10
	Qualitative/descriptive observations of intervention (e.g. vague estimate of time since introduction)	Population size, range, density, fecundity	5

Competing interests

The authors declare that they have no competing interests.

Authors' contributions

PR conceived of the study, coordinated development of the protocol and drafted the manuscript. HD, DH, MP, NW and HW participated in development of the protocol and drafting of the manuscript. All authors read and approved the final manuscript.

Funding disclosure

This proposal is submitted as a partnership between CABI, the National Marine Fisheries Service (NMFS), CBSG/IUCN and USDA's Invasive Species Coordination Program. The funding for this review comes from the USDA Invasive Species Coordination Program, managed by Ms. Hilda Diaz-Soltero, USDA Senior Invasive Species Coordinator, who is an advisor to the review team but not taking an active role in any phase of the selection and reviewing of the evidence captured by the CABI review team.

Author details

[1]CABI, Nosworthy Way, Wallingford, Oxfordshire, United Kingdom. [2]USDA, National Invasives Species Council, 1201 Eye Street, NW., 5th Floor, Washington, DC, USA.

References

1. Clavero M, Garcı-Berthou E: **Invasive species are a leading cause of animal extinctions.** *Trends Ecol Evol* 2005, **20**(3):110.
2. Gurevitch J, Padilla DK: **Are invasive species a major cause of extinctions?** *Trends Ecol Evol* 2004, **19**:470–474.
3. Miller RR, Williams JD, Williams JE: **Extinctions of North American fishes during the past century.** *Fisheries* 1989, **14**:22–38.
4. Burbidge AA, Manly FJ: **Mammal extinctions on Australian islands: causes and conservation implications.** *J Biogeogr* 2002, **29**:465–473.
5. Moher D, Cook DJ, Eastwood S, Olkin I, Rennie D, Stroup DF: **Improving the quality of reports of meta-analysis of randomised controlled trials: the QUOROM statement – Review.** *Lancet* 1999, **354**:1896–1900.
6. McNeely JA, Mooney HA, Neville LE, Schei PJ, Waage JK: *Global Strategy on Invasive Alien Species.* Gland, Switzerland and Cambridge UK: IUCN; 2001.
7. Stevens A, Milne R: **From The effectiveness revolution and public health.** In *Progress in Public Health.* Edited by Scalley G. London: Royal Society for Medicine Press; 1997.
8. Pullin AS, Knight TM: **Support for decision making in conservation practice: an evidence-based approach.** *J Nat Conserv* 2003, **11**(2):83–90.

How effective are on-farm conservation land management strategies for preserving ecosystem services in developing countries? A systematic map protocol

Jessica Thorn[1*], Jake Snaddon[2], Anthony Waldron[3], Kasper Kok[4], Wen Zhou[5], Shonil Bhagwat[6], Kathy Willis[1,7,8] and Gillian Petrokofsky[1,5]

Abstract

Background: An extensive body of literature in the field of agro-ecology claims to show the positive effects that maintenance of ecosystem services can have on sustainably meeting future food demand, by making farms more productive and resilient, and contributing to better nutrition and livelihoods of farmers. In Africa alone, some research has estimated a two-fold yield increase if food producers capitalize on new and existing knowledge from science and technology. Site-specific strategies adopted with the aim of improving ecosystem services may incorporate principles of multifunctional agriculture, sustainable intensification and conservation agriculture. However, a coherent synthesis and review of the evidence of these claims is largely absent, and the quality of much of this literature is questionable. Moreover, inconsistent effects have commonly been reported, while empirical evidence to support assumed improvements is largely lacking.

Objectives: This systematic map is stimulated by an interest to (1) collate evidence on the effectiveness of on-farm conservation land management for preserving and enhancing ecosystem services in agricultural landscapes, by drawing together the currently fragmented and multidisciplinary literature base, and (2) geographically map what indicators have been used to assess on-farm conservation land management. For both questions, we will focus on 74 low-income and developing countries, where much of the world's agricultural expansion is occurring, yet 80% of arable land is already used and croplands are yielding well below their potential.

Methods/Design: To this end, reviewers will systematically search bibliographic databases for peer-reviewed research from Web of Science, SCOPUS, AGRICOLA, AGRIS databases and CAB abstracts, and grey literature from Google Scholar, and 22 subject-specific or institutional websites. Boolean search operators will be used to create search strings where applicable. Ecosystem services included in the study are pollination services; pest-, carbon-, soil-, and water-regulation; nutrient cycling; medicinal and aromatic plants; fuel wood and cultural services. Outputs of the systematic map will include a database, technical report and an online interactive map, searchable by topic. The results of this map are expected to provide clarity about synergistic outcomes of conservation land management, which will help support local decision-making.

Keywords: Agro-ecology, Conservation agriculture, Sustainable intensification, Ecosystem services, In-field assessment, Site-specific management, Land sharing, Decision-making, Evidence-based environmental policy

* Correspondence: jessica.thorn@merton.ox.ac.uk
[1]Department of Zoology, Long-term Ecology and Resource Stewardship Laboratory, Biodiversity Institute, University of Oxford, Oxford, UK
Full list of author information is available at the end of the article

Background

Food production systems are threatened in the face of growing food demand, climate change and land cover changes [1]. Agriculture accounts for 70% of water withdrawals worldwide [2], one third of all available energy [3], 75% of all deforestation [4], 19-29% of global GHG emissions, and is the largest contributor of non-CO2 GHG emissions [5]. Declines in ecosystem health have consequences for agricultural production, such as soil salinization from over-irrigation and eutrophication of watercourses from fertilizer application. Forty percent of arable land worldwide is already degraded [6]. Moreover, these trends are increasing as agriculture intensifies and expands. For example, between 1961-2005 agricultural production doubled in Sub-Saharan Africa [7], and globally, was one of the main drivers of degradation of 65% of natural ecosystems [8]. In the last century, forest cover decreased from 170-100 million ha and every year, palm oil cultivation is responsible for c. 300 000 ha of forest cover loss [9]. In the next 25 years, food production and availability must increase by 50–70% to keep pace with the demands of a global population expected to reach 8-10 billion, income growth, and changing consumer preferences [6,10]. To address these challenges, recent international meetings have been convened, such as the UN Summit of 2014, where leaders from 20 governments and 30 organizations pledged their commitment to addressing food security through the formation of the Global Alliance for Climate-Smart Agriculture. Balancing the need to provide enough food for a growing population while maintaining healthy ecosystems and habitats is thus arguably one of the most pressing issues of the 21st century [1,6].

Ecosystem services as incentives for conservation agricultural land management

An emerging strategy being championed for conservation is the ecosystem service framework, which proponents consider more likely to be relevant to agricultural landscapes and their associated people than traditional biodiversity conservation [11,12]. The ecosystem services framework can be used to capture how human action both impacts and is affected by ecosystem responses to land use and land use changes [13].

Although various comprehensive frameworks and classifications refined and omitted categories [14,15], the framework for ecosystem services referred to is based on the Millennium Ecosystem Assessment (MA) [8], as this was the first large-scale ecosystem service assessment and categories are widely recognized [16]. This includes supporting services (e.g. carbon regulation, pest regulation, nutrient cycling), regulating services (e.g. water/soil regulation and supply, pollination services), provisioning services (e.g. fuel wood, medicinal and aromatic plants) and cultural services (e.g. education, recreational, spiritual, tourism, bequest or aesthetic value). Ecosystem goods and services are stocks or flows of materials that deliver welfare gains or losses that are material (e.g. fuel wood), as well as non-material (e.g. recreational services) [17]. Ecosystem elements are both biotic and abiotic and are generally described in terms of amounts (e.g. taxonomic, functional, chemical or physical units) [18-20]. Ecosystem processes, often used interchangeably with ecosystem functions, are the complex interactions (e.g. events, reactions or operations) among elements of ecosystems (e.g. events, reactions or operations), and are generally described in terms of rates [21].

Since the publication of the MA in 2005, the ecosystem services framework has gained traction - in terms of research, a spectrum of tools, and funding mechanisms [22,23]. Dedicated journals have been launched (e.g. International Journal of Biodiversity Science in 2005, Ecosystem Services and Management in 2005, Ecosystem Services in 2012), alongside graduate programs (e.g. MSc in Ecosystem Services, University of Edinburgh). Funding bodies are also prioritizing research into more comprehensive quantification of values of ecosystem services and the link with human health and wellbeing, such as the $65 m 7 year programme on Ecosystem Services and Poverty Alleviation (ESPA) and the $11 m + 6 year Valuing Nature programme led by the National Environment Research Council [24]. Moreover, ecosystem services projects attract on average more than four times as much funding as traditional biodiversity conservation projects, through greater corporate sponsorship and a wider variety of finance tools [11]. Supported by this research, there is a growing spectrum of ecosystem assessment tools, including computer-based platforms using national data (e.g. Integrated Valuation of Ecosystem Services and Trade-offs (InVEST), modelling and scenario driven tools (e.g. MIMES, ARIES), as well as efforts to integrate these frameworks (e.g. the Common International Classification on Ecosystem Services (CICES)). The ecosystem services framework has been used for international negotiation and collaboration in platforms, such as the Ecosystem Services Partnership in 2008, the International Panel on Biodiversity and Ecosystem Services in 2012, and the EU 2020 Biodiversity Strategy (e.g. Target 2) [1]. National governments have also incorporated ecosystem services frameworks to inform budget assignment and thematic planning prioritization, such as the UK's National Ecosystem Service Assessment [17] and Foresight Report [1], that relates ecosystem services to agriculture and food security. The approach has further gained traction in the private sector, and has been used to conduct economic valuations in carbon (e.g. Voluntary Carbon Standard in South Africa), timber (e.g. Reduced Emissions from Deforestation and Degradation (REDD+) in Nigeria) and watersheds (e.g. Payments for Ecosystem Services in Costa Rica) [25].

The role of conservation land management in maintaining ecosystem services

Given the importance of ecosystem services to the sustainability and security of agricultural systems, as well as the current rate at which those services are being degraded by agricultural systems, a key need has arisen to implement ecosystem service conservation strategies on farms. A variety of alternative practices to conventional or intensive agriculture have been proposed, which we group under the term "conservation land management" for the purposes of this study. Conservation land management strategies preserve or enhance ecosystem services without compromising farm production and may be adopted before, during or after cultivation [26]. Strategies may be active, such as surface crop residue management, or passive, such as the existence of native vegetative patches in fields. Practices may incorporate principles, amongst others, of multifunctional agriculture (producing food and non-food commodities, maintaining wild crop varieties, traditional landraces and local culture [27]), sustainable intensification (relieving pressure on land expansion and limiting forest encroachment [7]), and conservation agriculture (practices of no-tillage, permanent soil cover using crop residues or cover crops, and crop rotation [28]). Such practices often require minimal inputs with opportunities for enhancing small-holder production [26].

Specific examples of conservation land management strategies include growing leguminous cover crops to fix nitrogen, retain moisture, stimulate root-growth and encourage below-ground microbial activity [29]; no till or minimum till systems and crop rotation, to influence soil organic carbon sequestration [26,30] and yield [26]; mosaic or matrix management of natural vegetation within or adjacent to farmland (e.g. set aside areas, buffer strips, hedgerows or field margins), to encourage the presence of beneficial wild pollinator populations [31]; fallowing to suppress leaching and erosion of organic matter and nutrients, and increase soil cation exchange [32]; intercropping and the use of push-pull systems to regulate detrimental pest populations and enhance natural enemy populations [33]; water conservation techniques, such as drip irrigation, alternative wet and dry irrigation, raised beds, tied ridges and ditches, and growing grass filter strips, to influence water regulation and supply and control erosion [34,35]; and the intercropping of timber trees with shade tolerant crops, or multi-story cropping, to reduce the presence of weeds and promote nutrient cycling [27]. To conceptualize a theory of change, Figure 1 shows examples of conservation land management strategies (single programs or comprehensive community initiatives)

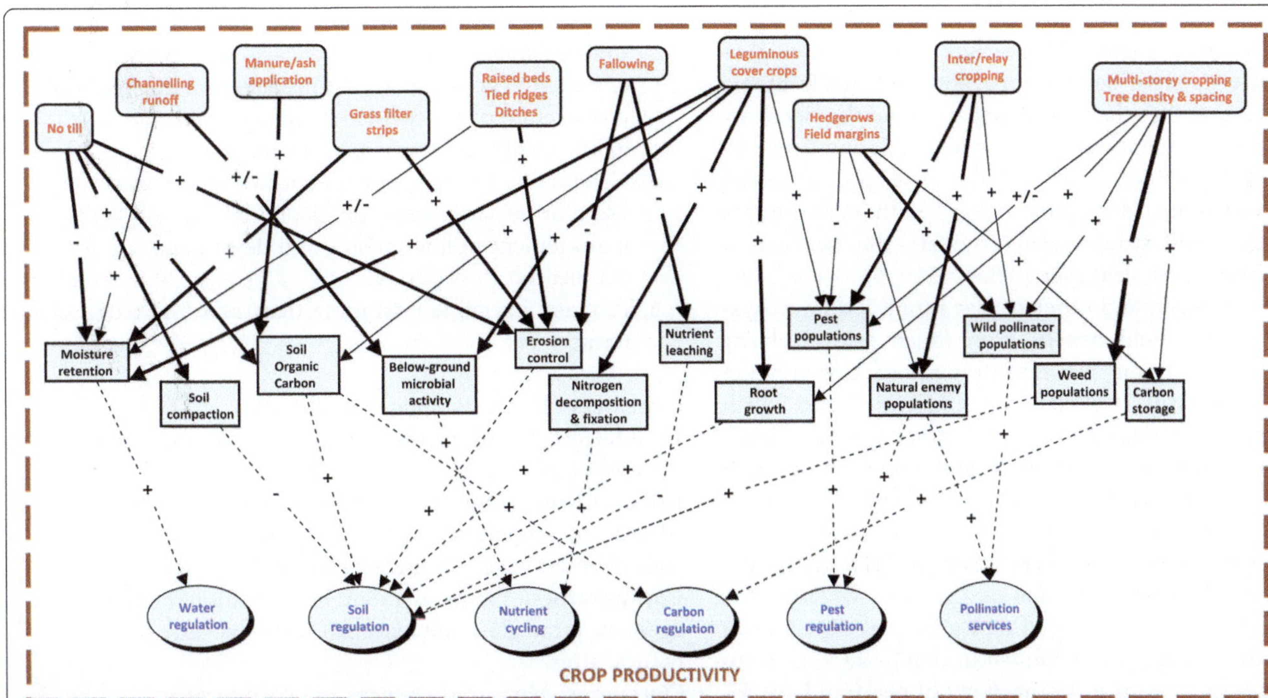

Figure 1 Illustrative theory of how conservation land management strategies may bring about change in ecosystem service provision. [Red] indicates conservation land management strategies; [Black] indicators; [Blue] ecosystem services; [+] indicates an increase; [-] indicates a decrease; thick solid lines are estimated relations referenced in the text; while dotted lines are proxies for ecosystem services. The box surrounding the figure indicates that all factors influence crop productivity.

(red) that may bring about outcomes on supporting or provisioning ecosystem services (blue), through key measurable indicators or proxies (black).

The figure illustrates the complex web of activity that is required to bring about change, while assumptions indicated in the flow arrows are not exclusive or exhaustive, and require varying degrees of research verification. We still lack a coherent evidence base showing how effectively these management strategies preserve or enhance ecosystem services overall.

Synthesizing evidence is complex for three main reasons. Firstly, change in conservation land management may affect various ecosystem services differently. For example, some studies report that long-term no-till can improve soil fertility, recovery and decrease erosion, but no-till can also lead to soil compaction, limit water infiltration and can hinder seed germination [36,37]. Other studies have reported that managing runoff can increase and stabilize crop production and deposit plant nutrients in soil, but runoff can adversely affect nutrient cycling [38]. The management of ecosystem services therefore requires making judgements about trade-offs, not least, the trade-off between agricultural production and environmental protection [23]. Secondly, impacts of land management on ecosystem services are often quantified by indicators or proxies of ecosystem processes, thought to subsequently impact ecosystem services. However, evidence for the adequacy of these proxies is often incomplete or inconsistently reported. For example, many studies suggest higher biodiversity allows for higher levels of ecosystem service provision [39], while others argue there is little hard evidence to show the necessity of a diversity of natural enemies in regulating pests on farms [40]. Thirdly, much of the evidence is spread across different disciplinary "silos", with very limited synthesis. Some studies also overstate the benefits of land management strategies [37].

Land managers, and other parties interested in ecosystem services, would benefit from much greater clarity and information on the effectiveness of conservation land management strategies, in order to decide which management strategies to implement at the farm level. When evidence is so extensive and disparate, a rapid first step in such an informational synthesis is a systematic map, a rigorous methodological tool of data extraction of peer-reviewed and grey literature [41]. Systematic maps have the same precision as a review, while no evidence synthesis is attempted and a critical appraisal of the quality of evidence is limited in depth [42,43]. Previous attempts to synthesize this body of research have focused on particular regions, such as Africa [44,45], a limited set of practices [28], or have evaluated management outcomes purely in terms of crop responses [45]. Our systematic map will build on this research, both geographically, and in terms of the management strategies and ecosystem services studied.

Against this background, the aim of this systematic map is to review the state of evidence that reports on the effectiveness of on-farm conservation land management for protecting or enhancing ecosystem services. First, we aim to provide a better summary of different strategies proposed and tested, in which crops, habitats and regions, and over what timeframes. Secondly, we will identify the pathways by which practices are assumed to influence ecosystem service provision by reporting on measurable indicators assessed in studies. We will differentiate between methodologies that are experimental, quasi-experimental and non-experimental and indicators that are physical, chemical, biological, social and/or economic. The spatial scale of the study is at the field level, as this is the scale at which most decisions for land management are made and need to be informed [46]. Our geographical coverage will be developing regions, as this is where much of agricultural expansion is occurring [1], yet 80% of arable land is already used [47] and croplands are yielding well below their potential [4]. In some cases, developing regions may also depend on ecosystem services rather than technological inputs to support agriculture, due to lower financial, technical and credit-borrowing capacity.

Objectives of the systematic map

1. Collate studies providing evidence on the effectiveness of on-farm conservation land management practices on ecosystem service provision in agricultural landscapes in low-income and developing countries.
2. Geographically map which indicators have been used for on-farm assessments of conservation land management in low-income and developing countries.
3. Produce an online interactive map, searchable by topic.

Elements of the systematic map question

Population: Farms in low/middle income and developing countries.

Intervention: Conservation land management strategies adopted to support productive agriculture, while simultaneously preserving or enhancing ecosystem services.

Comparators: Farms without conservation land management strategies, conventional/intensive agriculture or natural sites.

Outcomes: Measured changes in ecosystem services, including supporting services (e.g. carbon regulation, pest regulation, nutrient cycling), regulating services (e.g. water/soil regulation and supply, pollination services), provisioning services (e.g. fuel wood, medicinal and aromatic plants) and cultural services (e.g. education, recreational, spiritual, tourism, bequest or aesthetic value).

Method

Search strategy

The following search strategy and research question have been developed with stakeholders in two meetings in South Africa (February 2014) and UK (June 2014). Expertise of stakeholders span the fields of environment, conservation, biodiversity, development, agriculture, entomology, soil science, pollination, anthropology and ecology. Further comments on earlier drafts of this protocol were provided after the workshops and over email with other contributors (Additional file 1).

Language

The systematic map will be limited to studies published in English. This decision was made as the larger body of literature is in English, as well as this being the linguistic competency of the review team and also provides a mechanism for restricting the scale of the study [41,48]. Should a full systematic review be conducted arising from the map, French, Spanish or Portuguese would be considered to cover literature from regions in Africa, South East Asia and Latin America. Future assessments will create language-specific search strings associated with the research question.

Key search terms

A list of key terms, searched at levels of title, and abstract level is listed in Additional file 2. Each of the terms relate to the components of the research question and PICO (Population Intervention Comparator Outcome). The list was compiled by experts from invited institutes and universities, who met at the two stakeholder workshops. Terms were built into strings, used in preliminary scoping searches conducted in the Web of Science (WOS) CAB Abstracts, and Google Scholar. In WOS 27 search strings were tested, with the final string resulting in 7558 hits. The search strategy contains synonyms and near-synonyms, and does not make a distinction between definitions used in the primary literature. We will use these strings as the basis of the search, however an iterative approach to identifying search terms will be adopted to improve the strategy and help minimize bias. Details of the search logic and the development of the final strings are found in Additional file 3. Boolean search operators will be used to connect search terms in the usual way and subject to the specific rules of individual databases. In the search wildcards will be used with care and will vary slightly from database to database. Such variations between search strategies in each database and source will be documented and reported in the final map. The date of the search will be documented, allowing for updating of future mapping.

Sources of publications

Bibliographic databases

The following bibliographic databases will be considered, covering material from the natural and social sciences, from both peer-reviewed and grey literature:

1. Thomson Reuter's (formally ISI) Web of Science, New York, USA http://apps.webofknowledge.com/
2. Elseviers' SCOPUS http://www.elsevier.com/online-tools/scopus
3. CAB Abstracts published by CAB International, Wallingford, UK http://www.cabdirect.org/
4. AGRICOLA Agricultural Research Database http://www.ebscohost.com/academic/agricola
5. AGRIS Agricultural Science and Technology Information Systems http://agris.fao.org/agris-search/index.do

Search engine searching

In addition to the results from the above-mentioned agricultural and forestry databases, the meta-search engine platform Google Scholar (http://scholar.google.com) will be searched to access studies not captured in the above databases. The first 200 search results will be assessed at the level of title and abstract. Results, organized by relevance, will be compared to results of existing searches in bibliographic databases, and added to the reference list.

Targeted searches

Targeted searches will be conducted to source peer-reviewed and grey literature, including searching particular journals, specialist organizations, online databases and individual research papers. This will be restricted to .pdf or word documents available online. Results will be compared to searches from bibliographic databases and omitted papers included in the reference file.

Individual journals

First, we will hand search individual e-journals whose topic areas are closely aligned with our research question, based on the outcomes of searches using bibliographic databases. Journals include Ecological Indicators, Ecosystem Services, Integrated Environmental Assessment and Management and Agriculture, Ecosystems and Environment and Field Crop Research.

Key research papers and case studies through stakeholder engagement

Second, key research papers, reports and case studies relevant to the research question will further be identified by

(a) two stakeholder workshops with experts in key thematic areas, in Cape Town, South Africa and Oxford, UK; (b) online calls on the Oxford Long-Term Ecology Lab and CIFOR websites and (c) consultations with an advisory board of experts throughout 2014/5. We will rely on the advisory panel to recommend relevant dissertations, where the rich data has not been published in academic journal. Snowball sampling techniques will be used to identify further literature and will help minimize selection bias.

Grey literature for specialist searching

Third, a grey literature search will include organizational reports, conference papers or proceedings and policy briefs, station and annual reports, found on searches of subject-specific websites and institutional websites (Table 1). Specialist organizations listed include research organizations, national/bi-multi-lateral organizations, donors, and private industry that work the sector in agricultural land management and conservation. Key terms used to search grey literature will be amended as compared to strings used for peer-reviewed bibliographic databases. Literature will be searched through thematic (e.g. agricultural biodiversity) and geographic areas (e.g. Africa, Asia) and simple key terms based on a subset of search terms (e.g. (ecosystem OR ecological OR environmental) AND (function OR service OR good OR process)

Table 1 Websites of specialist organizations and online databases

Organization	Website
Centre for International Forestry (CIFOR)	http://www.cifor.org/library/
Consultative Group for International Agricultural Research (CGIAR)	www.cgiar.org/resources/cgiar-library
Alliance for a Green Revolution in Africa (AGRA)	http://agra-alliance.org
International Rice Research Institute (IRRI)	http://irri.org
International Centre for Tropical Agriculture (CIAT)	http://ciat-library.ciat.cgiar.org
Integrated Water Management Institute (IWMI)	www.iwmi.cgiar.org
International Institute of Tropical Agriculture (IITA)	www.iita.org/
International Potato Centre (CIP)	http://cipotato.org
Africa Rice Centre (AfricaRice)	www.africarice.org
World Resources Institute (WRI)	www.wri.org/
International Crops Research Institute for the Semi-Arid Tropics (ICRISAT)	http://oar.icrisat.org/cgi/search/advanced
Climate Change Agriculture and Food Security (CCAFS)	http://ccafs.cgiar.org/publications
World Agroforestry Centre (ICRAF)	http://outputs.worldagroforestry.org
International Center for Agricultural Research in the Dry Areas (ICARDA)	http://www.icarda.org/
Comprehensive Africa Agriculture Development Programme (CAADP)	http://www.caadp.net/
Institute of Environment and Agricultural Research (INRA)	www.inra.fr/en/liste/dossiers/76
Monsanto Agricultural	www.monsanto.com/products/pages/biotech-technical-publications.aspx
Syngenta Foundation	http://www.syngentafoundation.org/index.cfm
Department for International Development (DFID)	http://r4d.dfid.gov.uk/
Ecosystem Services Partnership (ESP)	www.es-partnership.org
Ecoagriculture Partners	www.ecoagriculture.org/publications.php
International Union for the Conservation of Nature (IUCN)	www.iucn.org/knowledge/publications_doc/publications/
United Nations Development Programme (UNDP)	www.undp.org/content/undp/en/home/librarypage.html
Overseas Development Institute (ODI)	www.odi.org/publications
International Maize and Wheat improvement (CIMMYT)	http://repository.cimmyt.org/xmlui/discover
The Economics of Ecosystems and Biodiversity (TEEB)	www.teebweb.org
Convention of Biological Diversity (CBD)	www.cbd.int
Bioversity International	www.bioversityinternational.org/e-library/publications/
Practical Action	http://practicalaction.org/publishing
International Institute on Environment and Development (IIED)	http://pubs.iied.org

AND (farm OR agriculture OR agro-ecosystems OR agro-ecology OR cultivation OR crops OR conservation agriculture) AND (impact OR assess OR method OR protocol)). To assist in the screening in the grey literature, text extraction software will be used to identify relevant key words and strings related to the research questions.

Comprehensiveness of search

Alternative terms for the search have been tested using a test library (Additional file 4), and disaggregated using wildcard and other variations of similar words to ensure key literature relevant to the research question is picked up in the search, as well as to ensure specificity.

Searches using strings in the test library adequately picked up studies assessing the effects of conservation land management on ecosystem processes, as well as ecosystem service impacts on agricultural production (e.g. pollination). Results identified studies that quantify specific ecosystem services or disservices, or assess how conservation agricultural land management strategies benefit or degrade ecosystem services.

Site-specific land management strategies included, amongst others, organic (e.g. ash, biochar, straw, urea, farmyard manure) and inorganic fertilizer application (NPK), mulching or surface crop residue retention, crop rotation, inter-/relay- cropping, tied/raised ridges and ditches, hedgerows, field margins, grass filter strips, integrated pest and weed management, terracing, hand pollination, cover cropping, mosaic/matrix management, fallowing, reforestation, multipurpose agroforestry crops, and various tillage regimes.

Studies employed methodologies from the social, economic and natural sciences. Amongst others, various studies assessing soil measured soil organic material, and chemical properties of pH, macronutrients and heavy metals, and physical properties such as soil bulk density and texture. Nutrient cycling was measured through proxies of microbial activity and biomass, while various studies measured biodiversity metrics looking at community composition of key taxonomic indicators. Assessments of carbon services assess above/below-ground biomass, growth and survival rate of trees, stem diameter at breast height and density. Pollination service assessments measured visitation rates, flower pulp analysis, fruit set, and arthropod diversity and abundance. Social indicators included yield, income, farm size, and amounts harvested. Table 2 presents some of the coding variables to be used in the systematic map, with results extracted from studies in the test library.

Study screening, inclusion and exclusion criteria

Search results from all databases will be exported into library folders in EndNote reference manager, and duplicates removed in a common folder. The final set of citations will be screened against the following eligibility criteria by five reviewers. Each study will be assessed for inclusion sequentially at title, abstract stages, and full text. At the beginning of each stage of screening, Kappa analysis will be performed by the four reviewers to confirm agreement between them on a random subset of 100 studies. An online calculator for free-marginal kappa will be used, available at http://justusrandolph.net/kappa/ [49]. If the kappa coefficient is below 0.6, then the kappa analysis will be repeated on additional sets of randomly selected studies until agreement reaches at least 0.6. The reviewers will use DateX systematic review evaluation software [50] to document their decisions. Exclusion of studies will be conservative, and to supplement the kappa analysis of agreement, random samples of rejected studies will be examined and any disagreements between the reviewers discussed and resolved through regular meetings.

Relevant population

- Relevant farming systems will include terrestrial food and cash crops, including cereals, grains, roots and tubers, pulses, vegetables oilseeds and tree oils and agroforestry crops that are multipurpose trees for food, fodder, fuel, medicine and shade. Crops are categorized according to the FAO 2012/13 major commodities list [51]. Mixed and mono-cropping systems will be included, as will organic, conventional, hybrid and genetically-modified crops. Marine biomes or mangroves with maricultural or aquacultural activities will be excluded (see Liquete et al [52] for a review on marine ecosystem services), as will livestock farming and pastures (land covered with grass or other plants suitable for grazing).
- Countries to be included in the study have been selected from three databases of globally-recognized organizations working to improve ecosystem services, reduce food insecurity and support economic development (based on FAO 2013, WB 2014, ESPA 2014 [53-55]). The list of 74 countries, details of the motivation of this classification, together with a description of calculations, is found in Additional file 5.

Relevant intervention

- Studies assessing a conservation land management strategy with the aim to improve ecosystem services will be included. Studies describing general biology, foraging/behavioural/spatial ecology, growth behaviour of crops, DNA markers that reveal genetic phylogeny or reporting plant-breeding

Table 2 Examples from the test library summarizing conservation land management strategies assessed, ecosystem services measured, indicators (non-exclusive), crops and country of assessment

Conservation land management	Ecosystem service	Properties studied	Indicators	Crop	Country	Publication	Year	Authors
Native vegetative patches in farmland	Pollination	Biological	Flower visitation networks, seed mass, honeybee and visitor abundance	Sunflower	South Africa	Ecology Letters	2011	Carvalheiro, L. G., Veldtman, R., Shenkute, A. G., Tesfay, G. B., Pirk, C. W. W., Donaldson, J. S., and Nicolson, S. W.
Reforestation, crop rotation	Soil, carbon	Chemical and biological	Soil organic carbon, microbial biomass, dissolved organic carbon	Rice	China	Journal of Plant Ecology	2013	Cui, J., Zhang, R., Bu, N., Zhang, H., Tang, B., Li, Z., Jiang, L., Chen, J. and Fang, C.
Surface crop residue retention, crop rotation, ridge and furrow systems, legume intercrop	Water, soil, carbon	Chemical and physical	Water infiltration rates, soil moisture, soil organic carbon, texture, NPK, heavy metals	Maize	Malawi	Field Crops Research	2013	Thierfelder, C., Chisui, J. L., Gama, M., Cheesman, S., Jere, F. D., Bunderson, W. T., Eash, N. S. and Rusinamhodzi, L.
Fallowing with pigeon pea and elephant grass	Soil, carbon, biodiversity	Chemical and biological	Soil organic carbon, pH, enzyme activity, microbial biomass	Maize	Ghana	Soil Biology and Biochemistry	2008	Asuming-Brempong, S., Gantner, S., Adiku, S.G.K., Archer, G., Edusei, V. and Tiedje, J. M.
Agroforestry with herbaceous, shrub and tree species strips	Carbon, Soil	Chemical and biological	Soil organic carbon, species diversity and composition, pH, soil exchangeable cations, NPK	Oil palm	Brazil	Agroforestry Systems	2014	de Carvalho, W. R., Vasconcelos, S. S., Kato, O. R., Capela, C. J., B., and Castellani, D. C.
No-till, fallowing, farmyard manure and crop residue mulching	Carbon, soil, water	Chemical and physical	Soil organic carbon, soil texture, water pH, total base cations, bulk density, nitrogen	Cotton	Burkina Faso	Nutrient Cycling in Agro-ecosystems	2006	Ouattara, B., Ouattara, K., Serpantié, G., Mando, A., Sédogo, M., and Bationo, A.
Fallowing over various time periods	Carbon, biodiversity	Biological	Tree and shrub structure, floristic species richness, diversity, stem density, biomass	Tobacco	Tanzania	African Journal of Ecology	2005	Mangora, M. M.

methods, the development of improved cultivars or the selection of genotypic traits in plants will be excluded. Studies assessing only land uses differentiated by crop type will be excluded.

Relevant outcomes

- Changes in ecosystem services measurable at the field level will be included, though some processes will also influence the state of the ecosystem service at broader spatial scales (e.g. carbon sequestration influenced by global processes, or water regulation and supply influenced by catchment level processes). Services may be referred to specifically (e.g. soil organic content in the rhizosphere under alternate tillage regimes [56]), or more broadly (e.g. cultural services of Bedouin in Jordan [57]). However, studies on disease or nutrition outcomes will be excluded, including vector-borne diseases (e.g. malaria or typhoid management in irrigated systems) or fungal pathogens, as well as studies measuring off-farm non-timber forest products. Studies assessing only outcomes on yield or income will be excluded.

Relevant types of study design

- The spatial scale of the study is at the field level. Studies to be included are grounded, empirical assessments, field-based, or correlative experiments in natural environments. Studies assessing the impacts of land uses off the farm will be excluded (e.g. acid mine drainage influencing yield) or climatic influences (e.g. temperature or rainfall changes influencing crop phenology). Purely lab-based manipulative experiments including microscopy, infrared reflectance spectroscopy, chromatography, pot-based/green-house experiments will be excluded. Ex-situ methods of assessing services using remote sensing or modelling (e.g. crop simulation modelling) will be excluded. We will not adopt a strict definition of 'farm' and include studies in agricultural mosaics. This is because service provision in farmland boundaries (e.g. pollination by arthropods) may be affected by adjacent protected areas, riparian strips, buffer zones and the wider farming community.

- Single year and chronosequences assessing inter-annual variation, seasonal changes or interdecadal studies will be included. Long-term paleoecology studies will be excluded. Studies assessing the impact of land management during or after cultivation are included.
- Studies on ecosystem services that provide conceptual frameworks, assess regulatory/legislative frameworks, market access or information for farmers, as well as scientific reviews, meta-analyses, literature reviews, studies validating earlier estimates, clinical trials, patents or datasets will be excluded. Records of local ecological knowledge for assessing ecosystem services, beyond that which is captured in the literature search (e.g. unpublished interview transcripts, video interviews) will be excluded.
- Studies using social and ecological data will be included, while studies that look at economic valuation only (e.g. stated preferences or willingness to pay) will be excluded, unless they provide some direct assessment of basic ecological or social units (See Laurans et al [25] for such a review). Studies that describe methodologies, such as protocols or manuals, or conservation spatial planning tools, with no reference to case study applications will also be excluded.

No time limitation to the search will be applied to the search. However, two factors may result in some bias towards studies more recently published, i.e. literature in the format of Word/PDF documents or digitized library prints, and the context in which the term of ecosystem service has been cited, i.e. being relatively new and taken up in the last 25 years.

Potential effect modifiers and reasons for heterogeneity

For the purpose of the systematic map, reasons for heterogeneity, will not be considered (i.e. non-intervention variables that might influence the outcome [58]). However, we will collect data on region, country, geographical co-ordinates, spatial unit of analysis, crop type, farm type and biome. Future studies may consider other likely reasons for heterogeneity, including climate and agro-ecological zone, soil type, mineral texture class, topography (altitude and slope), species mix in cropping systems, previous land uses, other sources of income/livelihoods or development projects, amongst others.

Study quality assessment

In the case of a systematic map, the study will not seek to assess the quality of different types of evidence (e.g. use of controls, treatments, or baselines, account for environmental variability, and standardized sampling designs), but data are extracted that reflect the quality of studies. We will differentiate between studies that are experimental (studies estimating the impact of farming strategy using controls/treatments), quasi-experimental (without all factors affecting outcome or groups controlled) and non-experimental (no explicit manipulation of groups or design, including correlational, comparative, or longitudinal studies) (see Additional file 6). The study will include recommendations for reviewing the quality of studies or particular methodologies for a future systematic review(s), should they arise.

Coding variables

A standard coded questionnaire (Additional file 6) will be used to select and extract a subset of data from studies that have passed the full text screening stage, extracted by five reviewers. Data extracted will fall under three categories:

(1) Bibliographic information (publication type, author, journal title, year of publication, duration of study)
(2) Basic information of the study (region, country, geographical co-ordinates, spatial unit of analysis, crop type, farm type, biome)
(3) Primary ecosystem service (provisioning/supporting/regulating/cultural) and subtype of ecosystem (e.g. nutrient cycling), site-specific conservation land management practice, methodology (experimental/quasi-experimental/non-experimental), properties (physical/chemical/biological/social/economic) and indicators.

Data mapping and presentation

The output of the systematic map will be a technical report and a map. Summary statistics of study characteristics will be presented, together with a narrative synthesis in the discussion of general trends found in the literature. Methods of data presentation will depend on the accepted studies, but may include which indicators have been used to measure which land management strategies, or trends in the frequency, location, or duration of studies. In addition, a flow diagram will show the number of articles retrieved in the search, numbers passing each subsequent stage of screening, following PRISMA - Preferred Reporting Items for Systematic Reviews and Meta-Analyses [59]. An online interactive map will indicate temporal, thematic and spatial knowledge gaps. To enable such visualisation, background maps powered by OpenLayers (http://openlayers.org) and Google Maps (https://developers.google.com/maps/) are routinely used

[60]. The database of sites can be created in PostgreSQL format (www.postgresql.org) and the mapping may be facilitated by PostGIS (http://postgis.refractions.net). The database of this nature normally contains geographical co-ordinates and background information about studies. This can enable the user to switch on and off the relevant fields of data, distinguished with coloured markers/shapes with a sliding timeline based on the year of publication.

Additional files will include:

(a) an Excel database of coded data for the systematic map,
(b) an EndNote database of all the studies included in the review, and
(c) an EndNote database of studies excluded at full text with reasons for exclusion.

Discussion

The study may illuminate the evolution of the study of conservation land management and ecosystem service provision in productive landscapes. The systematic map should contribute to a clearer understanding of what conservation land management strategies have been adopted on farms and what indicators have been used for their assessment. The results of the map will be used to critically reflect upon the state and scope of evidence and will help identify knowledge gaps to inform new research initiatives or future investments [41,61].

Should sufficient primary research exist, recommendations will be put forward for a full systematic review of a more narrowly defined question. Further development of the systematic map might include examination of the full text, or extension of the database coding and categorization.

A preliminary assessment has led us to some hypothesized trends of certain crops, regions, and ecosystem services that have been favoured. For example, systems commonly studied may include commodity crops of high global economic value and extent (e.g. coffee) or staple crops (e.g. rice or wheat). Regions commonly studied may be where there is high degradation and production potential, such as Brazil's cerrado, China's Loess Plateau, or Mexico's Chiapas. Positive effects of crop diversity may be more commonly studied in tropical than arid agro-ecosystems [62]. Certain ecosystem services may also have been favoured, such as those that are relatively simple to measure (e.g. soil), or those that provide direct goods for human consumption such as provisioning services (e.g. water) [63]. Other ecosystem services may have been discounted (e.g. cultural services), in part due to the subjective nature of perception and appreciation [15], and the need to apply anthropological methodologies that do not fall within the skillset of those typically conducting valuations, including ecologists and economists.

Additional files

Additional file 1: Expert workshops and list of participants.

Additional file 2: Search strings tested in Web of Science.

Additional file 3: Potential search strings tested in Web of Science.

Additional file 4: List of publications in test library.

Additional file 5: Rationale and classification process of countries included in the review.

Additional file 6: Categories (codebook) of data recorded for articles included in the database.

Abbreviations

ARIES: Artificial Intelligence for Ecosystem Services; CABI: CAB International (formally Commonwealth Agricultural Bureaux); CEE: Collaboration for Environmental Evidence; CCAFS: Climate Change Agriculture and Food Security; CICES: Common International Classification of Ecosystem Services; CIFOR: Centre for International Forestry Research; DFID: UK Department for International Foreign Development; DNA: Deoxyribonucleic acid; EBF: Evidence Based Forestry Initiative; ES: Ecosystem Service; EU: European Union; ESPA: Ecosystem Services and Poverty Alleviation; FAO: United Nations Food and Agricultural Organization; GHG: Green House Gas; GNI: Gross National Income; InVEST: Integrated Valuation of Ecosystem Services and Trade-offs; LIFDC: Low-Income Food Deficit Countries; MA: Millennium Ecosystem Assessment; MIMES: Multi-scale Integrated Models of Ecosystem Services; PRISMA: Preferred Reporting Items for Systematic Reviews and Meta-Analyses; REDD: Reduced Emissions from Deforestation and Degradation; OECD: Organization of Economic Co-operation and Development; PICO: Population Intervention Comparator Outcome; TEEB: The Economics of Ecosystems and Biodiversity; United Kingdom: UK; WB: World Bank; WOS: Web of Science.

Competing interests

The authors declare that they have no competing interests.

Authors' contributions

JT jointly with GP conceived the study and secured funding support. JT, JS, AW, KK, WZ, and GP co-wrote the manuscript. SB presented the mapping methodology. JT developed the data extraction strategy and built the test library, and will coordinate the mapping process, analysis and presentation of the results. JT will implement the search and screen the articles with four other reviewers. All authors read and approved the final manuscript.

Acknowledgements

This research is made possible through the financial and technical support of the Evidence-Based Forestry (EBF) Initiative of the Centre for International Forestry Research (CIFOR), supported through the United Kingdom's Department for International Development (DFID), Climate Change Agriculture and Food Security (CCAFS) and the University of Oxford Long-term Ecology Lab. The authors would like to thank the experts that attended the stakeholder workshop in Cape Town, South Africa and Oxford, UK at the launch of the review, and the advisory panel and anonymous reviewers for their feedback on earlier versions of the manuscript.

Author details

[1]Department of Zoology, Long-term Ecology and Resource Stewardship Laboratory, Biodiversity Institute, University of Oxford, Oxford, UK. [2]Centre for Biological Sciences, University of Southampton, Southampton, UK. [3]Edward Grey Institute, Department of Zoology, University of Oxford, Oxford, UK. [4]Land Dynamics Department, Wageningen University, Wageningen, Netherlands. [5]Centre for International Forestry Research, Bogor, Indonesia. [6]Department of Geography, Faculty of Social Sciences, Open University, Milton Keynes, UK. [7]Royal Botanical Gardens, Kew, UK. [8]Department of Biology, University of Bergen, Bergen, Norway.

References

1. Foresight. The future of food and farming executive summary. London, UK: The Government Office for Science; 2011.
2. UNESCO. UN World Water Development Report 4 – managing water under uncertainty and risk. Paris, France: UNESCO; 2012.
3. FAO, WFP and IFAD. State of food insecurity in the world 2012: economic growth is necessary but not sufficient to accelerate reduction of hunger and malnutrition. Rome, Italy: FAO; 2012.
4. Tilman D, Balzer C, Hill J, Befort BL. Global food demand and the sustainable intensification of agriculture. Proc Natl Acad Sci. 2011;108(50):20260–4.
5. CCAFS. Food emissions. 2014 [10 August 2014]; Available from: http://ccafs.cgiar.org/bigfacts/#.
6. Oxford Martin Commission. Now for the long term: report of the Oxford Martin Commission for Future Generations, Oxford Martin School. Oxford, UK: University of Oxford; 2013.
7. Dile YT, Karlberg L, Temesgen M, Rockström J. The role of water harvesting to achieve sustainable agricultural intensification and resilience against water related shocks in sub-Saharan Africa. Agr Ecosyst Environ. 2013;181:69–79.
8. Millennium Ecosystem Assessment. Ecosystems and human wellbeing: synthesis. Washington, DC: Island Press; 2005.
9. Steffan-Dewenter I, Kessler M, Barkmann J, Bos MM, Buchori D, Erasmi S, et al. Tradeoffs between income, biodiversity, and ecosystem functioning during tropical rainforest conversion and agroforestry intensification. Proc Natl Acad Sci. 2007;104(12):4973–8.
10. Cabell JF, Oelofse M. An indicator framework for assessing agro-ecosystem resilience. Ecol Soc. 2012;17(1):18.
11. Goldman RL, Tallis H, Kareiva P, Daily GC. Field evidence that ecosystem service projects support biodiversity and diversify options. Proc Natl Acad Sci. 2008;105(27):9445–8.
12. Poppy GM, Chiotha S, Eigenbrod F, Harvey CA, Honzák M, Hudson MD, et al. Food security in a perfect storm: using the ecosystem services framework to increase understanding. Phil Trans R Soc B. 2014;369:1–13.
13. Matzdorf B, Meyer C. The relevance of ecosystem services framework for developed countries' environmental policies a comparative case study of the US and EU. Land Use Policy. 2014;38:509–21.
14. Haines-Young R, Potchin M. Common International Classification for Ecosystem Service (CICES): Consultation on Version 4, August –December 2012 EEA Framework Contract No EEA/IEA/09/03; 2013.
15. TEEB, The Economics of Ecosystems and Biodiversity Ecological and Economic Foundations. Kumar P (ed). London, UK and Washington, USA: Earthscan; 2010
16. Maes J, Teller A, Erhard M, Liquete C, Braat L, Berry P, et al. Mapping and Assessment of Ecosystems and their Services. An analytical framework for ecosystem assessments under action 5 of the EU biodiversity strategy to 2020. Luxembourg: Publications office of the European Union; 2013.
17. UK NEA. United Kingdom National Ecosystem Assessment: understanding nature's value to society, Technical Report. Cambridge: UNEP-WCMC; 2011.
18. Schwartz MW, Brigham CA, Hoeksema JD, Lyons KG, Mills MH, Van Mantgem PJ. Linking biodiversity to ecosystem function: implications for conservation ecology. Oecologia. 2000;22:297–305.
19. De Bello F, Lavorel S, Dıaz S, Harrington R, Cornelissen JHC, Bardgett RD, et al. Towards an assessment of multiple ecosystem processes and services via functional traits. Biodivers Conserv. 2010;19:2873–93.
20. Mace GM, Norris K, Fitter AH. Biodiversity and ecosystem services: a multilayered relationship. Trends Ecol Evol. 2012;27(1):19–26.
21. Wallace KJ. Classification of ecosystem services: problems and solutions. Biol Conserv. 2007;139(3–4):235–46.
22. Pullin AS, Knight T. Time to build capacity for evidence synthesis in environmental management. Environ Evid. 2013;2(21):1–2.
23. Power AG. Ecosystem services and agriculture: tradeoffs and synergies. Philos Trans R Soc Lond B Biol Sci. 2010;365(1554):2959–71.
24. National Environment Research Council (NERC). Valuing Nature Programme. 2014 [12 June 2014]; Available from: http://www.valuing-nature.net/.
25. Laurans Y, Rankovic A, Bille R, Pirard R, Mermet L. Use of ecosystem services economic valuation for decision-making: questioning a literature blindspot. J Environ Manage. 2013;119:208–19.
26. Mupangwa W, Twomlow S, Walker S. Reduced tillage, mulching and rotational effects on maize (Zea mays L.), cowpea (Vigna unguiculata (Walp) L.) and sorghum (Sorghum bicolor L. (Moench)) yields under semi-arid conditions. Field Crop Res. 2012;132:139–48.
27. Leakey RB. Multifunctional Agriculture and Opportunities for Agroforestry: implications of IAASTD. In: Nair PKR, Garrity D, editors. Agroforestry-the future of global land use. Chapter 12. Netherlands: Springer; 2012. p. 203–14.
28. Friedrich T, Kienzle J, Kassam A. Conservation agriculture in developing countries: the role of mechanization. In: Innovation for sustainable agricultural mechanisation. Hanover, Germany: Food and Agricultural Organization; 2009.
29. Agamuthu P, Broughton WJ. Nutrient cycling within the developing oil palm-legume ecosystem. Agr Ecosyst Environ. 1985;13(2):111–23.
30. Ghimire R, Adhikari K, Chen Z, Shah SC, Dahal K. Soil organic carbon sequestration as affected by tillage, crop residue, and nitrogen application in rice–wheat rotation system. Paddy Water Environ. 2012;10(2):95–102.
31. Ricketts TH, Regetz J, Steffan-Dewenter I, Cunningham SA, Kremen C, Bogdanski A, et al. Landscape effects on crop pollination services: are there general patterns? Ecol Lett. 2008;11(5):499–515.
32. Aweto AO. Shifting cultivation and secondary succession in the tropics. Wallingford, UK: CABI International; 2013.
33. Trenbath BR. Intercropping for the management of pests and diseases. Field Crop Res. 1993;34(3–4):381–405.
34. Van der Hoek W, Sakthivadivel R, Renshaw M, Silver JB, Birley MH, Konradsen F. Alternate wet/dry irrigation in rice cultivation: a practical way to save water and control malaria and Japanese encephalitis? Colombo, Sri Lanka: International Water Management Institute; 2001.
35. Chapagain T, Riseman A, Yamaji E. Achieving more with less water: Alternate Wet and Dry Irrigation (AWDI) as an alternative to the conventional water management practices in rice farming. J Agric Sci. 2011;3(3):3–13.
36. Melero S, Panettieri M, Madejón E, Macpherson HG, Moreno F, Murillo JM. Implementation of chiselling and mouldboard ploughing in soil after 8 years of no-till management in SW, Spain: effect on soil quality. Soil Tillage Res. 2011;112(2):107–13.
37. Powlson DS, Stirling CM, Jat ML, Gerard BG, Palm CA, Sanchez PA, et al. Limited potential of no-till agriculture for climate change mitigation. Nature Clim Change. 2014;4(8):678–83.
38. Li C, Yue L, Kou Z, Zhang Z, Wang J, Cao C. Short-term effects of conservation management practices on soil labile organic carbon fractions under a rape–rice rotation in central China. Soil Tillage Res. 2012;119:31–7.
39. Carvalheiro LG, Veldtman R, Shenkute AG, Tesfay GB, Pirk CWW, Donaldson JS, et al. Natural and within-farmland biodiversity enhances crop productivity. Ecol Lett. 2011;14:251–9.
40. Mertz O, Ravnborg H, Lövei G, Nielsen I, Konijnendijk C. Ecosystem services and biodiversity in developing countries. Biodivers Conserv. 2007;16(10):2729–37.
41. Roe D, Fancourt M, Sandbrok C, Sibanda M, Giuliani A, Gordon-Maclean A. Which components or attributes of biodiversity influence which dimensions of poverty? Environ Evid. 2008;3(3):1–15.
42. Collaboration for Environmental Evidence. Guidelines for systematic reviews in environmental management. UK: Bangor University; 2013.
43. Ricketts TH, Daily GC, Ehrlich PR, Michener CD. Economic value of tropical forest to coffee production. Proc Natl Acad Sci U S A. 2004;101(34):12579–82.
44. Corbeels M, de Graaff J, Ndah TH, Penot E, Baudron F, Naudin K, et al. Understanding the impact and adoption of conservation agriculture in Africa: a multi-scale analysis. Agr Ecosyst Environ. 2014;187:155–70.
45. Corbeels M, Sakyi RK, Kühne RF and Whitbread A. Meta-analysis of crop responses to conservation agriculture in sub-Saharan Africa: Climate Change Agriculture and Food Security Report No. 12; 2014: Copenhagen, Denmark. p. 1-20.
46. Peh KSH, Balmford A, Bradbury RB, Brown C, Butchart SHM, Hughes FMR, et al. TESSA: a toolkit for rapid assessment of ecosystem services at sites of biodiversity conservation importance. Ecosystem Serv. 2013;5:51–7.
47. Glenn J, Gordon T, Florescu E. State of the future, millennium development project. Washington, DC: UN University; 2008.
48. van der Werf HMG, Petit J. Evaluation of the environmental impact of agriculture at the farm level: a comparison and analysis of 12 indicator-based methods. Agr Ecosyst Environ. 2002;93(1–3):131–45.
49. Randolph J. Online Kappa Calculator. 2008 [3 July 2014]; Available from: http://justusrandolph.net/kappa/.
50. Fancourt M. DateX: systematic review extraction software. 2015: Cambridge, UK.
51. FAO. Final 2012 data and preliminary 2013 for major commodities aggregated. Rome, Italy: Statistics Division; 2013.

52. Liquete C, Piroddi C, Drakou EG, Gurney L, Katsanevakis S, Chare A, et al. Current status and future prospects for the assessment of marine and coastal ecosystem services: a systematic review. PLoS One. 2013;8(7):1–15.
53. Ecosystem Services and Poverty Alleviation (ESPA). Regions of operation. 2014 [23 June 2014]; Available from: http://www.espa.ac.uk/our-work/regions.
54. FAO, Low-Income Food-Deficit Countries (LIFDC) - List for 2014. 2014. [2 April 2015]; Available from: http://www.fao.org/countryprofiles/lifdc/en/.
55. World Bank, Country and Lending Groups. 2014. [2 April 2015]; Available from: http://data.worldbank.org/about/country-and-lending-groups.
56. Zhang P, Wei T, Jia Z, Han Q, Ren X. Soil aggregate and crop yield changes with different rates of straw incorporation in semiarid areas of northwest China. Geoderma. 2014;230–231:41–9.
57. Sagie H, Morris A, Rofè Y, Orenstein DE, Groner E. Cross-cultural perceptions of ecosystem services: a social inquiry on both sides of the Israeli–Jordanian border of the Southern Arava Valley Desert. J Arid Environ. 2013;97:38–48.
58. Soderstrom B, Hedlund K, Jackson L, Katterer T, Lugato E, Thomsen I, et al. What are the effects of agricultural management on soil organic carbon (SOC) stocks? Environ Evid. 2014;3(2):1–8.
59. PRISMA. Preferred reporting items for systematic reviews and meta-analyses. 2009; Available from: http://www.prisma-statement.org/.
60. Peterson MP. Mapping the in cloud. New York, USA: The Guildford Press; 2014.
61. De Groot RS, Wilson MA, Boumans RMJ. A typology for the classification, description and valuation of ecosystem functions, goods and services. Ecol Econ. 2002;41:393–408.
62. Norfolk O, Eichhorn MP, Gilbert F. Traditional agricultural gardens conserve wild plants and functional richness in arid South Sinai. Basic Appl Ecol. 2013;14(8):659–69.
63. Boelee E, Chiramba T, Khaka E. An ecosystem services approach to water and food security in an ecosystem services approach to water and food security. Nairobi, Kenya, Colombo, Sri Lanka: International Water Management Institute and United Nations Environment Programme; 2011.

13

Comparative effectiveness of silvicultural interventions for increasing timber production and sustaining conservation values in natural tropical production forests. A systematic review protocol

Gillian Petrokofsky[1*], Plinio Sist[2], Lilian Blanc[2], Jean-Louis Doucet[3], Bryan Finegan[4], Sylvie Gourlet-Fleury[2], John R Healey[5], Barbara Livoreil[6], Robert Nasi[7], Marielos Peña-Claros[8], Francis E Putz[9] and Wen Zhou[7]

Abstract

Background: Currently, about 400 million hectares of tropical moist forests worldwide are designated production forests, about a quarter of which are managed by rural communities and indigenous peoples. There has been a gradual impoverishment of forest resources inside selectively logged forests in which the volume of timber extracted over the first cutting cycle was mostly from large, old trees that matured over a century or more and grew in the absence of strong anthropological pressures. In forests now being logged for a second and third time, that volume has not been reconstituted due in part to the lack of implementation of post-logging silvicultural treatments. This depletion of timber stocks renders the degraded forests prone to conversion to other land uses. Although it is essential to preserve undisturbed primary forests through the creation of protected areas, these areas alone will not be able to ensure the conservation of all species on a pan-tropical scale, for social, economic and political reasons. The conservation of tropical forests of tomorrow will mostly take place within human-modified (logged, domesticated) forests. In this context, silvicultural interventions are considered by many tropical foresters and forest ecologists as tools capable of effectively conserving tropical forest biodiversity and ecosystem services while stimulating forest production. This systematic review aims to assess past and current evidence of the impact of silviculture on tropical forests and to identify silvicultural practices appropriate for the current conditions in the forests and forestry sectors of the Congo Basin, Amazonia and Southeast Asia.

Methods: This systematic review will undertake an extensive search of literature to assess the relative effectiveness of different silvicultural interventions on timber production and the conservation value of forests, and to determine whether there is a relationship between sustainability of timber harvesting and the maintenance/conservation of other ecosystem services and biodiversity in production forests. Data will be extracted for meta-analysis of at least sub-sets of the review questions. Findings are expected to help inform policy and develop evidence-based practice guidelines on silvicultural practices in tropical forests.

Keywords: Silviculture, Tropical forests, Sustainable forest management, Conservation, Biodiversity

* Correspondence: gillian.petrokofsky@zoo.ox.ac.uk
[1]University of Oxford, Oxford, UK
Full list of author information is available at the end of the article

Background

While tropical silviculture still remains little-known or poorly acknowledged among decision-makers, there is also some strong antagonism even within the forestry scientific community, namely between forest ecologists and conservation ecologists. The former defend silvicultural interventions as tools to stimulate forest production whereas the latter accept only the most gentle interventions and promote biodiversity conservation through the creation of protected areas even within production forests. One source of confusion and conflict is that depending on how it is carried out, logging can represent either forest exploitation or a major silvicultural intervention designed to maintain forest productivity. From the outset, tropical silviculture faced the problem of reconciling timber production with long-term conservation of forest ecosystems, though the main objective in the early days was indeed to improve timber production [1]. The main challenge lies in finding thresholds of extraction intensity coupled with silvicultural treatments that are compatible with: (1) the maintenance of biodiversity and the main ecosystem services targeted for a given forest management unit, (2) the profitability for the actors involved.

Although it is essential to preserve undisturbed natural forests through the creation of protected areas, these areas alone will not be able to ensure the conservation of all species on a pan-tropical scale, for social, economic and political reasons. According to SEDAC (http://sedac.ciesin.columbia.edu/data/collection/wildareas-v2, cited by Hobbs *et al.* [2]), 83% of the land surface of the earth is impacted to some degree. Tropical forests are no exception, with (at most) only 1/2 of them being considered primary [3]. The conservation of tropical forests of tomorrow will mostly take place within human-modified (logged, domesticated) forests, but only if they are well managed. "Well managed" means that the implemented forest management must guarantee the perpetuation of timber resources, along with biodiversity and most of the other ecosystem services provided by forests. With advances in the agricultural sciences and increased demand for plantation crops and agricultural commodities (e.g., palm oil and beef), the opportunity costs of forest retention have also increased of late. Counterbalancing that trend is the emergence of markets for the payment of ecosystem services provided by forests which opens up economic incentives to support the implementation of silvicultural treatments.

Currently, about 400 million hectares of tropical moist forests worldwide are designated production forests, about a quarter of which are managed by rural communities and indigenous peoples [3]. There is now clear evidence that, with current logging practices based on a single diameter cutting limit and a rotation cycle of around 30 years, timber volumes will not recover within this rotation cycle [4]. The only way to keep an economically viable commercial volume is to expand the pool of species harvested and/or to reduce the minimum cutting diameter [4-6]. Today this challenge is further exacerbated because most tropical forests have already been logged, and because there is a greater diversity of stakeholders in the forestry sector (farmers, communities, urban society), which was long dominated by large logging companies exploiting forest concessions for timber production. In these forests, the volume of timber extracted over the first cutting cycle was mostly from large trees that grew over several centuries in the absence of strong anthropogenic pressures. When these forests are re-logged, that volume has not been recovered and, in many cases, the young trees expected to make up the next harvest are lacking in numbers or have not yet attained larger sizes as no silvicultural treatments were applied to favour them [7]. Figure 1 illustrates this reduced yield over time in contrast with sustained yield after silvicultural treatments.

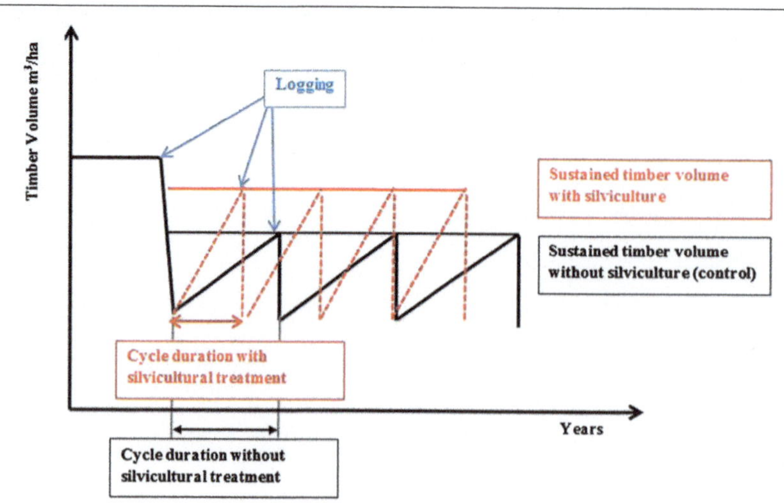

Figure 1 Expected impact of post-logging silviculture on both sustained timber volume and duration of rotation cycle.

This depletion of commercial stocks of timber reduces the financial value of standing forest and puts the forests at risk of being further severely degraded and ultimately converted to other uses. This problem was highlighted in Africa and South-East Asia in the 1960s by foresters who originated from temperate countries and who undertook a large variety of silvicultural trials with the goal of enriching tropical forests with commercial species, by fostering their regeneration either naturally or artificially. Those trials involved contrasting technical operations, from full planting to liberation of naturally regenerated future crop trees by poison-girdling competitors. Those trials generated a considerable amount of data and knowledge that has mostly remained unavailable, published in technical reports and/or non-English language journals, and fully under-valued – or even forgotten – as logging has developed into a vast profitable 'mining' operation in most forests dedicated to production.

Objective of the review

A systematic review of the impact of past silvicultural practices, and not just logging, can inform us about the available opportunities for managing and restoring the economic and environmental values of natural tropical production forests. The main objective of this systematic review is therefore to assess past and current knowledge of the impact of silviculture on the conservation values of tropical forests (its natural capital, including biodiversity, and the full range of ecosystem services that it underpins) and to identify silvicultural approaches that are appropriate for the forests and forestry sectors of the Congo Basin, Amazon Basin, and tropical regions of Southeast Asia (Table 1 gives a full list of countries and regions included in the review). One important goal of this systematic review is to evaluate the compatibility of silviculture for enhanced timber production with the maintenance of other ecosystem services. In so doing, we want to address the following primary and secondary questions:

Primary review question

How do silvicultural interventions affect timber production and conservation values in natural tropical production forests?

Secondary review questions

1. How do different silvicultural treatments affect both production and conservation values?
2. Where silviculture for sustained timber harvests and maintenance/conservation of other ecosystem

Table 1 List of terms used to build the search string and define the inclusion perimeter of the papers retrieved

A. Geographical terms (countries and regions) (using spelling variants as appropriate)	C. Intervention terms (using double quotes, and proximity operators, as appropriate)
Tropic* Or "Central America" Or "South America" Or Caribbean Or "Central Africa" Or "East Africa" Or "West Africa" Or "Southeast Asia" OR Mexico OR Belize OR "Costa Rica" OR "El Salvador" OR Guatemala OR Honduras OR Nicaragua OR Panama OR Bolivia OR Brazil OR Colombia OR Ecuador OR Guyana OR Galapagos OR Paraguay OR Peru OR Suriname OR Venezuela OR Anguilla OR "Antigua and Barbuda" OR Aruba OR Bahamas OR Barbados OR "Virgin Islands" OR Cayman Islands OR Cuba OR Dominica* OR Grenada OR Guadeloupe OR Haiti OR Jamaica OR Martinique OR Montserrat OR Antilles OR "Puerto Rico" OR "Saint Barthelemy" OR "Saint Kitts" OR Nevis OR "Saint Lucia" OR "Saint Martin" OR "Saint Vincent and the Grenadines" OR "Trinidad and Tobago" OR "Turks and Caicos" OR Angola OR Cameroon OR "Central African Republic" OR Chad OR Congo OR Zaire OR "Equatorial Guinea" OR Gabon OR Sudan OR Zambia OR Burundi OR Comoros OR Djibouti OR Eritrea OR Ethiopia OR Kenya OR Madagascar OR Malawi OR Mauritius OR Mayotte OR Mozambique OR Reunion OR Rwanda OR Seychelles OR Somalia OR Tanzania OR Uganda OR Benin OR "Burkina Faso" OR "Cote d'Ivoire" OR "Ivory Coast" OR Gambia OR Ghana OR Guinea* OR Liberia OR Mali OR Mauritania OR Niger* OR "Saint Helena" OR "Sao Tome and Principe" OR Senegal OR "Sierra Leone" OR Togo OR Brunei OR Burma OR Myanmar OR Cambodia OR "East Timor" OR Indonesia OR Laos OR "Lao PDR" OR Malaysia OR Philippines OR Thailand OR Vietnam OR India	*PRE-HARVEST/PLANNING* permanent sample plot OR minimum cutting diameter OR diameter cutting limit OR harvest plan OR forest management plan OR cutting cycle OR diameter limit OR liana/liane cutting OR climber cutting OR riparian buffers OR sustainable forest management OR sustained timber yields OR future crop trees OR potential crop trees OR advanced regeneration OR felling cycle OR rotation cycle OR skid trails OR logging roads OR snig tracks OR logging gaps OR planting OR sowing OR nursery OR criteria OR forest certification OR inventory OR stock inventory OR forest management unit HARVEST timber harvesting OR logging OR selective harvesting OR selective logging OR reduced- impact logging OR ril OR low-impact logging OR low-impact silviculture OR harvesting techniques OR logging damage OR yarding techniques OR seed tree retention OR shelterwood OR Malayan uniform system OR polycyclic OR monocyclic OR uneven-aged management OR even-aged management OR felling OR clear strip felling OR directional felling OR group felling OR tpti OR silin OR partial cutting OR patch cutting OR celos OR logging intensity OR selective management OR retention forestry OR annual coupe POST-HARVEST liberation thinning OR thinning OR girdling OR refinement OR poisoning OR poison-girdling OR enrichment planting OR line planting OR gap planting OR recovery rate OR reconstitution rate OR assisted natural regeneration OR topsoil removal OR regrowth method OR liana cutting
B. Population terms	**D. CAB Abstracts search variant**
forests OR rainforests OR woods OR woodlands OR kerangas OR cerrado OR wallaba OR varzea OR igapo OR terra firme OR mata atlantica	Set **A** AND (silvicultur*.ab. OR silvicultur*.ti. OR silvicultur*.de. OR KK110.cc. [a]OR logging.ab. OR logging.ti. OR logging.de. OR deforest*.ab. OR deforest*.ti. OR deforest*.de. OR forest management.de. OR forest inventories.de.) (See EndNote for explanation of field limits specified here)

services and biodiversity has been practiced, how do differences in these treatments affect socio-economic outcomes?

Methods

This systematic review will undertake an extensive search of literature to assess the impacts of different silvicultural interventions on timber production and the conservation value of natural tropical forests, and to further determine whether these outcomes can both be achieved by the same interventions and how different silvicultural interventions for combined sustained timber harvests and maintenance/conservation of other ecosystem services and biodiversity affect socio-economic outcomes. Data will be extracted for meta-analysis of at least a sub-set of the review questions. Findings are expected to help inform policy and develop evidence-based practice guidelines on the value (or otherwise) of silvicultural practices in tropical natural forests.

Search strategy

Given that the scope of this systematic review is broad and that it can be difficult to identify relevant studies from titles and abstracts, the authors are aware of the necessity of balancing comprehensiveness with specificity in the searches. "Forest management", for example, is in general a broader concept than the focus of this review, but "silviculture" alone is too narrow. A subgroup of authors tested various search strategies, using combinations of key words and phrases suggested in an initial two-day scoping meeting. These strategies were tested against a reference list of papers (see Additional file 1: Annexl) to refine the search and maintain a balance between, at the one end, extracting an unwieldy set of forestry references from the 1920s onwards which 'might' contain a table of useful data, and, at the other end, a narrow set of papers with known useful data, which is not representative of the overall body of evidence. Keywords and phrases for searches in the main bibliographic databases (indicated by an asterisk in Table 2) were derived from the following PICOs listed below. The terms and search logic are shown in Table 1. We recognise the importance of studies published in languages other than English. The search strategy will therefore be applied (and documented carefully for peer review) to the following languages: English, French, Spanish, Portuguese, German, and Dutch, subject to time and resources. Where time and resources permit, and recognizing the potentially high cost of translating relevant studies, we will also search in: Bahasa Indonesia, Japanese, and Thai.

PICOs

Population

Closed canopy natural forest used for timber production in the tropics (= natural tropical production forests). (A full list of countries and regions in the tropics is included in Table 1).

Interventions

Any silvicultural activity, including pre-harvest, harvesting (logging), post-harvest treatments, construction of secondary roads and skid trails (A full list of silvicultural activities is included in Table 1).

Comparator

Temporal or spatial comparisons of silvicultural practices.

Outcomes

Increase of timber yield or timber volume recovery, **or** changes in ecological conditions of the forest or socio-economic outcomes (latter only if studies also reporting timber yield and ecological condition outcomes).

Other

In addition to systematic searching (as above), a call for grey literature will be issued to stakeholders, including through the IUFRO subgroup on tropical silviculture, the International Society of Tropical Forests, LinkedIn and Research Gate and the personal contacts of authors. References in relevant books will also be searched as time and resources permit.

Table 2 Databases and websites

Bibliographic databases	Grey literature
*Web of Science	**Institutions**
*CAB Abstracts (with particular focus on the Forestry & Forest Products Abstracts subset)	Forest Research Institute of Malaysia
*Scopus	FAO
Agricola	USDA Forest Service
Agris	US Fish and Wildlife Service
BioONE	USAID
SciELO	GIZ
JSTOR	ICRAF
Index to Theses Online	CIFOR
Directory of Open Access Journal	CATIE
ProQuest	Agritrop (CIRAD)
	IITF
	Embrapa
Web searches	Royal Museum for Central Africa
Google	National Botanic Garden of Belgium
Google Scholar	University Departments with tropical forestry research collections

*Asterisks relate to sources which will be searched using Table 1 strategy.

Search terms & Boolean logic

Searches in CAB Abstracts, Web of Science and Scopus will be based on the logic A AND B AND C, where A-C are the sets of terms listed in Table 1. Within each set, alternative terms will be searched, using OR. CAB Abstracts, having a sub-database dedicated to forestry, will be searched using a modified strategy that makes use of CABICODES and descriptors from the CAB Thesaurus (Table 1 – section D). The search was devised in consultation with the Editor of Forestry Abstracts, and the former editor of Forest Products Abstracts (one of the review authors). In our scoping studies a simpler search performed better than the strategy used for Scopus and Web of Science. A much simpler search strategy, based on that devised for CAB Abstracts, but excluding the CABICODES, will be used to search resources that cannot accommodate complex search strings. These are NOT marked with an asterisk in Table 2.

Inclusion criteria

Studies will be included if they are from populations in the target countries or regions (Table 1), and include information on the silvicultural (and other) interventions and outcomes set out in the PICO list above. It is likely that a larger set of relevant Outcomes will be included once reviewers screen the full text of papers. Consequently, a complete list of silvicultural interventions and outcomes will be documented iteratively as these emerge. A subgroup of authors will select papers retrieved from the searches, following removal of duplicates. A preliminary inclusion/exclusion assessment of a randomly selected set of 100 articles will be conducted to test consistency between researchers' screening judgment. Cohen's Kappa metrics will be used to indicate a measure of consistency. A consistency co-efficient of 0.6 or greater is usually accepted as sufficient in the literature [8]. The first stage of inclusion/exclusion will entail screening for relevance of articles by title only. This stage will be followed by screening using titles and abstracts together; finally the assessment will be based on reading the full text. At each stage, studies will remain in the set to be included if they MAY contain relevant information about silvicultural interventions or outcomes. This will help reduce the possibility that we are overly-restrictive about relevant outcomes at the outset. Results of screening will be documented (including reasons for exclusion) and will be available as supplementary material to the systematic review paper.

Exclusion criteria

Papers will be excluded if they are exclusively about

- mangroves or woodlands (e.g. miombo) or pure conifer forests
- agroforestry systems
- tree plantations following clear-cutting

Effect modifiers

The following will be recorded, where available, as potential effect modifiers and sources of heterogeneity:

Geographical location
Altitude
Climate
Soil type
Time after logging or any other silvicultural intervention
Dominant tree species
Type, duration and intensity of silvicultural activity and logging intensity
Time elapsed from silvicultural intervention to sampling
Land use and history of land use in the neighbouring landscape
Proximity to urban centres (which have different sizes, socio-economic status, etc.)

Critical appraisal of studies

Studies included in the review after full text screening will be critically appraised for study design, the robustness of their study design and the extent to which authors have attempted to limit biases.

Variables outlined in Table 3 will be used to construct a list from which the quality of individual studies can be assessed. Studies will be categorised as having high or low susceptibility to bias on the basis of presence/absence of replication, measures of variance and potential effect modifiers. Studies with high susceptibility to bias will not be excluded from the narrative review, but they will be subjected to sensitivity analysis following Brooks *et al.* [9]. We are aware of the proposal by Bilotta *et al.* [10] that environmental systematic reviews should adapt the Environmental- Risk of Bias Tool and we will check the feasibility of this approach for our set of studies. We know that our review will include a number of older studies and we are aware of advances in study design in recent decades, and also of the general problem of pseudo-replication in published studies [11], and presumably also in unpublished studies that may be included in our review, and we will amend our proposed critical appraisal method if (a) it is too difficult to apply by reviewers (requiring subjective judgment that is difficult to standardise across reviewers, despite repeated kappa analysis), and (b) the criteria are too 'strict' (resulting in a very small set of studies that are considered to be of sufficiently high quality to be included for quantitative analysis in the review).

Table 3 Criteria for critical appraisal and data extraction from studies selected after full text screening

Main sections	Criteria for critical appraisal	Data extracted
Description of the study site		Latitude and longitude of sites
		Data on climate, topography, geology and soils
		Population size, main cities, main economic activities
Population		Type of forest indicated from which a standard list can be compiled for ease of comparison
		Basal area, timber volume before logging, biomass, number of trees, and other inventory metrics (to be developed iteratively)
Methods	Clear and repeatable methodology	Number of plots, treatments, blocks, replications (see also below)
	Adequate description of silvicultural treatments	Logging intensity (m^3/ha, basal area removed, number of trees removed, biomass removed)
	Replication - at least 3 plots or treatments	Number of plots/treatment
	Outcome measured after intervention	Number of years after silvicultural intervention
	Spatial autocorrelation between transects	
Results	Validity or reliability of data	Data for mean and SD
		Size of the sample
	Randomized sample	Y/N
	Variables measured allow the effect of the intervention to be assessed	
		% recovery of timber volume
		% recovery of biomass
		Biodiversity (floristic, faunistic, indicators)
		Other quantification or indicators of ecosystem services
		Measures of socio-economic outcome (including poverty, human health & well-being, livelihoods, social capital, social welfare, empowerment, equity, based on Pullin *et al.* (2013) [14])
		Type of models (matrix, individual based)

Data synthesis and presentation

A narrative synthesis of data from all studies included in the review will describe the quality of the results along with the study findings. Tables will be produced to summarise these results. Where studies report similar outcomes, meta-analysis may be possible. In these cases effect sizes will be standardised and weighted appropriately. Details of the quantitative analysis will emerge once full-text screening has yielded a set of studies that can be assessed for content and quality.

Subgroup analysis of categories of studies will also be performed where sufficient studies report common sources of heterogeneity. In addition to publication bias (see below), sensitivity analysis will be carried out where possible following Brooks *et al.* [9]. Silvicultural effects will be presented visually in plots of mean effect sizes and variance. If meta-analysis of effect sizes is possible, it will take the form of random-effects models, and meta-regression will be performed where effect modifiers cause significant heterogeneity between studies.

Publication bias

There is evidence that in many research areas papers are more likely to be published if they demonstrate clear, positive results (or strong negative effects), and that papers that show little or no effect are less likely to be published. To assess the possibility of such publication bias, we will compare studies in the unpublished 'grey' literature with studies in academic journals [12] to assess whether there is evidence of publication bias. If data allow, we will assess bias using funnel plots, which show effect sizes and standard error or sample sizes [13].

Endnote

[a]For the CAB Abstracts database, searching is limited to Title, Abstract and keyword fields using the field limiters shown in Table 1. Web of Science and Scopus have a default search mode that interrogates only those fields, which will be used. This eliminates papers that contain any of our keywords in, for example, journal titles, institution addresses, etc.

KK110 is one of a suite of CABICODES, which are applied as additional indexing categories by abstractors compiling CAB Abstracts. It relates to Forest management & Silviculture. For a full list see http://www.cabi.org/Uploads/CABI/publishing/training-materials/additional-resources/cabi-codeguide.pdf.

Additional file

Additional file 1: Annex1. Reference list to test search comprehensiveness.

Competing interests

The authors declare that they have no competing interest.

Authors' contributions

All authors discussed the review title in a 2-day workshop, agreed the scope and extent of the review, and contributed to the draft. All have read and agreed the Protocol. The full systematic review will be under the leadership of Plinio Sist.

Acknowledgements

We acknowledge the financial support of the Center for International Forestry Research (CIFOR) from their Evidence-Based Forestry programme that is funded primarily by the UK Department for International Development (DFID) through their KNOW-FOR grant to CIFOR.

Author details

[1]University of Oxford, Oxford, UK. [2]Cirad, Montpellier, France. [3]Université de Liège, Gembloux Agro-Bio Tech, Liège, Belgium. [4]Catie, Turrialba, Costa Rica. [5]Bangor University, Bangor, UK. [6]Fondation pour la Recherche sur la Biodiversité, Paris, France. [7]CIFOR, Bogor, Indonesia. [8]Wageningen University, Wageningen, Netherlands. [9]University of Florida, Gainesville, FL, USA and CIFOR.

References

1. Dawkins HC, Philip MS. Tropical moist forest silviculture and management: a history of success and failure. Wallingford, UK: CAB International; 1998. p. 359. ISBN 0-85199-255-2.
2. Hobbs RJ, Higgs E, Harris JA. Novel ecosystems: implications for conservation and restoration. Trends Ecol Evol. 2009;24:599–605.
3. Blaser J, Sarre A, Poore D, Johnson SL. Status of tropical management. ITTO Technical Series, Yokohama, Japan. No. 38 2011p. 460.
4. Putz FE, Zuidema PA, Synnott T, Pena-Claros M, Pinard MA, Sheil D, et al. Sustaining conservation values in selectively logged tropical forests: the attained and the attainable. Conservation Letters. 2012;5:296–303.
5. Sist P, Gourlet-Fleury S, Nasi R. IUFRO international conference report: What future is there for tropical forest silviculture? Bois et Forêts des Tropiques. 2011;310:3–6.
6. Gourlet-Fleury S, Mortier F, Fayolle A, Baya F, Ouedraogo D, Benedet F, et al. Tropical forest recovery from logging: a 24 year silvicultural experiment from Central Africa. Philos Trans R Soc Lond B Biol Sci. 2013;368:20120302.
7. Catinot R. Sylviculture tropicale en forêt dense africaine [Tropical silviculture in African moist forest]. Bois et Forêts des Tropiques. 1965;100:5–18.
8. Cohen J. A coefficient of agreement for nominal scales. Educ Psychol Meas. 1960;20:37–46.
9. Brooks J, Waylen KA, Mulder MB. Assessing community-based conservation projects: A systematic review and multilevel analysis of attitudinal, behavioral, ecological, and economic outcomes. Environ Evid. 2013;2:2.
10. Bilotta GS, Milner AM, Boyd IL. Quality assessment tools for evidence from environmental science. Environ Evid. 2014;3:14.
11. Ramage BS, Sheil D, Salim HM, Fletcher C, Mustafa NZ, Luruthusamay JC, et al. Pseudoreplication in tropical forests and the resulting effects on biodiversity conservation. Conserv Biol. 2013;27:364–72.
12. Pullin A, Bangpan M, Dalrymple S, Dickson K, Haddaway NR, Healey JR, et al. Human well-being impacts of terrestrial protected areas. Environ Evid. 2013;2:19.
13. Leimu R, Koricheva J. What determines the citation frequency of ecological papers? Trends Ecol Evol. 2005;20:28.
14. Sterne JA, Jüni P, Schulz KF, Altman DG, Bartlett C, Egger M. Statistical methods for assessing the influence of study characteristics on treatment effects in 'meta-epidemiological' research. Stat Med. 2002;21:1513–24.

14

The fitness consequences of inbreeding in natural populations and their implications for species conservation – a systematic map

Linda E Neaves[1], Jacqualyn Eales[2], Raj Whitlock[3], Peter M Hollingsworth[1], Terry Burke[4*] and Andrew S Pullin[2]

Abstract

Background: Threatened species often have small and isolated populations where mating among relatives can result in inbreeding depression increasing extinction risk. Effective management is hampered by a lack of syntheses summarising the magnitude of, and variation in inbreeding depression. Here we describe the nature and scope of the literature examining phenotypic/fitness consequences of inbreeding, to provide a foundation for future syntheses and management.

Methods: We searched the literature for articles documenting the impact of inbreeding in natural populations. Article titles, abstracts and full-texts were assessed against *a priori* defined criteria, and information relating to study design, quality and other factors that may influence inbreeding responses (e.g. population size) was extracted from relevant articles.

Results: The searches identified 11457 articles, of which 614 were assessed as relevant and included in the systematic map (corresponding to 703 distinct studies). Most studies (663) assessed within-population inbreeding resulting from self-fertilisation or consanguineous pairings, while 118 studies assessed among-population inbreeding due to drift load. Plants were the most studied taxon (469 studies) followed by insects (52 studies) and birds (43 studies). Most studies investigated the effects of inbreeding on components of fitness (e.g. survival or fecundity; 648 studies) but measurements were typically under laboratory/greenhouse conditions (486 studies). Observations were also often restricted to the first inbred generation (607 studies) and studies frequently lacked contextual information (e.g. population size).

Conclusions: Our systematic map describes the scope and quality of the evidence describing the phenotypic consequences of inbreeding. The map reveals substantial evidence relating to inbreeding responses exists, but highlights information is still limited for some aspects, including the effects of multiple generations of inbreeding. The systematic map allowed us to define several conservation-relevant questions, where sufficient data exists to support systematic reviews, e.g. How do inbreeding responses vary with population size? However, we found that such syntheses are likely to be constrained by incomplete reporting of critical contextual information. Our systematic map employed the same rigorous literature assessment methods as systematic review, including a novel survey of study quality and thus provides a robust foundation to guide future research and syntheses seeking to inform conservation decision-making.

Keywords: Genetic variation, Conservation genetics, Systematic review, Inbreeding depression, Outbreeding, Heterosis, Genetic load, Isolated populations, Small populations

* Correspondence: t.a.burke@sheffield.ac.uk
[4]Department of Animal and Plant Sciences, University of Sheffield, Sheffield S10 2TN, UK
Full list of author information is available at the end of the article

Background

Natural populations of many species are faced with an increasing number of pressures from human-driven environmental changes, such as habitat degradation and fragmentation. These pressures can lead to local reductions in population size, and increases in the isolation of populations. It is widely accepted that small isolated populations are at risk of extinction from demographic processes and environmental stochasticity, but they also face two genetic threats [1-3]. The erosion of genetic diversity and the reduced effectiveness of selection (and increased effects of chance) in small populations can limit opportunities for adaptation to changing conditions. The effects of these evolutionary limitations, however, may only be realised in the long-term and hence may not represent a significant short-term threat. In contrast, breeding amongst close relatives in small isolated populations may result in an immediate loss of fitness in their offspring, known as inbreeding depression [2]. Even though the detrimental consequences of inbreeding are expressed in the short-term, inbreeding depression is infrequently considered in the conservation of natural populations [4].

Inbreeding refers to the mating among related individuals and results in an increase in homozygosity. The fitness costs associated with inbreeding are due to recessive deleterious alleles that confer a disadvantage on the individual possessing them. In heterozygous individuals the effects of these alleles are masked by a functional allele, but in homozygous individuals the deleterious effects are exposed as there is no functional counterpart [5,6]. In addition, the fitness of homozygous individuals may also be reduced relative to heterozygous individuals where possessing two different and complimentary alleles confers a benefit (e.g. disease/pathogen resistance) [6]. Where populations remain small, drift tends to lead to the random fixation of alleles throughout the population, even if they are deleterious. Termed drift load, the resultant loss of fitness can only be restored through the introduction of new genetic diversity from elsewhere (i.e. outcrossing with another population). Thus, inbreeding can occur at two levels, within populations via self-fertilisation and consanguineous mating and among populations through drift [2].

In each generation the loss of fitness due to inbreeding depression impacts on the demography of the population, as fewer individuals are recruited, and this reinforces inbreeding as population size declines. Ultimately this negative feedback of reduced population size on fitness can result in an "extinction vortex" [4]. The detrimental consequences of inbreeding depression can be alleviated through the introduction of genes from another population [7,8]. Outcrossing typically enhances fitness by masking recessive deleterious alleles and restoring heterozygosity in overdominant loci [6]. This 'genetic rescue' has been used effectively to restore fitness in small inbred populations of a range of threatened taxa (e.g. [9-12]). Madsen et al [11], for example reported the recovery of a small inbred adder population, where breeding amongst relatives had resulted in reduced fitness and survival and subsequent population decline, following the introduction of eight unrelated males from a large nearby population. The enhanced population growth associated with the introduction of new genetic material continued past the first generation [10].

The introduction of individuals from another population, however, is not always beneficial. Isolated populations can accumulate genetic differences through selection and drift. Where inter-mating between these populations occurs the fitness of hybrid offspring can be reduced due to differential adaptations or the presence of intrinsic genetic incompatibilities (i.e. the breakup of co-adapted gene complexes) [13,14]. This reduction in the fitness of hybrid offspring is termed outbreeding depression. Recent reviews assessing the relative costs and benefits indicate that, in general, outcrossing is not detrimental [15]. Rather, outcrossing in small isolated populations may be beneficial, and the greatest risks of outbreeding depression are associated with populations inhabiting highly divergent environmental conditions or that have been isolated for over 500 years [16]. Thus, it appears that in many instances, particularly in the small populations where inbreeding is likely, concern about outbreeding depression should not prevent actions to alleviate inbreeding depression.

The phenotypic outcomes of inbreeding have been the focus of several reviews, in both captive and natural populations. These reviews have demonstrated that inbreeding frequently leads to significant fitness costs, and hence is relevant to conservation [2,17,18]. O'Grady et al. [19], for instance, quantified the average overall effect of inbreeding on fitness and then, using computer simulations showed this led to an average 37% decrease in the median time to extinction. Many of these reviews, however are now dated and more recent reviews have been restricted to particular taxonomic groups (e.g. [20]) or focused on particular aspects of inbreeding depression (e.g. [21]) and hence only represent a subset of the available evidence. In addition, while some of these reviews are meta-analytic [18,20,21], none are systematic, and thus may represent a biased view of the available evidence. Therefore there is a need to synthesise the evidence base describing the phenotypic/fitness consequences of inbreeding in a transparent way to inform conservation strategies.

Systematic mapping approaches provide a structured overview of the evidence base for a particular research question. The systematic mapping technique is still relatively new in conservation science but it applies many of the robust, repeatable and transparent scientific methods utilised within systematic reviews. However, unlike systematic reviews, which use meta-analyses to answer a specific question, systematic maps aim to categorise the literature according to study aims, design or other factors that may be of interest. In addition, maps can be applied to broader questions, synthesising potentially disparate studies that could not be directly compared via meta-analysis [22]. Here, we employed systematic mapping techniques to synthesise studies describing the fitness consequences of inbreeding in natural populations to provide an objective assessment of the extent, nature and quality of the evidence. Systematic maps do not usually attempt data extraction or critical appraisal of the evidence base. However, this can limit our understanding of the evidence base. Therefore we undertook a more detailed assessment, coding data that both described the study and assessed its quality from the full-text. We also extracted information on factors that may influence the costs of inbreeding (e.g. life history traits, population size), and hence are informative for decision-making in conservation. By collating the evidence base in this way we aimed to generate a searchable map (database) that can be used by conservation practitioners to identify relevant studies, as well as to highlight gaps in research, and pinpoint areas where there is sufficient evidence to conduct systematic reviews.

Objective of the systematic map

Question formulation

Understanding the impact of, and factors influencing, inbreeding in populations of conservation concern has been identified as an important issue for conservation practitioners [23-25]. The need for a systematic review to assess the impacts of inbreeding on wild populations was discussed by the UK Conservation Genetics Working Group, which includes representatives from UK-based government and non-government conservation organisations and scientists working in the field of conservation genetics. Feedback from these meetings was used to shape the scope of review and develop a draft systematic review protocol. To maximise relevance to stakeholders this review protocol was peer-reviewed and published on-line in accordance with the guidelines for systematic reviews in environmental management [26].

A systematic review protocol was established in 2010 [27]. The initial searches retrieved a large volume of relevant but disparate literature. Given this, it was deemed more appropriate to synthesise this body of evidence as a systematic map, rather than a systematic review. Thus, the primary question posed in the review was used to define the scope of this systematic map.

Primary question

What are the fitness consequences of inbreeding in natural populations?

Primary objective

- To describe and map research that investigates the phenotypic consequences of inbreeding within natural populations.

Secondary objectives

- Assess and describe the extent to which studies report attributes that could influence the consequences of inbreeding depression (e.g. population size, mating system, higher-level taxonomic group).
- Quantify published research that assesses the relationship between the phenotypic consequences of inbreeding and levels of neutral genetic variation within populations.
- Provide a searchable map, in the form of a database.
- Identify knowledge gaps to guide future primary research.
- Inform and direct future research syntheses, by identifying areas where sufficient studies exist to support systematic review.

Methods

The methods for the systematic map followed those of the systematic review protocol [27] for search strategy, study inclusion criteria and study quality assessment. The coding criteria for the map were adapted from the Collaboration for Environmental Evidence (CEE) Systematic Review and Evidence Synthesis Guidelines [26] and from existing systematic map reports (e.g. [28]). In addition, the sources of heterogeneity described in the review protocol were used in the development of the coding criteria.

Search strategy

Scope of search

The literature search was conducted in two phases. Searches in the first phase were run in November 2010; these interrogated all literature indexed prior to this search date. We subsequently updated these searches on 12th August 2013. In this second phase we limited the search to all articles published during or after 2010. Both phases of literature searching used the same search terms, and interrogated the same databases.

The following online databases were searched for relevant literature and data:

- ISI Web of Science
- Scopus
- JSTOR

In addition, web searches were performed in Google scholar (scholar.google.com) using the search term 'inbreeding depression' and the first 100 hits were assessed for relevance. Further searches of the grey literature were not performed, as our previous experience in reviewing phenotypic responses to intraspecific outbreeding indicated that this would be inefficient, since the relevant data was published primarily in the peer-reviewed literature [15]. In this previous review, searches within the libraries of UK-based conservation agencies and non-governmental organisations (NGO) yielded no relevant hits, while the relevant data from theses were almost always also published in the primary literature. Bibliographies of reviews that were identified as relevant were searched for relevant primary studies.

Search terms

We used the search strings listed in Table 1 to retrieve articles indexed in the above online databases. Search terms were identified by reference to articles cited in traditional reviews [2,18], consultation with subject experts within the review group and simplified trial searches. Compound search strings were generated by combining each of the search strings. The sensitivity of the search strings was tested and refined using 15 'control' papers, selected to cover a range of publication dates and whose titles contained a variety of terminology to describe responses to inbreeding (Additional file 1). This ensured that the specificity of the searches was maximised, minimising the number of papers retrieved without losing sensitivity. Records of the search strategies used were maintained to ensure repeatability and transparency.

The results from each database were combined in a single ENDNOTE X7 library file in order to create a database containing the raw search results for this systematic map. A separate file was generated for each search. When each stage of the searching was completed (in 2010 and 2013) all the database libraries were incorporated into a single library and any duplicates were removed using the automatic function in ENDNOTE X7 (and during the review process). The process was repeated when the libraries from the two search stages were combined (Figure 1).

Table 1 Search strings used to query online databases

Group	Search string
i. Inbreeding related strings	"In*breeding coefficient$"
	"Cost$ of in*breeding"
	(inbred SAME mating*) NOT (("Quantitative trait loc*")OR(QTL*))
	(inbred SAME (offspring OR progeny))
	Selfed SAME out*
	"Optimal outcrossing" OR "Outcrossing distance"
	"Benefit* of dispersal"
	"Cost* of dispersal"
	("Natal dispersal" AND (inbred OR in*breeding OR heterosis OR self* OR fitness))
	(Philopatr* AND (inbred OR in*breeding OR heterosis OR self* OR fitness))
ii. Fitness related strings	(Heterosis AND in*breeding)
	(Depression SAME in*bre*)
	(Depression SAME fitness)
	"Genetic load"

Boolean syntax follows the ISI Web of Knowledge template, and was adapted as necessary for use in databases using other syntax.

Study inclusion criteria

Relevant subject(s)

Studies that assessed the outcomes of inbreeding in natural populations of wild species, at any location globally, and experimental individuals and progeny derived from these were included. Natural populations were defined as those that have been founded by natural/spontaneous colonisation, or naturalised or (re-) introduced populations that persist in the absence of human intervention. No inter-specific studies or studies on hybrid swarms were considered. Studies involving species that were agricultural cultivars or strains, or whose populations are under captive management were excluded (e.g. zoo populations). The inbreeding responses in these populations may differ from those in natural populations as they may have been subject to artificial bottlenecks, differing selection regimes (e.g. selection for increased yield or captive conditions), controlled breeding or stock movement. Similarly, human-maintained inbred lines, lab strains, and artificially selected populations were excluded. For each study, the possibility of using a subset of populations was considered, where this subset fitted with the guidelines set out in this section. Studies that used terms such as "provenance", "land-race" or "cultivar" to describe study populations were retained until it could be ascertained whether they fitted the criteria.

Species mating system reports, dispersal studies and studies on philopatry were included only if there was evidence that appropriate crosses (inbred/non-inbred) were carried out or observed/inferred and appropriate progeny traits were measured (e.g. progeny traits that are not parental traits; germination, hatching rate, fitness components of progeny).

Types of intervention/exposure

The intervention/exposure were "inbred crosses" resulting from:

- Experimental or observed natural crosses among related individuals within the study population;
- Natural or experimentally induced self-fertilisation;
- Experimental crosses between individuals separated by known distances within the population;
- Experimental or observed natural crosses among apparently random individuals within the study population.

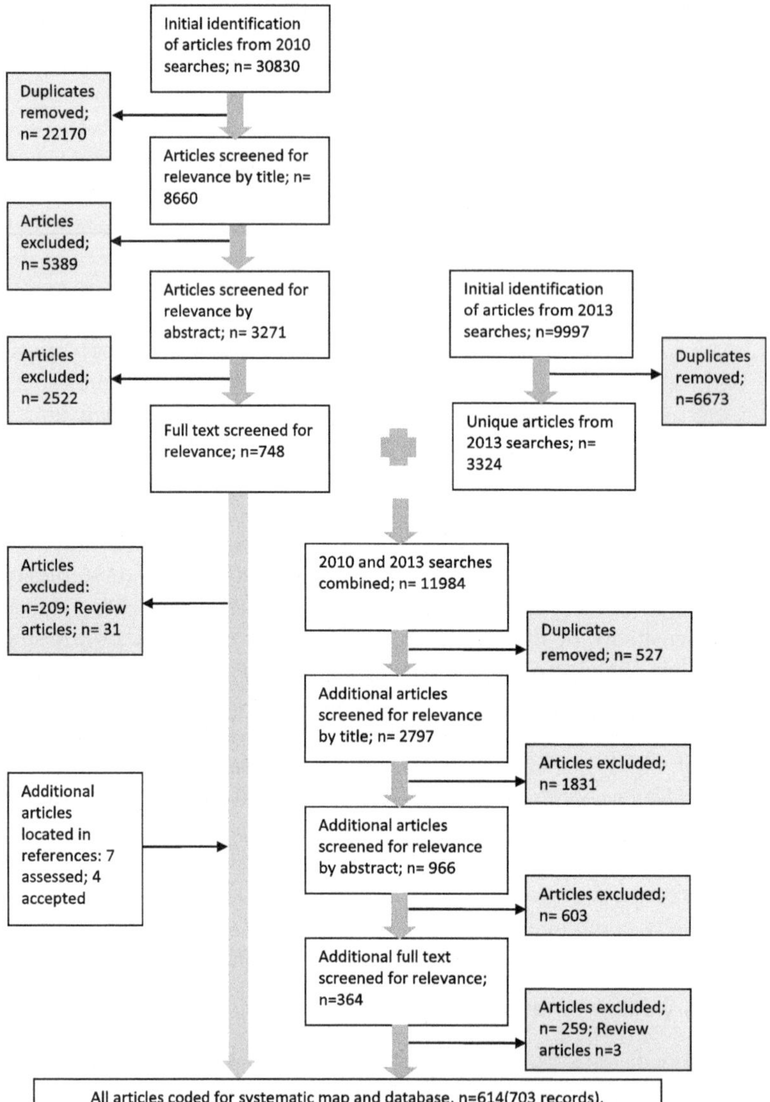

Figure 1 Articles included and excluded at each stage of the systematic mapping process. Flow chart describing the article assessment process applied to the literature searches undertaken to construct the systematic map. Numeric values indicate the number of articles included or excluded at each stage.

Types of comparator

The comparator takes the form of observed or experimental non-inbred crosses. The comparator and its corresponding exposure/intervention cross were defined according to the information presented in the study. The comparator is the offspring stemming from relatively less related parent individuals, and included crosses between randomly selected or unrelated individuals within the population or crosses between individuals separated by larger physical distances within the population. In addition, crosses between geographically separated populations, or crosses following the translocation of individuals from a donor population were included, where studies provided data allowing inbreeding to be evaluated in sub-populations relative to the total population. This meant that cases of genetic rescue and heterosis were included in the map. It should be noted this was done within the scope of studies retrieved using the searches described above and so does not constitute a synthesis of studies assessing phenotypic responses to outbreeding.

Types of outcome

Relevant outcomes were measures of traits of the progeny arising from inbred and non-inbred crosses. This included measurements of fitness components (life history traits) such as survival or mortality, reproductive effort or success, or early-acting components of viability that are unambiguously traits of the offspring, and traits more distantly related to fitness such as size, growth rate and responses to stressors.

Types of study

We considered empirical studies where comparator and intervention/exposure could be identified on the basis of pedigrees, or where physical distances between individuals paired and mated could be used as a proxy, and where appropriate outcome measures were also reported. Pedigrees could be based on controlled experimental crosses or through well-resolved marker-based or observational pedigrees where mating was not constrained.

Articles that were book sections were assessed for relevance. Meeting abstracts with no data were not considered for this review. Review articles were retained if their subject was congruent with the review subject and/or were likely to contain relevant data. Other reviews were rejected. Articles that were *errata*, commentaries, or that contained no empirical data, or that were QTL or genetic map studies without inbreeding measurements or that focussed on humans were excluded. Studies that did not obviously deal with whole organism biology (e.g. molecular biology, medical or biochemistry studies) were also excluded.

Article screening

Studies were assessed for inclusion in the systematic map based on a hierarchical assessment of relevance by scanning article titles, followed by reading the abstract of articles with relevant titles, followed by reading the full-text of articles with relevant titles and abstracts (Figure 1). Studies were deemed relevant based on the presence of the relevant subject, intervention and comparator (control and inbred mating or crosses within natural populations) and outcome (fitness) measurements, as described in the previous sections. Decisions were inclusive at the title and abstract stages when the relevance of the study was unclear.

The objectivity and repeatability of the article selection process was determined during title, abstract and full text appraisal. Two investigators independently assessed the same randomly selected subsets of articles at each stage and the consistency of these assessments was determined via kappa analysis [29]. Kappa analysis measures the difference in the frequency of the observed agreement between investigators and that expected by chance. The kappa coefficient ranges from -1 to 1, where 1 is total agreement, 0 is equivalent to chance and negative values indicate agreement is less than chance, suggesting systematic disagreement between investigators (the results of the kappa analysis are presented in Table 2). Reviewers discussed the inclusion criteria to optimise the methods and ensure the application of the criteria was consistent.

Study quality assessment

The quality of studies was assessed at the full text level. The procedure for quality assessment was based on that proposed in the original protocol (Table 3). However, four design features (relating to comparator and exposure crosses, study populations, selection of individuals crossed, and status of the study population) were excluded from the list of criteria due to limitations in applying criteria at the article level or a lack of information provided by authors (Table 3). The quality score was not used to inform the inclusion or exclusion of studies but to provide information on the quality of relevant studies.

Coding system for the systematic map

Articles were described and categorised using a combination of keywords or continuous values generated from the information reported in the studies and expert knowledge in the review team (Additional file 2). Articles were considered to contain multiple studies in cases where inbreeding outcomes were measured in more than one species, or where the results of divergent experimental designs were reported. These studies were maintained as separate entries in our database.

Articles were categorised according to; the full reference, year of publication, study species, types of intervention crosses, types of comparator crosses, the evidence-type defining the comparator and intervention crosses, and types of outcomes (i.e. phenotypic traits measured). In addition, we recorded, where possible, the following potential sources of heterogeneity to address the secondary objective of documenting the reporting frequency for factors that could influence responses to inbreeding:

- *Taxonomic group*: Higher-level taxonomic category for species; amphibian, annelid, arachnid, bird, bryozoan, cestode, crustacean, diatom, echinoderm, fish, fungi, insect, mammal, mollusc, plant, platyhelminth, reptile, rotifer and tunicate.
- *Mode of reproduction*: Primary means of reproduction in the study organism.
- *Indices of genetic diversity*: Measures of genetic variation reported and the types of variation/markers examined (when reported).
- *Physical distance*: Geographic distance separating intervention and comparator crosses.
- *Population size*: The number of individuals in the study population.
- *Population history*: Any information on the population included such as past population size, number of founding individuals or the rate of inbreeding.
- *Period of inbreeding*: Maximum progeny generation where phenotypic consequences of inbreeding were measured.

Table 2 Summary of the article assessment procedure

Stage of article assessment	Total number assessed	Number passing assessment	Agreement level, κ (interpretation according to [30])		Subset assessed in duplicate (% of total)	
			RW vs. JE (2010 searches)	JE vs. LEN (2013 searches)	RW vs. JE (2010 searches)	JE vs. LEN (2013 searches)
Title assessment	11457	4244	1. 0.43 (moderate)	0.61 (substantial)	100 per trial (total 500; 5.75%)	279 (9.84%)
			2. 0.62 (substantial)			
			3. 0.74 (substantial)			
			4. 0.56 (moderate)			
			5. 0.74 (substantial)			
Abstract assessment	4244	1119	1. 0.20 (slight)	0.56 (moderate)	1. 50	75 (7.76%)
			2. 0.43 (moderate)		2. 49	
			3. 0.66 (substantial)		3. 50	
			4. 0.72 (substantial)		4. 50	
			5. 0.59 (moderate)		5. 330	
			5b. 0.70 (substantial)		5b. 330	
					(total 859; 26.22%)	
Full text assessment	1119	614*	N/A	0.96 (almost perfect)	N/A	50 (4.48%)

Objectivity of article assessment was determined using the Kappa coefficient of agreement via independent assessments of a subset of articles [29]. Articles were assessed by three reviewers (RW, JE and LEN).
*79 articles could not be assessed and were excluded at this stage due to being written in a language other than English or required additional information from the authors before inclusion due to insufficient/unclear reporting.

- *Fitness trait class*: Categorical description of whether the trait measured was a direct or indirect measure of fitness.
- *Coefficient of inbreeding*: Within comparator (non-inbred) and/or exposure (inbred) group.
- *Environmental conditions*: Categorical description of the environment the outcomes of inbreeding were measured in.
- *Ploidy*: The number of set of chromosomes contained within an individual.
- *Dispersal ability*: Categorical description of the potential for gene flow between populations.

All articles relevant at the full-text assessment stage were coded by a single reviewer (LN) to ensure consistency. Coding of keywords was revised following discussions with the review group to ensure relevance and clarity of coding. Any articles already coded were re-coded using the revised terms/values. The keywords used for each category recorded are reported in Additional file 2.

Systematic map database

Following coding of relevant articles and assessment of quality (Table 3) we created a searchable systematic map (database) to describe the scope of available data on inbreeding responses and to identify knowledge gaps. This database can be used for further analysis of the data, as it enables articles to be searched and ordered according to keywords and permits more complex cross tabulations.

Results

Review descriptive statistics

The primary literature search yielded 11457 articles, which were screened against the inclusion criteria to yield a final total of 614 articles (Figure 1; Table 2). The studies included in the map and studies excluded at each level are listed in Additional file 3. The earliest study identified as relevant under our search strategy and inclusion criteria was published in 1963 (Figure 2). Subsequently, the number of articles reporting phenotypic responses to inbreeding remained relatively constant until 1993, after which the rate of publication increased linearly with time (Figure 2). Following coding, the 614 articles were separated into 703 studies/records in the database. This database is included as a Microsoft Access file (Additional file 4).

Quantity of relevant papers

Taxonomic group

Studies documenting the effects of inbreeding included a range of high-level taxonomic groups, specifically, amphibian, annelid, arachnid, bird, bryozoan, cestode, crustacean, diatom, echinoderm, fish, fungi, insect, mammal, mollusc, plant, platyhelminth, reptile, rotifer and tunicate (Figure 3). Of the 703 studies included in the map, 66.7%

Table 3 Study quality assessment criteria

	Design feature	Study attribute
Internal validity of study	Comparator and exposure crosses	* Did exposure (inbred) crosses take place contemporaneously with control (non-inbred) crosses? Yes: 1 point, otherwise: 0 points
		1. Were the pedigrees underpinning all the crosses known (because of experimental manipulation; 1 points), or unknown (0 points)?
	Scale of evidence for relative level of inbreeding between exposure and comparator groups	2. Was the coefficient of inbreeding known for both the exposure and comparator crosses? No pedigree information or physical distance used as proxy, for relative inbreeding level (0 points). Inbreeding coefficient or pedigree known for exposure (1 point). Known for both (2 points)
	Outcome measure	3. Were the outcome measures components of fitness (survival, fecundity, viability; 1 point), or indirect measures of these (e.g. growth rate, body mass, size; 0 points)?
	Environment for trait measurements	4. Were the traits measured in the field (2 points), under experimental conditions that closely approximate field conditions (1 point), or under non-native experimental conditions (0 points)?
	Study populations	* Were populations selected randomly or using sampling that was stratified over variation in population location, distribution or size (1 point, otherwise 0 points)?
		5. Was the cost of inbreeding observed in multiple study populations (1 point; 0 points otherwise)?
	Selection of individuals for crossing experiments, or selection of natural observed crosses	* Were crosses selected randomly from a population of possible crosses, or was selection stratified over variability in location or timing of crosses (1 point; otherwise 0 points)?
External validity of study	Observation window for progeny individuals	6. Were the fitness consequences of inbreeding within progeny tracked until at least the F_2 generation (1 point; otherwise 0 points)?
	Status of populations studied	* Did the study include re-introduced, or naturalized population(s)? 1 point if all populations were naturally founded, otherwise 0 points

Those study attributes marked by an asterisk (*) were removed from the quality assessment due to insufficient information.

(469 studies) assessed the consequences of inbreeding in plants. This was approximately 9 times greater than the next most frequently studied animal taxon (insects: 52 studies). Among animal taxa the most frequently studied were insects, followed by birds, mammals and molluscs.

Types of crosses

Relevant studies documented a range of different cross types, with most studies using more than one type of inbred cross. The inbred group (the intervention) was typically based on self-fertilisation (472 studies, 66.8% of the total). It should be noted that self-fertilisation was more commonly employed for plant studies (409 studies, 87.2% of plant studies) than for animals (63 studies, 26.8% of animal studies), which typically employed consanguineous crosses (e.g. sib-mating; 162 studies, 68.9% of animal studies). The non-inbred control crosses (the comparator) were usually crosses between randomly selected parent individuals within the study population that were known, or presumed to be unrelated (593 studies, 83.9%; Figure 4a). We did not distinguish between unrelated and randomly-selected crosses within populations as there was often insufficient information reported to enable unambiguous classification. A small proportion of studies assessed among-population inbreeding, using within-population crosses as the inbred cross and between-population crosses as the non-inbred cross (112 studies; 15.5%). 81 studies evaluated inbreeding responses using a combination of self-fertilisation/consanguineous crosses, within-population crosses and among-populations crosses. It should be noted that information on cross-types is reported at the study level and hence multiple crosses were often recorded for each study. This means that interrogation of the database can produce results that include uninformative comparisons such as 'within population'/'within population' crosses (since this could be classed as both inbred and non-inbred cross). It was not possible to overcome this coding artefact whilst maintaining description of the full range of potential combinations of cross types used to evaluate inbreeding responses evaluated in each study.

Responses to inbreeding

Studies measured the phenotypic outcomes of inbreeding in a range of traits. Survival was the most frequently reported (81.9%; 576 studies), followed by size (52.0%, 366 studies). We classified 12.5% of studies (88) as reporting "other responses", which included resistance to pathogens, parasites or herbivory, responses to stress, attractiveness or mating success. The majority of studies (92.0%; 647 studies) reported the cost of inbreeding in at least one component of fitness (i.e. survival, viability or fecundity; Figure 5). Other measures more distantly related to fitness (e.g. size) were also highly reported (63.8%, 449 studies).

Inbreeding responses were measured under laboratory/greenhouse/common garden (controlled) conditions in 62.1% of studies (439 studies), while 26.7% of studies included measurements under natural conditions (189 studies; Figure 6a). Most studies focusing on plants observed phenotypic consequences to inbreeding under controlled

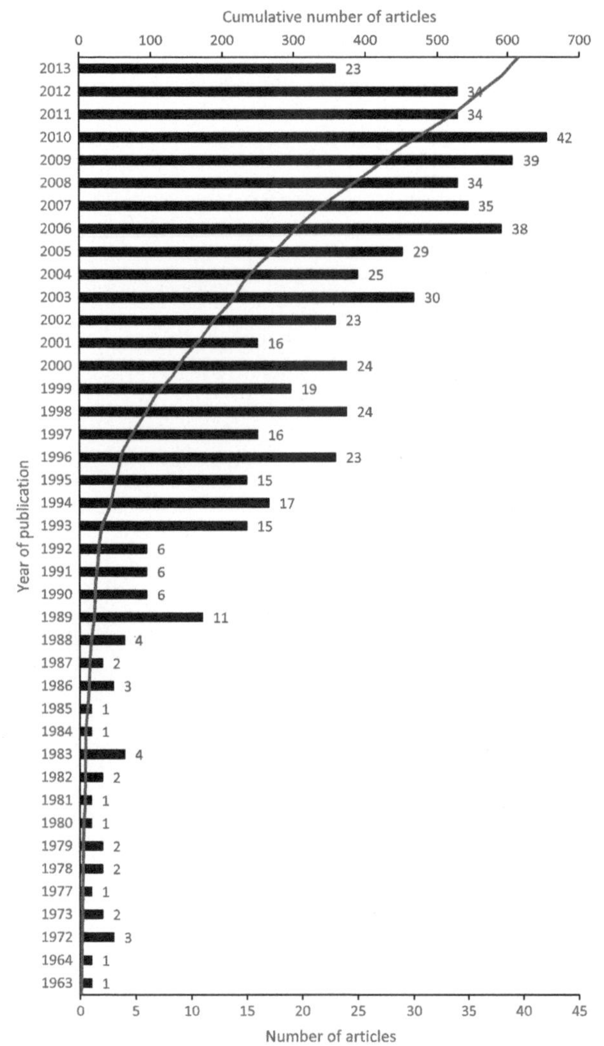

Figure 2 Publication frequency of articles presenting data on inbreeding responses over time. The 614 articles shown are limited to those included in this systematic map, deemed as relevant to our review question at the full text level. Bars indicate the number of articles published annually and the line represents the cumulative number of articles over time.

environmental conditions. In contrast, the environmental contexts for inbreeding studies focusing on animals were more variable. Observations of inbreeding costs in invertebrates (e.g. insects and molluscs) were largely confined to controlled conditions, while responses in vertebrates (e.g. birds and mammals) were usually observed in the field (Figure 6b).

Sources of heterogeneity

A subset of the relevant studies reported information on potential sources of heterogeneity that could modify inbreeding responses (Figure 7). Factors associated with experimental design, such as test environment (703 studies) and generations of inbreeding (654 studies) were well reported, but reporting of the coefficient of inbreeding was less frequent (140 studies). In contrast, sources of heterogeneity relating to the study population or study species were, in general, poorly reported. Mode of reproduction was frequently reported or could be determined based on the taxonomic group (e.g. mammals; 235 studies), while 120 studies reported population size. Only 38 studies reported levels of within-population genetic diversity assessed by neutral markers. Information on population history was occasionally reported, although the information presented was highly variable and included qualitative and quantitative information on past population size, changes in levels of fragmentation/isolation and bottlenecks. Other potential sources of heterogeneity that we had intended to record, such as dispersal ability and ploidy, were insufficiently reported to enable inclusion in the database.

Quality of relevant papers

We used six study quality assessment criteria to describe the internal and external validity of each study. The maximum quality score possible was 8 points (Table 3).

Overall, the average quality score for literature investigating the costs of inbreeding was 4.3 (range: 1-7, Figure 8). Total quality did not vary substantially over time, nor did it vary with taxonomic group, although a greater proportion of studies in birds and mammals possessed higher quality scores.

Evidence used to distinguish (at least one) possible inbred/non-inbred comparisons was primarily based on experimental manipulation (controlled crosses), followed by pedigrees estimated using molecular markers (Table 4). 59 studies employed other methods such as observational pedigrees (22 studies), several of which contained potential inaccuracies or errors; physical distance between crosses (8 studies); or anatomical structure of flowers (e.g. chasmogamous vs. cleistogamous flowers; 6 studies). In 666 cases (94.7% of studies) the coefficient of inbreeding was reported or could be determined for the inbred cross, but only 194 (27.6% of the total number) of these also provided information for the non-inbred cross (Table 4).

Phenotypic outcome measures typically included at least one component of fitness (92.2%, 648 studies; Table 4). However, in order to be maximally informative inbreeding studies must be carried out under natural conditions. In this respect the evidence base is poor, with inbreeding responses more frequently assessed in controlled (non-natural) conditions (54.9%, 386 studies). For instance, only 25.0% (176) of the studies that measured a direct component of fitness did so under natural conditions.

Studies assessing inbreeding responses are most robust where multiple populations are evaluated across multiple generations. However, most studies only investigated inbreeding responses within a single population (59.9%;

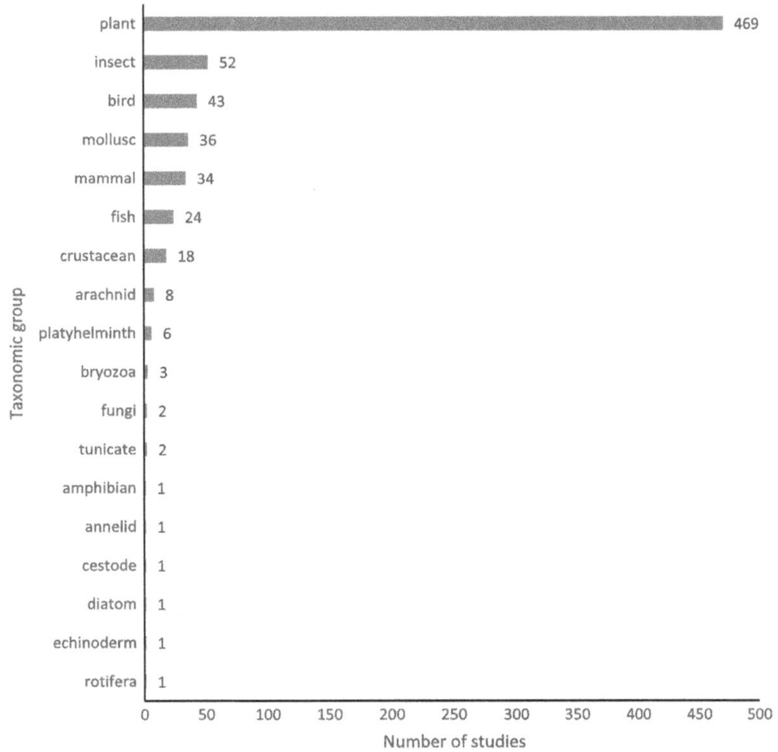

Figure 3 Number of studies by high-level taxonomic categories within the systematic map.

421 studies) and the phenotypic consequences of inbreeding were usually observed only within the first generation following inbreeding (607 studies; 86.3%; Table 4). 28 of the studies replicated the study in the same population in multiple years.

Discussion

Our systematic map shows there is a substantial amount of data available on the phenotypic consequences of inbreeding in wild populations. Studies typically reported factors associated with experimental design, such as cross types and pedigrees. In contrast, reporting of potential sources of variation in inbreeding responses, such as population size or mode of reproduction was relatively poor. The map shows key limitations in understanding the consequences of inbreeding in the wild are due to studies frequently measuring responses under laboratory/ greenhouse (controlled) conditions, only investigating a single generation of inbreeding or only assessing a single population.

The amount of primary literature documenting phenotypic consequences of inbreeding has increased in most years since 1993. The importance (and even the existence) of inbreeding depression in natural populations was questioned by some authors in the 1990s [31,32]. This corresponds to, and likely resulted in the initial increase in the rate of publications documented in this systematic map. Subsequent key studies and reviews (e.g. [18,19,33]) that demonstrated fitness costs associated with inbreeding in natural populations and validated its relevance to conservation have resulted in a continuous, and increasing number of publications on the topic. In addition, the subsequent recognition of conservation genetics as a distinct discipline within conservation biology and establishment of several journals devoted to this field are also likely to have facilitated the increasing rate of publication.

The cross types employed by studies in this map broadly fall into two categories, those that focus on mating among related individuals within populations, and those that study inbreeding effects due to drift load among geographically separated populations. The systematic map shows the effects of inbreeding within populations are more commonly investigated. These effects are relatively well understood and accepted [2] but these studies frequently employ crosses with extreme levels of inbreeding. The inclusion of crosses reflecting lower levels of inbreeding can refine our understanding of how the costs increase with the level of inbreeding, which can inform risk assessment in conservation strategies. In contrast to within-population inbreeding, the map shows that only a relatively small proportion of studies used crossing designs that allowed the quantification of drift load, despite the fact that these studies have provided some of the most dramatic

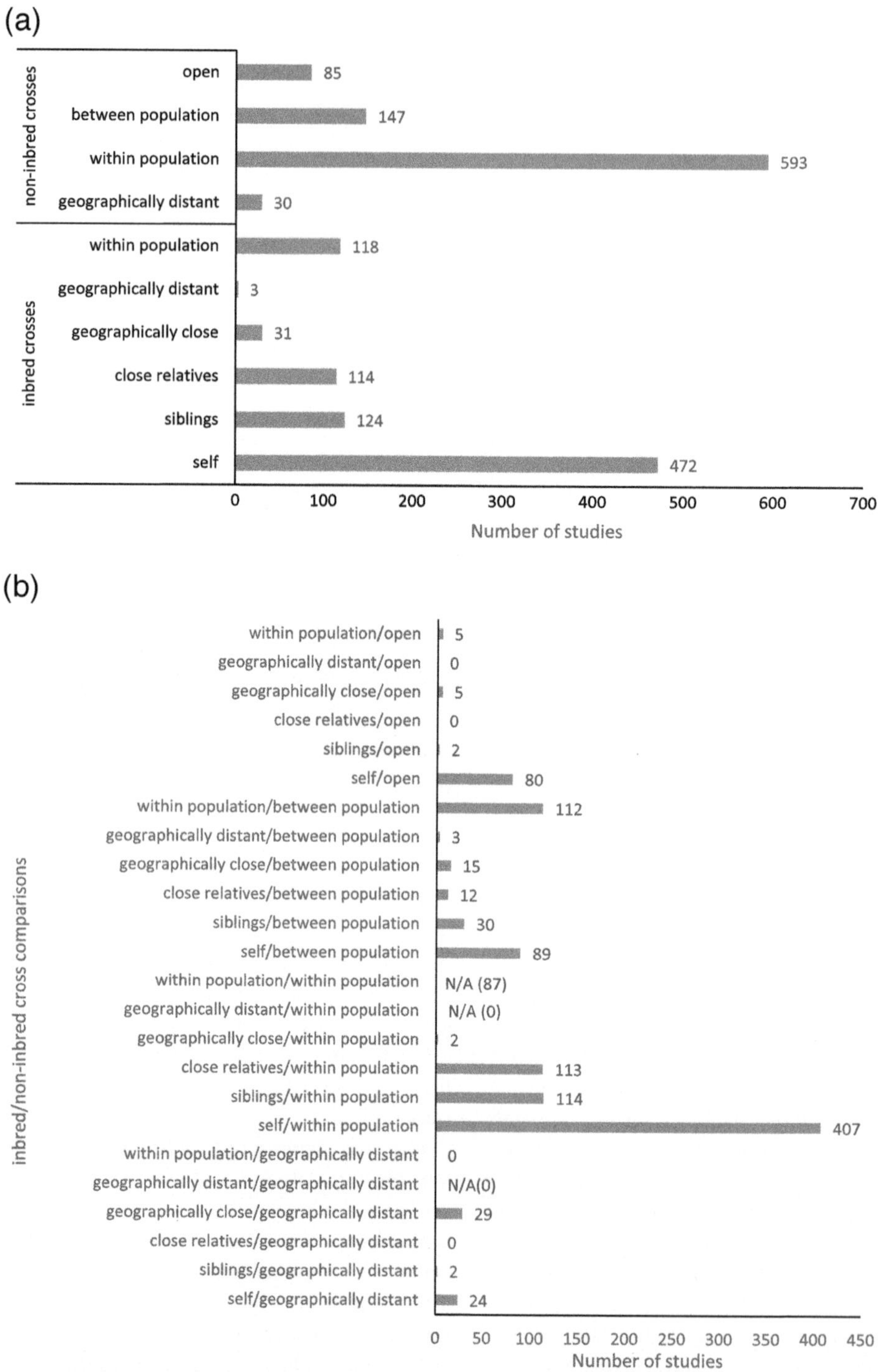

Figure 4 Types of inbred and non-inbred crosses used to assess the costs of inbreeding. (a) Number of studies by cross type for inbred and non-inbred crosses **(b)** Frequencies of combinations of inbred and non-inbred crosses comprising study designs. N/A denotes uninformative comparisons for inbreeding responses with bracketed numbers showing the number of studies.

examples of inbreeding depression (e.g. [10,11]). There are several potential reasons for this. The effects of drift load are only expected to be important in small populations where there has been sufficient time for genetic drift to act and the investigation of drift load requires more complex crossing designs that can distinguish between the effects of inbreeding and outbreeding between populations. In addition, for many threatened species such comparisons may not be possible, as no additional large populations exist. It is also possible that our search strategy may

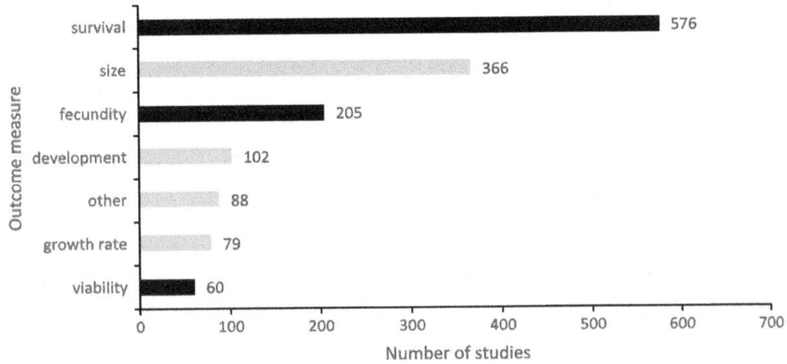

Figure 5 Number of studies by trait types within the systematic map. Black bars reflect direct measures of fitness (viability, fecundity and survival) and grey bars refer to indirect measures.

not have retrieved studies focused on outbreeding responses (but see Limitations of the map section). The inclusion of between-population crosses, is however, necessary to ascertain the effects of inbreeding where drift has led to an increase in baseline levels of inbreeding within the population, as within-population comparisons alone can underestimate the costs of inbreeding. A recent review of the effect of outcrossing indicates the fitness of small populations benefits from outcrossing with other populations, suggesting these populations suffered inbreeding depression [15]. In contrast, a meta-analysis focusing on inbreeding effects suggested that costs were reduced in small populations [20]. This latter study however, could not determine if this was the result of purging of deleterious alleles or higher baseline levels of inbreeding masking the effects of inbreeding depression. Thus in small populations, where drift load can increase the baseline level of inbreeding, crossing designs that enable inbreeding effects to be assessed at both levels are necessary to determine the overall cost of inbreeding and accurately inform conservation.

Recessive alleles with large deleterious effects may be purged by selection when they are exposed by inbreeding, reducing the genetic load and hence the costs of inbreeding. In theory, purging of deleterious alleles will reduce the costs of inbreeding over time and fitness may be recovered, or even enhanced [34]. While some studies suggest the relationship between population fitness and inbreeding can be influenced by purging [35,36], several reviews indicate the effectiveness of purging is highly variable [37,38]. The investigation of the impact of purging requires inbreeding to be followed across multiple generations. The map shows that the primary research in natural populations is limited, with most studies only assessing one generation of inbreeding. Similar limitations on studies of outbreeding responses have also been noted [15]. In addition, information on factors that can influence the effectiveness of purging, such as the rate of inbreeding (classified here as population history; [36]), are poorly reported in natural populations. Measuring the costs of inbreeding across multiple generations, and the potential for purging, is not straightforward, particularly in natural populations, but it may be critical to elucidating the long-term consequences of inbreeding in natural populations.

To be maximally informative for conservation, studies need to assess the effects of inbreeding using scenarios that mimic those found in nature. However, this aim typically conflicts with obtaining robust and repeatable results that control for (and disentangle) potentially confounding variables. The map shows that many studies sacrifice realism for control in the environmental conditions the effects of inbreeding are measured in, with the costs typically assessed in controlled, often relatively benign laboratory/greenhouse conditions. Several studies indicate that the costs of inbreeding increase, or may only become apparent under more stressful conditions [21,39]. Thus, while measures under controlled conditions may increase accuracy, it does not necessarily reflect the costs of inbreeding that would occur under natural conditions, which may be much greater. The map does, however, show that while numerous studies continue to assess inbreeding under controlled environments, the number of studies assessing the costs under natural conditions has increased (e.g. [40,41]).

Substantial variation in the costs of inbreeding has been documented (e.g. [21,37]). This map shows that individual studies rarely assessed multiple species or multiple populations within a species in an attempt to document variation and fewer attempted to test potential causes. Variation in inbreeding responses could be related to species- or population- specific attributes, or other study specific effects, which need to be considered when interpreting the phenotypic effects of inbreeding reported. Information on the potential sources of heterogeneity we were interested in was reported sporadically. Population level effects such as size, levels of genetic diversity and demographic history were relatively poorly reported

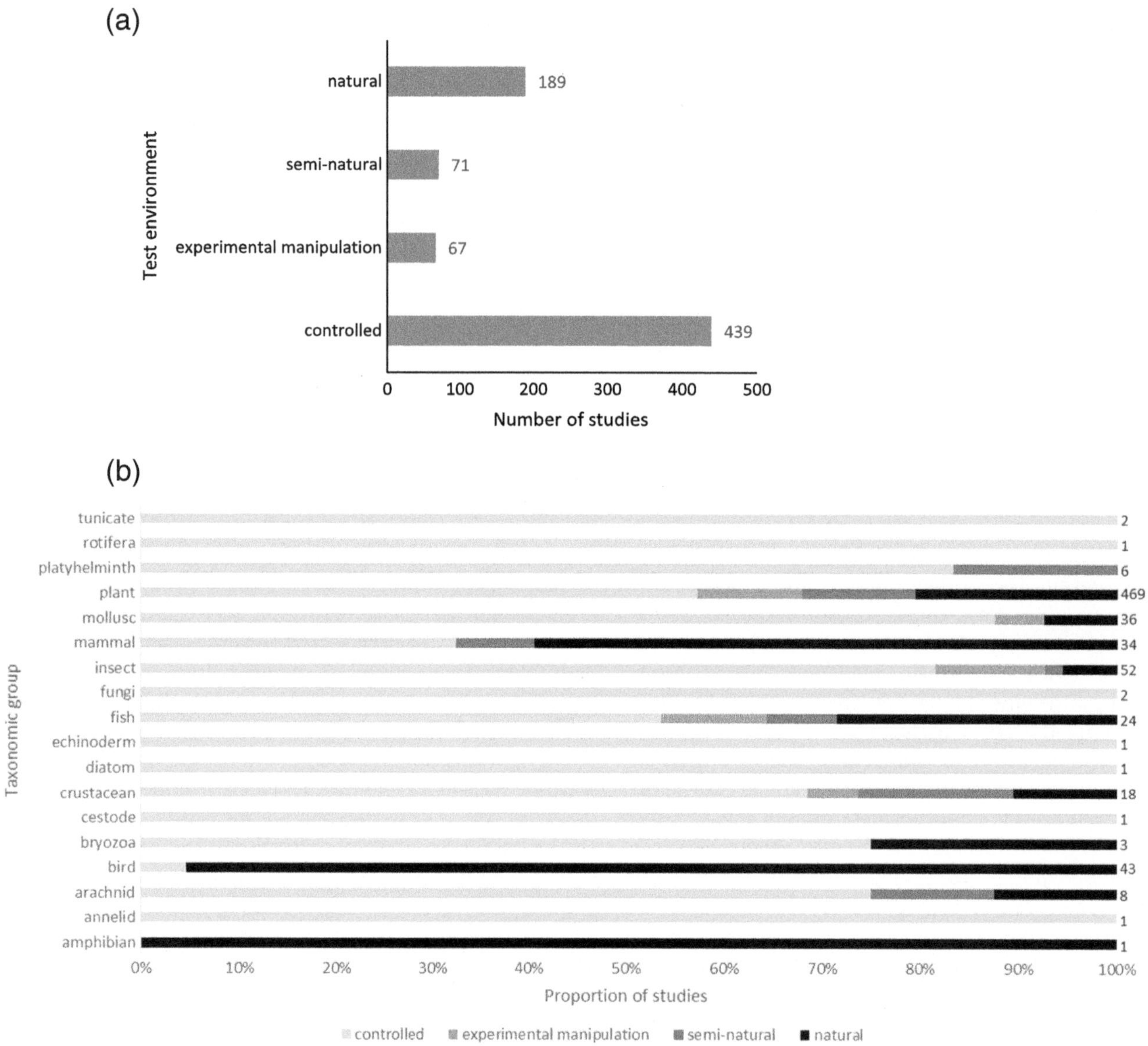

Figure 6 Inbreeding studies categorised by test environment. (a) Number of studies by test environment within the systematic map. **(b)** Frequency of studies by test environment for each taxonomic category. Natural conditions are black, semi-natural conditions are dark grey, experimentally manipulated conditions are light grey and controlled laboratory, greenhouse or common garden conditions are white. Number of studies is shown to the right.

compared with potential experimental effects such as the environmental context traits were measured under. In many instances, particularly for rare or threatened species this reflects limitations on the availability of individuals/ populations for sampling. It is also likely the map underestimates the number of studies where relevant additional information on species and population level effects is available, as it may be reported in separate articles or it may be available through consultation with authors. Nonetheless, the lack of reporting of potential sources of variation is likely to represent a pervasive problem within the inbreeding literature and is a common criticism of primary research synthesised in other reviews [15,42].

Understanding the influence of potential sources of variation on the effects of inbreeding can inform decision-making in conservation and enable the prediction of high-risk situations where intervention may be required. In general, the consequences of inbreeding are based on outcrossing diploids species [4]. Thus, the first step in assessing risk is an understanding of species traits. However, the map shows there is substantial variation in reporting of many traits, while sufficient data on the

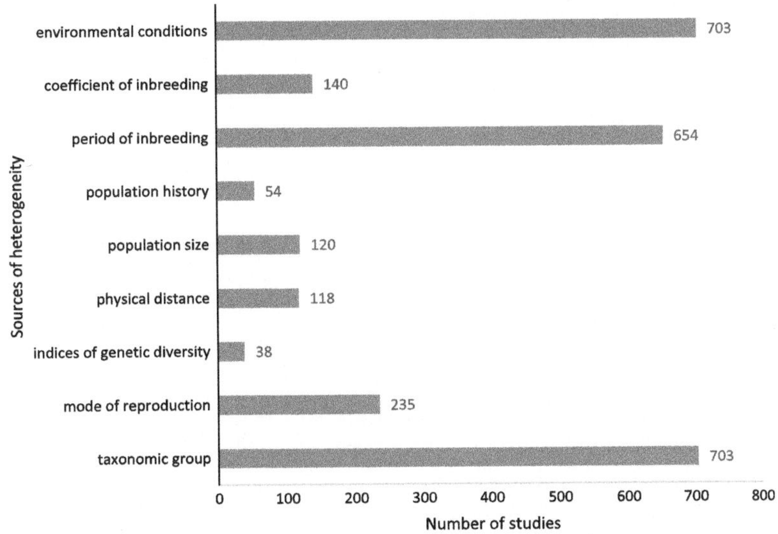

Figure 7 Number of studies within the systematic map reporting potential sources of heterogeneity.

mode of reproduction appears to exist; information on the ploidy of study species appears insufficient to enable review.

Among populations within species potentially useful proxies for identifying high-risk situations include levels of genetic diversity within populations (e.g. heterozygosity) or population size, as both have been shown to correlate with inbreeding [43,44]. Genetic diversity is typically measured at neutral loci, which do not impact on fitness, and as a result measures of neutral and quantitative variation do not always correlate [45]. Thus, the underlying cause of the correlation between levels of neutral genetic diversity measures such as heterozygosity and fitness is not well resolved [43]. The map shows that while not extensive, sufficient studies exist to support syntheses to further elucidate the extent to which measures of neutral genetic diversity can reflect genetic load and hence the costs of inbreeding. From a practical standpoint, estimates of population size may be more easily obtained and frequently form part of monitoring and risk assessment. Hence determining the relationship between census size and inbreeding is likely to be useful for managers. Although the effective population size, the equivalent number of individuals contributing to the next generation is more closely related to inbreeding risk, it is difficult to determine and previous studies have shown census population size correlates with both effective population size and genetic diversity [43,46]. The relationship between inbreeding depression and census size is not straightforward as shown in a recent meta-analysis in plants [20]. Thus, broader

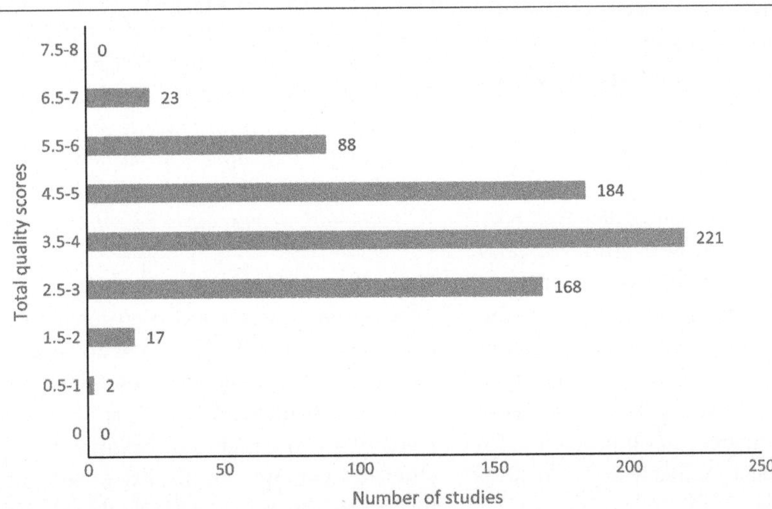

Figure 8 The total quality score for studies. The numbers of studies with each total quality score for the internal/external study validity. Number of studies is shown.

Table 4 Percentage (and number) of studies by quality scores for individual attributes

Quality attribute	Quality score			
	0	0.5	1	2
1. Evidence underpinning pedigrees	8.39%(59)	6.40%(45)	85.21%(599)	n/a
2. Coefficient of inbreeding known	5.26%(37)	n/a	67.14%(472)	27.60%(194)
3. Outcome measures	7.82%(55)	n/a	92.18%(648)	n/a
4. Environment for measurements	54.91%(386)	n/a	18.21%(128)	26.88%(189)
5. Number of replicate populations	59.89%(421)	n/a	40.11%(282)	n/a
6. Number of generations assessed	86.34%(607)	n/a	13.66%(96)	n/a

Quality attribute numbers correspond to those in Table 3.

reviews are required to provide insights into this relationship and refine predictions. The systematic map shows sufficient studies exist for meta-analysis to further elucidate the potential for census size to predict inbreeding depression (but see Possible systematic review topics below).

Research gaps

The systematic map highlights several areas in the primary research that are less well explored in natural populations.

- The long-term, multigenerational consequences of inbreeding, especially the potential for the restoration of fitness via purging of genetic load.
- Costs of moderate-low levels of bi-parental inbreeding.
- The relative contribution of within- and among-population inbreeding to the loss of fitness.
- The severity of inbreeding depression in natural conditions.
- The extent and causes of variation in inbreeding responses across populations of the same species, and particularly in response to differences in population size or levels of within-population neutral genetic diversity.

Possible systematic review topics

The systematic map highlights several subtopics that contain suitable primary research for synthesis into a systematic review. This systematic map revealed two distinct lines of investigation into inbreeding responses; the consequences of inbreeding resulting from consanguineous matings within populations and the impacts due to isolation and drift. The latter is likely encompassed by a recent systematic review of the consequences of outbreeding [15] and further review at this time is unlikely to yield additional information.

The primary research for responses to self-fertilisation/consanguineous mating is extensive and there are several subtopics which are likely to yield a suitable amount of data for meta-analysis. Specific questions could include:

- *How do species' traits influence the cost of inbreeding in natural plant populations?* Plants were clearly the focus of the primary research with 469 studies in the map and hence a review could provide relatively robust conclusions, although the high proportion of studies employing self-fertilisation may limit the relevance of these conclusions to other taxonomic groups. A subset of these data, relating to population size has already been meta-analysed (but see the following point). The primary research for invertebrates and vertebrates is sparser but could still be informative, particularly in vertebrates where studies often assess inbreeding under natural conditions.
- *How does population size influence the costs of inbreeding in natural populations?* Information on the relationship between inbreeding costs and population size has the potential to inform conservation strategy as a proxy for risks assessment. This area has recently been meta-analysed for plants [20]. This review was not systematic but future meta-analyses would need to ascertain potential overlap. The primary research will support a similar review on animals.
- *How do levels of neutral genetic variation within populations affect the costs of inbreeding?*
- *How does the environmental context influence phenotypic responses to inbreeding?*

Limitations of the map

The search strategy and the repeatability of the inclusion criteria were relatively robust to missing relevant studies based on sensitivity analyses of the searches and kappa analyses of inclusion criteria. However there may be some cases in which authors may publish inbreeding coefficients/genetic variation data separately to fitness data relating to the same individuals. In these cases, it is possible that neither of the two published articles may be covered by the search strategy. The obvious approach to address the problem would be to widen the search strategy to include all inbreeding coefficient/genetic variation studies and all fitness studies and then compare the author lists for matches but this would be expected to substantially increase the number of studies to be screened for relevance. In addition, although we employed a broad definition of inbreeding designed to capture primary literature investigating both within- and among-population inbreeding [47], it is possible that some studies assessing among-population inbreeding were published as outcrossing studies, and so may not have been covered by our searches.

There are also limitations associated with interpreting the results of queries in the database, due to the independent coding of attributes for each study. This occurs because each study may measure multiple variables or cross types, but not all the potential combinations of these may have been assessed. For instance, a study may report survival under controlled conditions but growth rate under both natural and controlled conditions. In the database the study will be recorded as measuring the traits survival and growth rate and that traits were measured under natural and controlled conditions. Thus, the record would imply that both traits were measured under natural conditions even though survival was only measure in controlled conditions. This means that a small number of false relationships between variables/attributes may be indicated by some queries in the database. This was limited, to some extent, by the separation of articles into studies where multiple species or experimental designs were employed but could not be overcome without loss of detail or further dissection of records into individual observations.

Conclusions

This systematic map provides a substantial database of research relevant to the primary question: 'What are the fitness consequences of inbreeding in natural populations?' The map is up-to-date as of 12 August 2013. Searches and screening procedures were carried out according to the original systematic review protocol, including kappa analyses. Articles were categorised according to the experimental design, sources of heterogeneity and the quality of the study.

Implications for policy and conservation management

The map provides a research tool for managers interested in the potential consequences of inbreeding and can be used to gather data for a range of subtopics. The information provided will enable users to identify relevant publications and assess the amount of information and, importantly, the quality of this information for a given topic. The map also highlights the potential for secondary syntheses to generate information that could be incorporated into conservation planning.

Implications for research

The map highlights several areas where information is limited or lacking and suggests future primary research should aim to assess the longer term multi-generational impacts of inbreeding, across multiple populations and under natural conditions. In addition, studies that simultaneously investigate within-population and among-population inbreeding are required to strengthen and refine our understanding of situations where inbreeding depression may not be detected due to the fixation of genetic load.

Finally, the potential for meta-analysis to be used to investigate the factors influencing variation in the phenotypic consequences of inbreeding appears limited, with the map showing a lack of reporting about potential sources of heterogeneity. This is particularly important from a conservation perspective where these patterns can lead to the identification of useful proxies for predicting the risks posed by inbreeding depression.

Additional files

Additional file 1: Record of searches.

Additional file 2: Coding categories and keywords used in the database.

Additional file 3: All articles retrieved by the searches and their inclusion/exclusion at each stage of assessment.

Additional file 4: Searchable database of inbreeding research. The database includes all articles accepted at the full-text level and is coded according to the keywords outlined in Additional file 2.

Competing interests

The authors declare that they have no competing interests.

Authors' contributions

LEN led the literature survey and assessment, carried out the data extraction and coding for the map. JE and RW designed and participated in the searches, and extraction of data. PMH, ASP and TB conceived of the study, formulation of the question and contributed to the design. LEN and JE drafted the manuscript with input from all the authors. All authors read and approved the final manuscript.

Acknowledgements

LEN, JE and RW were supported by a NERC knowledge exchange grant (NE/H001824/1) to PMH, ASP & TB. We thank Marie Chadburn for assistance in conducting literature searches and assessment of search results. We also thank the project review group, Mike Bruford, Richard Ennos, Mike Fay, Stuart Piertney and the representatives from UK conservation organisations, Barbara Jones (representing Natural Resources Wales), Ian Taylor (Natural England), Robin Payne (Scottish Natural Heritage), Barbara Smith (Game and Wildlife Trust) and Richard Smithers for assistance with question formulation and critical feedback on this review.

Author details

[1]Royal Botanic Garden Edinburgh, 20A Inverleith Row, Edinburgh EH3 5LR, UK. [2]Centre for Evidence-Based Conservation, School of Environment and Natural Resources and Geography, Bangor University, Bangor LL57 2UW, UK. [3]Institute of Integrative Biology, University of Liverpool, The Biosciences Building, Crown Street, Liverpool L69 7ZB, UK. [4]Department of Animal and Plant Sciences, University of Sheffield, Sheffield S10 2TN, UK.

References

1. Frankham R. Inbreeding in the wild really does matter. Heredity. 2010;104:124.
2. Keller LF, Waller DM. Inbreeding effects in wild populations. Trends Ecol Evol. 2002;17:230–41.
3. Spielman D, Brook BW, Frankham R. Most species are not driven to extinction before genetic factors impact them. Proc Natl Acad Sci U S A. 2004;101:15261–4.
4. Frankham R, Ballou JD, Briscoe DA. Introduction to Conservation Genetics. Cambridge: Cambridge University Press; 2010.
5. Charlesworth D, Willis JH. The genetics of inbreeding depression. Nat Rev Genet. 2009;10:783–96.
6. Charlesworth B, Charlesworth D. The genetic basis of inbreeding depression. Genet Res. 1999;74:329–40.
7. Hedrick PW, Adams JR, Vucetich JA. Reevaluating and broadening the definition of genetic rescue. Conserv Biol. 2011;25:1069–70.
8. Hedrick P, Fredrickson R. Genetic rescue guidelines with examples from Mexican wolves and Florida panthers. Conserv Genet. 2010;11:615–26.
9. Johnson WE, Onorato DP, Roelke ME, Land ED, Cunningham M, Belden RC, et al. Genetic restoration of the Florida panther. Science. 2010;329:1641–5.
10. Madsen T, Ujvari B, Olsson M. Novel genes continue to enhance population growth in adders (*Vipera berus*). Biol Conserv. 2004;120:145–7.
11. Madsen T, Shine R, Olsson M, Wittzell H. Conservation Biology: restoration of an inbred adder population. Nature. 1999;402:34–5.
12. Willi Y, van Kleunen M, Dietrich S, Fischer M. Genetic rescue persists beyond first-generation outbreeding in small populations of a rare plant. Proc R Soc B Biol Sci. 2007;274:2357–64.
13. Edmands S. Between a rock and a hard place: evaluating the relative risks of inbreeding and outbreeding for conservation and management. Mol Ecol. 2007;16:463–75.
14. Lynch M. The genetic interpretation of inbreeding depression and outbreeding depression. Evolution. 1991;45:622–9.
15. Whitlock R, Stewart GB, Goodman SJ, Piertney SB, Butlin RK, Pullin A, et al. A systematic review of phenotypic responses to between-population outbreeding. Environ Evid. 2013;2:13.
16. Frankham R, Ballou JD, Eldridge MDB, Lacy RC, Ralls K, Dudash MR, et al. Predicting the probability of outbreeding depression. Conserv Biol. 2011;25:465–75.
17. Hedrick PW, Kalinowski ST. Inbreeding depression in conservation biology. Annu Rev Ecol Syst. 2000;31:139–62.
18. Crnokrak P, Roff DA. Inbreeding depression in the wild. Heredity (Edinb). 1999;83:260–70.
19. O'Grady JJ, Brook BW, Reed DH, Ballou JD, Tonkyn DW, Frankham R. Realistic levels of inbreeding depression strongly affect extinction risk in wild populations. Biol Conserv. 2006;133:42–51.
20. Angeloni F, Ouborg NJ, Leimu R. Meta-analysis on the association of population size and life history with inbreeding depression in plants. Biol Conserv. 2011;144:35–43.
21. Fox CW, Reed DH. Inbreeding depression increases with environmental stress: an experimental study and meta-analysis. Evolution. 2011;65:246–58.
22. Grant MJ, Booth A. A typology of reviews: an analysis of 14 review types and associated methodologies. Heal Inf Libr J. 2009;26:91–108.
23. Fay MF, Sutcliffe J, Jones B, Taylor I. Proceedings of a Conservation Genetics Workshop Held at the Royal Botanic Gardens, Kew. 2001.
24. Ferris R. Research Needs for UK Biodiversity. London: DEFRA; 2007.
25. Gregory A, Burke T, Ferris T, Robson J, Smithers R, Whitlock R. The conservation of genetic diversity: science and policy needs in a changing world. 2006. JNCC report, No. 383.
26. Collaboration for Environmental Evidence. Guidelines for systematic review and evidence synthesis in environmental management. version 4.2. 2013. Environmental Evidence: www.environmentalevidence.org/wp-content/uploads/2014/06/Review-guidelines-version-4.2-final.pdf.
27. Whitlock R, Eales J, Chadburn M, Neaves LE, Hollingsworth PM, Burke T, Pullin AS. How do species' characteristics influence the cost of inbreeding? CEE Protocol 10-014. Collaboration for Environmental Evidence. 2012. http://www.environmentalevidence.org/wp-content/uploads/2014/06/Protocol10-014.pdf.
28. Randall N, James K. The effectiveness of integrated farm management, organic farming and agri-environment schemes for conserving biodiversity in temperate Europe - a systematic map. Environ Evid. 2012;1:4.
29. Cohen JE. A coefficient of agreement for nominal scales. Educ Psychol Meas. 1960;30:37–46.
30. Landis JR, Koch GG. The measurement of observer agreement for categorical data. Biometrics. 1977;33:159–74.
31. Lande R. Genetics and demography in biological conservation. Science. 1988;241:1455–60.
32. Caughley G. Directions in conservation biology. J Anim Ecol. 1994;63:215–44.
33. Ralls K, Ballou JD, Templeton A. Estimates of lethal equivalents and the cost of inbreeding in mammals. Conserv Biol. 1988;2:185–93.
34. Theodorou K, Couvet D. On the expected relationship between inbreeding, fitness, and extinction. Genet Sel Evol. 2006;38:371–87.
35. Larsen LK, PeLabon C, Bolstad GH, Viken Å, Fleming IA, Rosenqvist G. Temporal change in inbreeding depression in life-history traits in captive populations of guppy (*Poecilia reticulata*): evidence for purging? J Evol Biol. 2011;24:823–34.
36. Pekkala N, Emily Knott K, Kotiaho JS, Puurtinen M. Inbreeding rate modifies the dynamics of genetic load in small populations. Ecol Evol. 2012;2:1791–804.
37. Crnokrak P, Barrett SCH. Perspective: purging the genetic load: a review of the experimental evidence. Evolution. 2002;56:2347–58.
38. Byers DL, Waller DM. Do plant populations purge their genetic load? Effects of population size and mating history on inbreeding depression. Annu Rev Ecol Syst. 1999;30:479–513.
39. Reed DH, Fox CW, Enders LS, Kristensen TN. Inbreeding–stress interactions: evolutionary and conservation consequences. Ann N Y Acad Sci. 2012;1256:33–48.
40. Nielsen JF, English S, Goodall-Copestake WP, Wang J, Walling CA, Bateman AW, et al. Inbreeding and inbreeding depression of early life traits in a cooperative mammal. Mol Ecol. 2012;21:2788–804.
41. Brekke P, Bennett PM, Wang J, Pettorelli N, Ewen JG. Sensitive males: inbreeding depression in an endangered bird. Proc R Soc B Biol Sci. 2010;277:3677–84.
42. Castellanos M, Verdú M. Meta-analysis of meta-analyses in plant evolutionary ecology. Evol Ecol. 2012;26:1187–96.
43. Reed DH, Frankham R. Correlation between fitness and genetic diversity. Conserv Biol. 2003;17:230–7.
44. Leimu R, Mutikainen PIA, Koricheva J, Fischer M. How general are positive relationships between plant population size, fitness and genetic variation? J Ecol. 2006;94:942–52.
45. Reed DH, Frankham R. How closely correlated are molecular and quantitative measures of genetic variation? A meta-analysis. Evolution. 2001;55:1095–103.
46. Frankham R. Effective population size/adult population size ratios in wildlife: a review. Genet Res. 1995;66:95–107.
47. Wright S. The interpretation of population structure by F-statistics with special regard to systems of mating. Evolution. 1965;19:395–420.

Permissions

All chapters in this book were first published in EE, by BioMed Central; hereby published with permission under the Creative Commons Attribution License or equivalent. Every chapter published in this book has been scrutinized by our experts. Their significance has been extensively debated. The topics covered herein carry significant findings which will fuel the growth of the discipline. They may even be implemented as practical applications or may be referred to as a beginning point for another development.

The contributors of this book come from diverse backgrounds, making this book a truly international effort. This book will bring forth new frontiers with its revolutionizing research information and detailed analysis of the nascent developments around the world.

We would like to thank all the contributing authors for lending their expertise to make the book truly unique. They have played a crucial role in the development of this book. Without their invaluable contributions this book wouldn't have been possible. They have made vital efforts to compile up to date information on the varied aspects of this subject to make this book a valuable addition to the collection of many professionals and students.

This book was conceptualized with the vision of imparting up-to-date information and advanced data in this field. To ensure the same, a matchless editorial board was set up. Every individual on the board went through rigorous rounds of assessment to prove their worth. After which they invested a large part of their time researching and compiling the most relevant data for our readers.

The editorial board has been involved in producing this book since its inception. They have spent rigorous hours researching and exploring the diverse topics which have resulted in the successful publishing of this book. They have passed on their knowledge of decades through this book. To expedite this challenging task, the publisher supported the team at every step. A small team of assistant editors was also appointed to further simplify the editing procedure and attain best results for the readers.

Apart from the editorial board, the designing team has also invested a significant amount of their time in understanding the subject and creating the most relevant covers. They scrutinized every image to scout for the most suitable representation of the subject and create an appropriate cover for the book.

The publishing team has been an ardent support to the editorial, designing and production team. Their endless efforts to recruit the best for this project, has resulted in the accomplishment of this book. They are a veteran in the field of academics and their pool of knowledge is as vast as their experience in printing. Their expertise and guidance has proved useful at every step. Their uncompromising quality standards have made this book an exceptional effort. Their encouragement from time to time has been an inspiration for everyone.

The publisher and the editorial board hope that this book will prove to be a valuable piece of knowledge for researchers, students, practitioners and scholars across the globe.

List of Contributors

Nicola P Randall
Harper Adams University College, Newport, Shropshire TF10 8NB, UK

Katy L James
Harper Adams University College, Newport, Shropshire TF10 8NB, UK

Jeremy Brooks
The Ohio State University School of Environment and Natural Resources, 210 Kottman Hall, 2021 Coffey Rd, Columbus, OH 43212, USA

Kerry Ann Waylen
The James Hutton Institute, Social, Economic, and Geographic Sciences Group Cragiebuckler, Aberdeen, Scotland AB15 8QH, UK

Monique Borgerhoff Mulder
Anthropology Department, University of California, Davis, 210 Young Hall, One Shields Ave., Davis, CA 95616, USA

Raj Whitlock
Institute of Integrative Biology, University of Liverpool, The Biosciences Building, Crown Street, Liverpool L69 7ZB, UK

Gavin B Stewart
Centre for Reviews and Dissemination, University of York, York YO10 5DD, UK

Simon J Goodman
Faculty of Biological Sciences, University of Leeds, Leeds LS2 9JT, UK

Stuart B Piertney
Zoology Building, School of Biological Sciences, University of Aberdeen, Aberdeen AB24 2TZ, UK

Roger K Butlin
UK Population Biology Network, Department of Animal and Plant Sciences, University of Sheffield, Sheffield S10 2TN, UK

Andrew S Pullin
Centre for Evidence-Based Conservation, School of Environment and Natural Resources and Geography, Bangor University, Bangor LL57 2UW, UK

Terry Burke
UK Population Biology Network, Department of Animal and Plant Sciences, University of Sheffield, Sheffield S10 2TN, UK

Karissa M Ramstead
School of Forestry, Northern Arizona University, P.O. Box 15018, Flagstaff, AZ 86011, USA

James A Allen
School of Forestry, Northern Arizona University, P.O. Box 15018, Flagstaff, AZ 86011, USA

Abraham E Springer
School of Earth Sciences and Environmental Sustainability, Northern Arizona University, P.O. Box 4099, Flagstaff, AZ86011, USA

Dilys Roe
International Institute for Environment and Development (IIED), 80-86 Grays Inn Road, London WC1X 8NH, UK

Mike Day
International Institute for Environment and Development (IIED), 80-86 Grays Inn Road, London WC1X 8NH, UK
Centre for International Forestry Research (CIFOR), Jalan CIFOR, Situ Gede, Sindang Barang, Bogor 16115, Indonesia

Francesca Booker
International Institute for Environment and Development (IIED), 80-86 Grays Inn Road, London WC1X 8NH, UK
Centre for International Forestry Research (CIFOR), Jalan CIFOR, Situ Gede, Sindang Barang, Bogor 16115, Indonesia

Wen Zhou
Centre for International Forestry Research (CIFOR), Jalan CIFOR, Situ Gede, Sindang Barang, Bogor 16115, Indonesia

Sophie Allebone-Webb
Zoological Society of London (ZSL), Regent's Park, London NW1 4RY, UK

Noëlle Kümpel
Zoological Society of London (ZSL), Regent's Park, London NW1 4RY, UK

Nicholas A O Hill
Zoological Society of London (ZSL), Regent's Park, London NW1 4RY, UK
Conservation Science Group, Department of Zoology, University of Cambridge, Cambridge CB2 3EJ, UK

Juliet Wright
Zoological Society of London (ZSL), Regent's Park, London NW1 4RY, UK
Department of Life Sciences, Imperial College London, Silwood Park Campus, Buckhurst Road, Ascot SL5 7PY, UK

Niki Rust
International Institute for Environment and Development (IIED), 80-86 Grays Inn Road, London WC1X 8NH, UK
Durrell Institute of Conservation and Ecology, University of Kent, Canterbury, Kent CT2 7NZ, UK

Terry CH Sunderland
Centre for International Forestry Research (CIFOR), Jalan CIFOR, Situ Gede, Sindang Barang, Bogor 16115, Indonesia

Kent Redford
Archipelago Consulting, P.O. Box 4750, Portland, Maine 04112-4750, USA

Gillian Petrokofsky
Centre for International Forestry Research (CIFOR), Jalan CIFOR, Situ Gede, Sindang Barang, Bogor 16115, Indonesia
Department of Zoology, University of Oxford, The Tinbergen Building, South Parks Road, Oxford OX1 3PS, UK

Dominic A Andradi-Brown
Department of Life Sciences, Imperial College London, Silwood Park Campus, Buckhurst Road, Ascot, Berkshire SL5 7PY, UK

Caroline Howe
Centre for Biodiversity and Environment Research, Department of Genetics, Evolution and Environment, University College London, London WC1E 6BT, UK

Georgina M Mace
Centre for Biodiversity and Environment Research, Department of Genetics, Evolution and Environment, University College London, London WC1E 6BT, UK

Andrew T Knight
Department of Life Sciences, Imperial College London, Silwood Park Campus, Buckhurst Road, Ascot, Berkshire SL5 7PY, UK
Department of Botany, Nelson Mandela Metropolitan University, P.O. Box 77000, Port Elizabeth 6031, South Africa

Andrew S Pullin
Centre for Evidence-Based Conservation, School of Environment, Natural Resources and Geography, Bangor University, LL57 2UW Bangor, Gwynedd, UK

Mukdarut Bangpan
Evidence-informed Policy and Practice Information and Coordinating Centre (EPPI-Centre), Institute of Education, University of London, 20 Bedford Way, WC1H 0NR London, UK

Sarah Dalrymple
Centre for Evidence-Based Conservation, School of Environment, Natural Resources and Geography, Bangor University, LL57 2UW Bangor, Gwynedd, UK

Kelly Dickson
Evidence-informed Policy and Practice Information and Coordinating Centre (EPPI-Centre), Institute of Education, University of London, 20 Bedford Way, WC1H 0NR London, UK

Neal R Haddaway
Centre for Evidence-Based Conservation, School of Environment, Natural Resources and Geography, Bangor University, LL57 2UW Bangor, Gwynedd, UK

John R Healey
School of Environment, Natural Resources and Geography, Bangor University, LL57 2UW Bangor, Gwynedd, UK

Hanan Hauari
Evidence-informed Policy and Practice Information and Coordinating Centre (EPPI-Centre), Institute of Education, University of London, 20 Bedford Way, WC1H 0NR London, UK

Neal Hockley
School of Environment, Natural Resources and Geography, Bangor University, LL57 2UW Bangor, Gwynedd, UK

Julia P G Jones
School of Environment, Natural Resources and Geography, Bangor University, LL57 2UW Bangor, Gwynedd, UK

Teri Knight
Centre for Evidence-Based Conservation, School of Environment, Natural Resources and Geography, Bangor University, LL57 2UW Bangor, Gwynedd, UK

Carol Vigurs
Evidence-informed Policy and Practice Information and Coordinating Centre (EPPI-Centre), Institute of Education, University of London, 20 Bedford Way, WC1H 0NR London, UK

Sandy Oliver
Evidence-informed Policy and Practice Information and Coordinating Centre (EPPI-Centre), Institute of Education, University of London, 20 Bedford Way, WC1H 0NR London, UK

Biljana Macura
Department of Land, Environment, Agriculture and Forestry (TESAF), University of Padua, Viale dell'Università 16, 35020, Agripolis, Legnaro (PD), Italy
Centre for Evidence-Based Conservation, School of Environment, Natural Resources and Geography, Bangor University, Gwynedd, LL57 2UW, UK

Laura Secco
Department of Land, Environment, Agriculture and Forestry (TESAF), University of Padua, Viale dell'Università 16, 35020, Agripolis, Legnaro (PD), Italy

Andrew S Pullin
Centre for Evidence-Based Conservation, School of Environment, Natural Resources and Geography, Bangor University, Gwynedd, LL57 2UW, UK

Claes Bernes
Mistra Council for Evidence-Based Environmental Management Royal Swedish Academy of Sciences, SE-104 05 Stockholm, Sweden

Bengt Gunnar Jonsson
Department of Natural Sciences, Mid-Sweden University, SE-851 70 Sundsvall, Sweden

Kaisa Junninen
Metsähallitus Parks & Wildlife Finland, c/o UEF, P.O. Box 111FI-80101 Joensuu, Finland

Asko Lõhmu
Institution of Ecology and Earth Sciences, Tartu University, Vanemuise 46, 51014 Tartu, Estonia

Ellen Macdonald
Department of Renewable Resources, University of Alberta, 751 General Services Building, Edmonton, AB T6G 2H1, Canada

Jörg Müller
Department of Conservation and Research, Bavarian Forest National Park, Freyunger Str. 2, 94481 Grafenau, Germany

Jennie Sandström
Department of Natural Sciences, Mid-Sweden University, SE-851 70 Sundsvall, Sweden

Dilys Roe
International Institute for Environment and Development (IIED), 80-86 Grays Inn Road, London WC1X 8NH, UK

Max Fancourt
United Nations Environment Programme - World Conservation Monitoring Centre (UNEP-WCMC), 219 Huntingdon Road, Cambridge CB3 0DL, UK

Chris Sandbrook
United Nations Environment Programme - World Conservation Monitoring Centre (UNEP-WCMC), 219 Huntingdon Road, Cambridge CB3 0DL, UK

Mxolisi Sibanda
International Institute for Environment and Development (IIED), 80-86 Grays Inn Road, London WC1X 8NH, UK

Alessandra Giuliani
International Institute for Environment and Development (IIED), 80-86 Grays Inn Road, London WC1X 8NH, UK

Andrew Gordon-Maclean
International Institute for Environment and Development (IIED), 80-86 Grays Inn Road, London WC1X 8NH, UK

Philip D Roberts
CABI, Nosworthy Way, Wallingford, Oxfordshire, United Kingdom

Hilda Diaz-Soltero
USDA, National Invasives Species Council, 1201 Eye Street, NW., 5th Floor, Washington, DC, USA

David J Hemming
CABI, Nosworthy Way, Wallingford, Oxfordshire, United Kingdom

Martin J Parr
CABI, Nosworthy Way, Wallingford, Oxfordshire, United Kingdom

Nicola H Wakefield
CABI, Nosworthy Way, Wallingford, Oxfordshire, United Kingdom

Holly J Wright
CABI, Nosworthy Way, Wallingford, Oxfordshire, United Kingdom

Jessica Thorn
Department of Zoology, Long-term Ecology and Resource Stewardship Laboratory, Biodiversity Institute, University of Oxford, Oxford, UK

Jake Snaddon
Centre for Biological Sciences, University of Southampton, Southampton, UK

Anthony Waldron
Edward Grey Institute, Department of Zoology, University of Oxford, Oxford, UK

Kasper Kok
Land Dynamics Department, Wageningen University, Wageningen, Netherlands

Wen Zhou
Centre for International Forestry Research, Bogor, Indonesia

Shonil Bhagwat
Department of Geography, Faculty of Social Sciences, Open University, Milton Keynes, UK

Kathy Willis
Department of Zoology, Long-term Ecology and Resource Stewardship Laboratory, Biodiversity Institute, University of Oxford, Oxford, UK

Gillian Petrokofsky
Department of Zoology, Long-term Ecology and Resource Stewardship Laboratory, Biodiversity Institute, University of Oxford, Oxford, UK
Centre for International Forestry Research, Bogor, Indonesia

Gillian Petrokofsky
University of Oxford, Oxford, UK

Plinio Sist
Cirad, Montpellier, France

Lilian Blanc
Cirad, Montpellier, France

Jean-Louis Doucet
Université deLiège, Gembloux Agro-Bio Tech, Liège, Belgium

Bryan Finegan
Catie, Turrialba, Costa Rica

Sylvie Gourlet-Fleury
Cirad, Montpellier, France

John R Healey
Bangor University, Bangor, UK

Barbara Livoreil
Fondation pour la Recherche sur la Biodiversité, Paris, France

Robert Nasi
CIFOR, Bogor, Indonesia

Marielos Peña-Claros
Wageningen University,Wageningen, Netherlands

Francis E Putz
University of Florida, Gainesville, FL, USA and CIFOR

Wen Zhou
CIFOR, Bogor, Indonesia

Linda E Neaves
Royal Botanic Garden Edinburgh, 20A Inverleith Row, Edinburgh EH3 5LR, UK

Jacqualyn Eales
Centre for Evidence-Based Conservation, School of Environment and Natural Resources and Geography, Bangor University, Bangor LL57 2UW, UK

Raj Whitlock
Institute of Integrative Biology, University of Liverpool, The Biosciences Building, Crown Street, Liverpool L69 7ZB, UK

Peter M Hollingsworth
Royal Botanic Garden Edinburgh, 20A Inverleith Row, Edinburgh EH3 5LR, UK

Terry Burke
Department of Animal and Plant Sciences, University of Sheffield, Sheffield S10 2TN, UK

Andrew S Pullin
Centre for Evidence-Based Conservation, School of Environment and Natural Resources and Geography, Bangor University, Bangor LL57 2UW, UK